MW00581786

ADVANCES IN
COASTAL AND OCEAN ENGINEERING

ADVANCES IN COASTAL AND OCEAN ENGINEERING

Volume 1

Editor

Philip L.-F. Liu
Cornell University

World Scientific
Singapore • New Jersey • London • Hong Kong

Published by

World Scientific Publishing Co. Pte. Ltd.

P O Box 128, Farrer Road, Singapore 9128

USA office: Suite 1B, 1060 Main Street, River Edge, NJ 07661

UK office: 57 Shelton Street, London WC2H 9HE

Library of Congress Cataloging-in-Publication Data

Advances in coastal and ocean engineering / editor Philip L.-F. Liu

p. cm.

ISBN 9810218249 (v. 1)

1. Coastal engineering. 2. Ocean engineering. I. Liu, Philip L.-F.

TC205.A32 1995

627'.58--dc20 94-42318

CIP

Printed in Singapore by Uto-Print

PREFACE TO THE REVIEW SERIES

The rapid flow of new literature has confronted scientists and engineers of all branches with a very acute dilemma: How to keep up with new knowledge without becoming too narrowly specialized. Collections of review articles covering broad sectors of science and engineering are still the best way of sifting new knowledge critically. Comprehensive review articles written by discerning scientists and engineers not only separate lasting knowledge from the ephemeral, but also serve as guides to the literature and as stimuli to thought and to future research.

The aim of this review series is to present critical commentaries of the state-of-the-art knowledge in the field of coastal and ocean engineering. Each article will review and illuminate the development of scientific understanding of a specific engineering topic. Our plans for this series include articles on sediment transport, ocean waves, coastal and offshore structures, air-sea interactions, engineering materials, and seafloor dynamics. Critical reviews on engineering designs and practices in different countries will also be included.

P. L.-F. Liu, 1994

PREFACE TO THE FIRST VOLUME

This volume of the review series is a collection of five papers reviewing a wide range of research topics in coastal engineering.

The first paper, written by Yeh, discusses one of the fundamental issues concerning many fluid flow problems, namely, free-surface boundary conditions. The free surface is defined as the sharp interface between air and water. Yeh reviewed the effects of the air flow, surface tension, as well as the viscosity on the free-surface boundary conditions. The highlights of the paper are the discussion on the vorticity condition and the vortical behavior associated with the dynamics of the free surface.

Foda presents a survey on another boundary dynamics: the seafloor dynamics. In coastal water, fluid motions under water waves mobilize sediment on the seafloor and consequently wave energy is dissipated. The process is dependent on wave characteristics and sediment properties. Foda reviews recent studies on the nonlinear wave energy transfer into the seabed and different modes of wave-induced sediment fluidizaiton processes in cohesive as well as in noncohesive seabeds. Several issues concerning the interactions between sediment deposits and marine structures, such as pipeline and breakwater, are also briefly reviewed.

One of the active research areas in modeling wave propagation is the construction of a unified model which is valid from deep water to shallow water. Liu discusses several existing models. When the wave amplitude is small in deep water, a model based on the Boussinesq approximation seems to be performing well. However, if the nonlinearity is not negligible in deep water, none of the existing models can propagate waves from deep water to shallow water.

The sediment movement in the surf zone is a complex system. It can usually be decomposed into the longshore and the cross-shore components. Dean focuses his discussion on the cross-shore sediment transport process. In particular, the characteristics of the equilibrium profile and the dynamics of the beach profile evolution are reviewed. Several cross-shore transport models are compared and evaluated.

In the last paper, van der Meer presents a comprehensive review on the design consideration for a rubble mound breakwater. Both hydraulic and structural responses are discussed. Design formulas and graphs are presented, which can be used for a conceptual design of a rubble mound breakwater.

P. L.-F. Liu, 1994

CONTRIBUTORS

Robert G. Dean
Coastal and Oceanographic Engineering Department,
University of Florida,
336 Weil Hall,
Gainesville, FL 32611-2083
USA.

Mostafa A. Foda
Hydraulic and Coastal Engineering,
Department of Civil Engineering,
412 O'Brien Hall,
University of California at Berkeley,
Berkeley, CA 94720,
USA.

Philip L.-F. Liu
School of Civil and Environmental Engineering,
Hollister Hall,
Cornell University,
Ithaca, NY 14853,
USA.

Harry Yeh
Department of Civil Engineering,
University of Washington,
Wilcox FX-10,
Seattle, WA 98190,
USA.

Jentsje W. van der Meer,
Delft Hydraulics,
P. O. Box 152,
8300 AD Emmeloord,
THE NETHERLANDS.

CONTENTS

FREE-SURFACE DYNAMICS

HARRY YEH

There are two different restoring forces involved in a water-wave problem: gravitational and surface-tension forces. Both effects are discussed from basic fluid-mechanics view points. Maintaining the analyses as rigorous as possible but still within the basis of the continuum hypothesis, the free surface is treated as the air-water (two-fluid) interface, which creates a material discontinuity in fluid properties. The jump discontinuity condition of the general form of conservation equation is derived. At a material boundary such as the air-water interface, the jump condition of conservation of mass yields the kinematic free-surface boundary condition, the jump condition of conservation of linear momentum yields the dynamic free-surface boundary condition. The no-slip boundary condition arises at a material boundary as a consequence of the jump condition for mechanical-energy conservation; note that the no-slip condition is usually an assumed condition, but our anaylsis demonstrates that it is a necessary condition for the mechanical energy consideration at the interface.

A constant magnitude of pressure along the free surface is often imposed in traditional water-wave problems for irrotational fluid motions. It is demonstrated that, if this condition is imposed, water must be considered to be inviscid. Nevertheless, the resulting potential-flow solution requires the fictitious energy input from the surroundings, i.e., air, through the free surface. Under a more rigorous free-surface condition such as vanishing stresses on the interface, irrotational flows cannot exist at the free surface and must form a boundary layer. It is also demonstrated that the existence of air and variations in surface tension due to surfactant affect significantly wave-attenuation behavior.

In a real-fluid environment, flows at the free surface must be vortical. Because of this observation, the jump condition for the vorticity equation is examined. The resulting condition requires that the tangential component of fluid acceleration must be continuous across the fluid-fluid interface, which is not trivial since fluids are allowed to deform. It is demonstrated that the jump condition for three-dimensional flows is grossly different from that of two-dimensional flows; as far as behaviors of vortical flows at the free surface are concerned, two-dimensional flow approximation is inadequate. Vorticity normal to the interface and vorticity bending caused by the interface curvature appear to play important roles in determining free-surface vortex motions. It is found that the consideration of air above the free surface is essential to characterize vortical flows correctly at the air-water interface. Even though the air may not be important for water-flow dynamics, the air condition is crucial for the creation of fluid rotation at the interface of water.

Throughout this article, the air-water interface behaviors are examined based on the basic fluid mechanics view point. This will provide clear physical interpretations of free-surface dynamics, which hopefully leads to improvement of models.

1. Introduction

Traditionally, water-wave motions are formulated with the assumption of irrotational flows of incompressible, homogenous, and inviscid fluids. The formulated problems as such have the following characteristics:

1) the governing equation is the Laplace equation of the velocity potential which is a linear partial-differential equation,
2) the location of the free-surface boundary is not known *a priori*, hence two independent equations are necessary to describe the free-surface boundary conditions, and
3) these boundary conditions at the free surface are nonlinear, which make exact analytical solutions formidable.

The nonlinearity of the problem is associated with wave steepness, which is proportional to the ratio of wave amplitude to wave length. Since wave breaking prevents wave steepness from exceeding a certain limit, the nonlinear effects associated with surface-wave problems are weak and limited. Taking this physical limitation to be an advantage, solutions to water-wave problems are often approximated in the form of asymptotic expansions. Such an approach accurately yields the solutions for nonlinear waves as well as the predictions of long-time propagation behaviors of wave trains and wave packets.

Due to advances in computational capabilities with computers, much progress has been made in a variety of water-wave problems in recent years. As long as irrotational and two-dimensional flows of an incompressible inviscid fluid are assumed for water-wave motions, fully nonlinear motions can be numerically predicted without approximations. For example, using the boundary-integral method, overturning wave-breaking processes were modeled by many researchers, e.g., Longuet-Higgins and Cokelet (1976). Numerical simulation efforts for fully nonlinear water-wave problems are now focused on three-dimensional motions and involvement of vortical flows. An attempt was made to simulate a three-dimensional overturning motion of a breaking wave by Xü (1992), and numerical three-dimensional wave runup onto a beach was demonstrated by Liu *et al.* (1991). Vortical flows were incorporated with the irrotational base flows by Nadaoka *et al.* (1991).

Almost all of the theoretical and numerical developments are based on the condition of constant pressure at the water-surface boundary. Even when some modifications associated with the surface boundary layer were made, the water surface is usually modeled by the free-surface boundary conditions, i.e., no stress exerted on the water surface. Basically, the effects of air on the water surface are not considered, neglecting the presence of air above the water surface altogether. On the other hand, the dynamics of air flows are essential for the predictions associated with wind-wave growth and air-sea gas exchange. A question can be addressed as to which circumstances the presence of air becomes important for the determination of water-wave dynamics. This type of question is particularly important since our predictions of water-wave problems become so refined that the negligence of the air effects can cause significant errors in comparison with the higher-order effects in the traditional water-wave problems.

What I attempt to present here are careful and systematic analyses of the free-surface dynamics. As discussed, the traditional water-wave problems are based on the assumption of constant pressure along the surface, or when viscous effects are included, it is assumed that stresses exerted on the surface are nil. In both cases, the presence of air is neglected because the air density is much smaller than that of water so that the air conditions are dynamically unimportant for the water flows beneath the air. But this is an approximation. In a more rigorous description, the water surface should be viewed as the air-water two-layer-fluid interface. All three surface models (1. constant-pressure surface; 2. stress-free surface; 3. two-layer-fluid interface) will be discussed in this article. It is noted, however, that the discussion of a more fundamental model is excluded, such as the treatment of the surface as a phase transition region between gas and liquid, instead of the water surface being a plane.

In Sec. 2, some mathematical identities and manipulations are reviewed, which will be used for the rest of the article. An emphasis is made on the mathematical presentation of a curved surface, which is essential to describe the air-water interface. In Sec. 3, selected topics of fundamental fluid mechanics are reviewed. The review is not intended to be complete but to remind the readers of detailed physical interpretations associated with the theories of fluid mechanics. The free-surface boundary conditions are derived rigorously in Sec. 4 by considering the interface to be a material surface at which fluid properties have a discontinuous jump. The fundamental conservation laws are applied to the discontinuous interface. Surface-tension effects including the effects of surfactant are also discussed. In Sec. 5, characteristics and effects of constant-pressure water-surface conditions are identified. The stress-free surface conditions, effects of air, and surface-tension effects are also discussed. Vorticity conditions and vortical-flow behaviors associated with the dynamics of the water surface are presented in Sec. 6.

2. Mathematics Preliminaries

Some of the mathematical notations used here are defined in this section. Throughout this article, unless stated otherwise, Cartesian coordinates (x, y, z) are used. The y-axis is taken in the direction opposite the gravity direction, the x- and z-axes point horizontally to form a right-hand system. This choice of coordinates gives an advantage when we consider a two-dimensional problem in a vertical plane, deleting the z-axis yields a convenient (x, y) coordinate system. Also, unless otherwise stated, a "two-dimensional" problem means that all the fluid properties and flow characteristics are uniform in the z-direction.

All the scalar quantities will be expressed with plain symbols, e.g., V (volume), ρ (density), T (temperature). Vector quantities are expressed with a bold-faced symbol or with a letter subscript, e.g., $\mathbf{u}$ or u_j (velocity), $\mathbf{x}$ or x_j (position), $\boldsymbol{\omega}$ or ω_j (vorticity). Other higher-order tensors are represented with the letter subscripts, e.g., τ_{ij} (stress tensor), e_{ij} (rate-of-strain tensor), δ_{ij} (Kronecker delta), and ε_{ijk} (alternating unit tensor). The standard index notation is often used for mathematical manipulations. An index subscript j refers to $(j = 1, 2, 3)$ in the three-dimensional space, e.g., $\mathbf{x}_j = (x_1, x_2, x_3)$. A repeated subscript of index implies summation over all values the subscript can take, e.g., $\mathbf{u} \cdot \mathbf{u} = u_j u_j = (u_1 u_1 + u_2 u_2 + u_3 u_3)$. The magnitude of a vector $\mathbf{a}$ is $a = |\mathbf{a}| = \sqrt{\mathbf{a} \cdot \mathbf{a}}$.

Note that if a vector product of two finite vectors vanishes, i.e., $\mathbf{a} \times \mathbf{b} = 0$, then $\mathbf{a}$ and $\mathbf{b}$ must be parallel to each other. On the other hand, if a scalar product of two finite vectors vanishes, i.e., $\mathbf{a} \cdot \mathbf{b} = 0$, then $\mathbf{a}$ and $\mathbf{b}$ must be perpendicular to each other. The symbol $\circ$ represents a dyadic product, i.e., $\mathbf{a} \circ \mathbf{b} = a_i b_j$.

Any tensor of the second order can be written as a sum of symmetric (e.g., $B_{ij} = B_{ji}$) and skew-symmetric (e.g., $C_{ij} = -C_{ji}$) tensors, i.e., $A_{ij} = B_{ij} + C_{ij} = \frac{1}{2}(A_{ij} + A_{ji}) + \frac{1}{2}(A_{ij} - A_{ji})$. A product of a symmetric tensor and a skew-symmetric tensor vanishes identically. Suppose B_{ij} is symmetric and C_{ij} is skew-symmetric. Then, $B_{ij}C_{ij} = B_{ji}C_{ij} = -B_{ji}C_{ji} = -B_{ij}C_{ij}$. Therefore, $B_{ij}C_{ij} = 0$. For a three-dimensional space, a symmetric tensor has six independent entries, while a skew-symmetric tensor has three independent entries. Hence, the three independent entries of a skew-symmetric tensor C_{jk}

are equivalent to the three components of the dual vector $\mathbf{a}$:

$$\mathbf{a} = \varepsilon_{ijk} C_{jk} \ . \tag{2.1}$$

Kronecker delta is defined as $\delta_{ij} = 1$ if $i = j; \delta_{ij} = 0$ if $i \neq j$. The alternating unit tensor, ε_{ijk}, is defined as $\varepsilon_{ijk} = 1$ if ijk is a cyclic order, i.e., 123, 231, or 312; $\varepsilon_{ijk} = -1$ if ijk is a noncyclic order, i.e., 321, 213, or 132; $\varepsilon_{ijk} = 0$ if any two indices are alike, e.g., 121, 331, or 333. Note that $\varepsilon_{ijk} = \varepsilon_{jki} = \varepsilon_{kij}$; $\varepsilon_{ijk} = -\varepsilon_{jik}$; $\varepsilon_{ijk} = -\varepsilon_{ikj}$; $\varepsilon_{ijk}\varepsilon_{ilm} = \delta_{jl}\delta_{km} - \delta_{jm}\delta_{kl}$.

Gradient is expressed with the del operator, $\nabla(\bullet)$, and divergence is expressed by $\nabla \cdot (\bullet)$, and curl with $\nabla \times (\bullet)$.

A general form of integral transformation from volumetric integral to surface integral is

$$\int_D \nabla \cdot F dV = \int_S F \cdot \mathbf{n} \, ds \ , \tag{2.2}$$

where F is any scalar, vector, or tensor function but continuously differentiable in a domain D, and S denotes the surface surrounding the domain D, and $\mathbf{n}$ is the unit vector normal to the surface S pointing outward from the domain D. Hence, the divergence operator $\nabla \cdot (\bullet)$ can be physically interpreted as a net effect of F acting on an infinitesimal volume, i.e.,

$$\nabla \cdot F \equiv \underset{\mathbf{v} \to 0}{Lim} \frac{1}{V} \int_S F \cdot \mathbf{n} \, ds \, . \tag{2.3}$$

If F is a vector, by replacing F by $\mathbf{a}$,

$$\int_D \nabla \cdot \mathbf{a} \, dV = \int_S \mathbf{a} \cdot \mathbf{n} \, ds \ , \tag{2.4}$$

and if F is a tensor, by replacing F by A_{ij},

$$\int_D \frac{\partial A_{ij}}{\partial x_j} dV = \int_S A_{ij} n_j ds \ , \tag{2.5}$$

also, if $F = \varepsilon_{ijk} a_j$, then

$$\int_D \nabla \times \mathbf{a} \, dV = \int_S \mathbf{n} \times \mathbf{a} \, ds \ . \tag{2.6}$$

Other useful integral transformations are the Stokes theorem:

$$\int_s (\nabla \times \mathbf{a}) \cdot \mathbf{n}\, ds = \int_c \mathbf{a} \cdot d\mathbf{x}\,, \tag{2.7}$$

where c is the boundary of the surface S, and the transport theorem:

$$\frac{D}{Dt}\int_D F dV = \int_D \left(\frac{DF}{Dt} + F\nabla \cdot \mathbf{u}\right) dV = \frac{\partial}{\partial t}\int_D F dV + \int_S F\mathbf{u} \cdot \mathbf{n}\, ds\,. \tag{2.8}$$

In (2.8), F is a scalar or vector function of $\mathbf{x}$ and t, D/Dt is the material derivative ($D/Dt = \partial/\partial t + \mathbf{u} \cdot \nabla$), and $\mathbf{u}$ is the fluid particle velocity. The transport theorem represents that the rate of change of F over a material volume D is equal to the rate of change of F over the fixed volume which instantaneously coincides with D plus the flux of F outward from the fixed domain.

In order to describe a surface (e.g., the air-water interface), the following notations and methods are used. Suppose the location of a surface is expressed by a function F as shown in Fig. 1.

$$F(x, y, z, t) = 0\,. \tag{2.9}$$

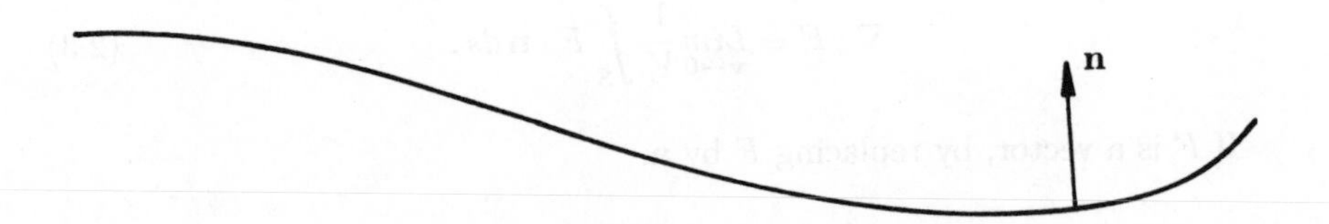

Fig. 1. The position of a surface, $F(x, y, z, t) = 0$. When the surface is concave in the direction of the unit vector normal to the surface, the curvature tensor is positive and the radius of the curvature is negative, i.e. $\kappa_{ij} > 0, R < 0$.

Let $\mathbf{x}$ be a position vector from an arbitrary fixed origin to a point on the surface $F = 0$. Since the increment $d\mathbf{x}$ lies in the plane tangent to the surface at the location $\mathbf{x}$, and the value of the function F is constant on the surface at a given instant of time, then by definition,

$$dF = d\mathbf{x} \cdot \nabla F = 0\,, \tag{2.10}$$

i.e., $d\mathbf{x}$ and ∇F are perpendicular to each other. Therefore, ∇F is pointing normal to the surface, and a unit normal vector on the surface is conveniently calculated by

$$\mathbf{n} = \frac{\nabla F}{\sqrt{\nabla F \cdot \nabla F}} = \frac{\nabla F}{n} \, , \tag{2.11}$$

where $n = |\nabla F|$.

The variation $d\mathbf{n}$ of the unit normal to a surface plays an important role in the theory of curvature. There are two different ways to treat the curvature of a surface: 1) use the expression of a surface in parametric form (e.g., Schwartz, Green and Rutledge, 1960) and 2) use the expression in a single equation (2.9) (e.g., Gibbs and Wilson, 1929). We adopt the latter approach here.

Observe that

$$d\mathbf{n} = d\left(\frac{\nabla F}{n}\right) = \frac{1}{n} d\mathbf{x} \cdot \nabla(\nabla F) - \frac{1}{n^2}(d\mathbf{x} \cdot \nabla n)\nabla F \, . \tag{2.12}$$

The variation $d\mathbf{n}$ of the unit normal vector is perpendicular to $\mathbf{n}$ because $\mathbf{n}$ is a unit vector; the last term in (2.12) points parallel to $\mathbf{n}$, hence it does not contribute to the evaluation of $d\mathbf{n}$. To show this more explicitly, observe that

$$d\mathbf{n} \cdot (\delta_{ij} - n_i n_j) = d\mathbf{n} \, , \tag{2.13}$$

and

$$\nabla F \cdot (\delta_{ij} - n_i n_j) = 0 \, , \tag{2.14}$$

where δ_{ij} is the Kronecker delta function and $n_i = \mathbf{n}$. Note that $(\delta_{ij} - n_i n_j)$ is an idemfactor for all vectors perpendicular to $\mathbf{n}$ and an annihilator for vectors parallel to $\mathbf{n}$. Therefore, (2.12) can be expressed by

$$d\mathbf{n} = d\mathbf{n} \cdot (\delta_{ij} - n_i n_j) = \frac{1}{n} d\mathbf{x} \cdot \nabla(\nabla F) \cdot (\delta_{ij} - n_i n_j) \, . \tag{2.15}$$

The last term of (2.12) vanishes because of (2.14). Since $d\mathbf{x} = d\mathbf{x} \cdot (\delta_{ij} - n_i n_j)$, (2.15) can be expressed by

$$d\mathbf{n} = d\mathbf{x} \cdot \left\{ \frac{(\delta_{ik} - n_i n_k) \cdot \dfrac{\partial^2 F}{\partial x_k \partial x_l} \cdot (\delta_{lj} - n_l n_j)}{n} \right\} = -d\mathbf{x} \cdot \kappa_{ij} \, . \tag{2.16}$$

Evidently, the quantity in the parenthesis is a symmetric tensor, which can be represented by κ_{ij} — the two-dimensional curvature tensor (dyadic) of the surface which has no component normal to the surface. Alternatively, κ_{ij} can be written as $\kappa_{ij} = -\nabla_\lambda \mathbf{n}$ where ∇_λ represents the gradient in the directions tangential to the surface. The surface curvature tensor is positive-definite when the surface is concave in the direction of the unit surface normal $\mathbf{n}$ (see Fig. 1). Since κ_{ij} is symmetric, it can be diagonalized by rotating the orthogonal axes to the principal directions to yield

$$\kappa_{ij} = \begin{bmatrix} \kappa_{11} & 0 \\ 0 & \kappa_{22} \end{bmatrix} = \begin{bmatrix} a_1\mathbf{e}_1\mathbf{e}_1 & 0 \\ 0 & a_2\mathbf{e}_2\mathbf{e}_2 \end{bmatrix}, \tag{2.17}$$

where a_1 and a_2 are scalars and $\mathbf{e}_1$ and $\mathbf{e}_2$ are two orthogonal unit vectors pointing in the principal direction in the plane tangent to the surface. Each component, described as a_1 and a_2 in (2.17) is termed the principal curvature. The average of the principal curvatures are termed the mean curvature, i.e., $\frac{1}{2}(a_1 + a_2)$. The principal curvature can be expressed with the corresponding principal radius of curvature, i.e., $a_1 = -1/R_1; a_2 = -1/R_2$. This correspondence can be demonstrated in Fig. 2, as $\Delta\mathbf{n}/\Delta s = 1/R$.

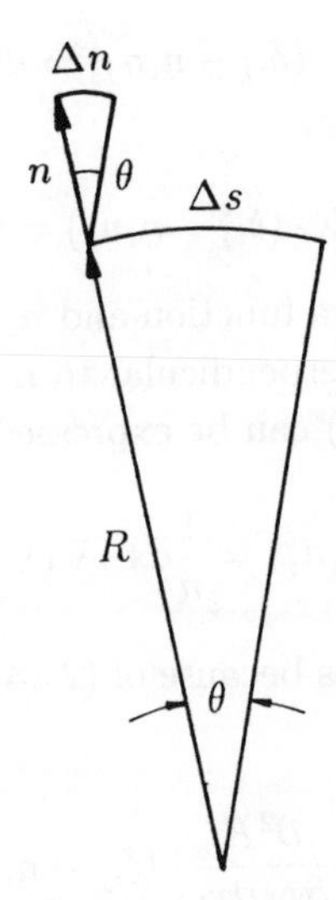

Fig. 2. A schematic relation between the curvature and radius of a surface.

To prove that the value of the mean curvature is invariant to the orientation of two orthogonal axes on the surface, first consider a given point Q on a surface $F = 0$ and $\mathbf{n}$ is the unit normal vector to the surface at Q. Let us take a plane surface q which passes through $\mathbf{n}$ at Q (see Fig. 3). Consider the curvature of the curve where the surface $F = 0$ and the plane q intersect. Let $\hat{\mathbf{n}}$ be the unit vector normal to that curve on the plane q. Note that the unit normal vector $\hat{\mathbf{n}}$ in the plane q, coincides with $\mathbf{n}$ at the point Q but it is not necessarily the case at other points unless the plane q intersects perpendicularly at every point with the surface $F = 0$.

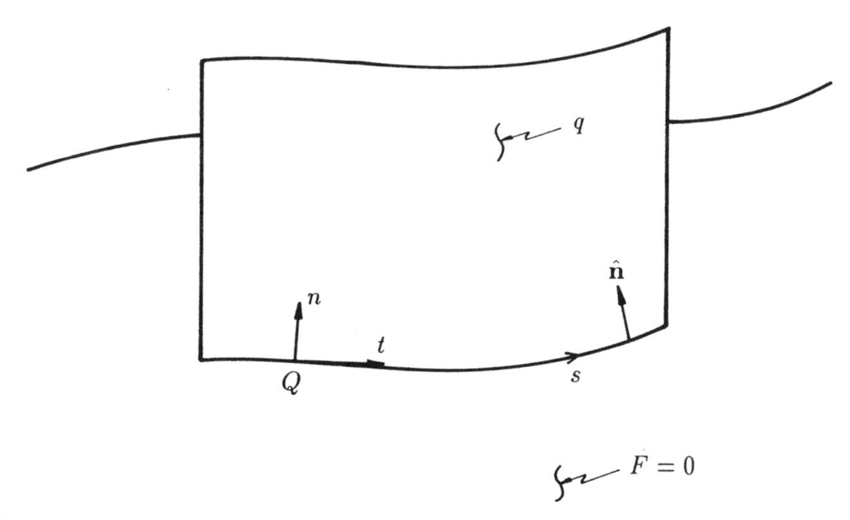

Fig. 3. A definition sketch for the proof of that the mean curvature is invariant to the orientation of two orthogonal axes on the surface.

The curvature of the intersecting curve, κ, is the limit of the ratio of the angle through which the tangent vector turns to the length of the arc; hence, the curvature of the intersection curve in the plane q is

$$\kappa = \left|\frac{d\mathbf{t}}{ds}\right| = \left|\frac{d^2\mathbf{x}}{ds^2}\right| , \tag{2.18}$$

where $\mathbf{t}$ is a unit vector tangent to the curve (see Fig. 3) and s is the length of arc. Since $\mathbf{t}$ and $\hat{\mathbf{n}}$ are unit vectors and perpendicular to each other, $|\frac{d\mathbf{t}}{ds}| = |\frac{d\hat{\mathbf{n}}}{ds}|$

and

$$\kappa = \left|\frac{d\mathbf{t}}{ds}\right| = -\frac{d\hat{\mathbf{n}}}{ds} \cdot \frac{d\mathbf{x}}{ds} = -\frac{d\hat{\mathbf{n}} \cdot d\mathbf{x}}{ds^2} \,. \tag{2.19}$$

The minus sign in (2.19) makes the curvature κ consistent with the sign convention for the surface curvature tensor κ_{ij}; κ is positive when the line is concave in the direction of $\hat{\mathbf{n}}$. Since $\mathbf{n} \perp d\mathbf{x}$ and $\hat{\mathbf{n}} \perp d\mathbf{x}$,

$$d(\mathbf{n} \cdot d\mathbf{x}) = 0 = d\mathbf{n} \cdot d\mathbf{x} + \mathbf{n} \cdot d^2\mathbf{x} \,,$$

and

$$d(\hat{\mathbf{n}} \cdot d\mathbf{x}) = 0 = d\hat{\mathbf{n}} \cdot d\mathbf{x} + \hat{\mathbf{n}} \cdot d^2\mathbf{x} \,. \tag{2.20}$$

Hence,

$$d\hat{\mathbf{n}} \cdot d\mathbf{x} + \hat{\mathbf{n}} \cdot d^2\mathbf{x} = d\mathbf{n} \cdot d\mathbf{x} + \mathbf{n} \cdot d^2\mathbf{x} \,, \tag{2.21}$$

and since $\mathbf{n}$ and $\hat{\mathbf{n}}$ are equal at Q,

$$d\hat{\mathbf{n}} \cdot d\mathbf{x} = d\mathbf{n} \cdot d\mathbf{x} \tag{2.22}$$

at Q. Therefore, (2.19) becomes

$$\kappa = -\frac{d\mathbf{n} \cdot d\mathbf{x}}{ds^2} = \frac{d\mathbf{x} \cdot \kappa_{ij} \cdot d\mathbf{x}}{d\mathbf{x} \cdot d\mathbf{x}} = a_1 \frac{(\mathbf{e}_1 \cdot d\mathbf{x})^2}{d\mathbf{x} \cdot d\mathbf{x}} + a_2 \frac{(\mathbf{e}_2 \cdot d\mathbf{x})^2}{d\mathbf{x} \cdot d\mathbf{x}} \,, \tag{2.23}$$

by using (2.17). Equation (2.23) can be also written as

$$\kappa = a_1 \cos^2 \alpha_1 + a_2 \cos^2 \alpha_2 \,, \tag{2.24}$$

where α_1 is the angle formed between $\mathbf{e}_1$ and $d\mathbf{x}$, and α_2 is between $\mathbf{e}_2$ and $d\mathbf{x}$.

Now, consider another plane q' that passes through $\mathbf{n}$ at Q but intersects perpendicularly to the plane q. On the new plane, the curvature κ' of the curve formed by the intersection of the surface (defined by $F = 0$) and the plane q' is

$$\kappa' = a_1 \cos^2 \beta_1 + a_2 \cos^2 \beta_2 = a_1 \sin^2 \alpha_1 + a_2 \sin^2 \alpha_2 \ , \tag{2.25}$$

where β_1 is the angle formed between $\mathbf{e}_1$ and $d\mathbf{x}$ on the plane q', and β_2 is between $\mathbf{e}_2$ and $d\mathbf{x}$. The mean curvature of κ and κ' is therefore,

$$\begin{aligned} \frac{\kappa + \kappa'}{2} &= \frac{a_1(\cos^2 \alpha_1 + \sin^2 \alpha_1) + a_2(\cos^2 \alpha_2 + \sin^2 \alpha_2)}{2} \\ &= \frac{a_1 + a_2}{2} \ . \end{aligned} \tag{2.26}$$

This proves that the mean value of the curvature in two normal sections at right angles to one another is constant and independent of the actual orientation of those sections as long as they are orthogonal.

Some other important properties of the unit normal vector on a surface are 1) $\nabla \mathbf{n} = \nabla_\lambda \mathbf{n}$ is symmetric; hence, 2) $\nabla \times \mathbf{n} = 0$, and 3) $\mathbf{n} \cdot \nabla \mathbf{n} = 0$.

3. Background

Prior to the discussions for the governing equations to describe the dynamics of fluid flows, we first define a few kinematic descriptions. First, the streamlines are defined as the curves tangent to velocity vectors at every point at a given instant. Mathematically, the streamlines can be defined by $\mathbf{u} \times d\mathbf{x} = 0$, where $\mathbf{u}$ is the fluid-velocity vector and $\mathbf{x}$ is the position vector of each streamline. A surface of streamlines connected by a closed curve identifies a streamtube. The strength of a streamtube is represented by the volume flux through the surface defined by the closed curve, and the strength is constant along a streamtube if the fluid is incompressible. A vortexline is an integral curve where its direction is tangent to the vorticity vector at every point at a given instant of time, and mathematically defined by $\boldsymbol{\omega} \times d\mathbf{x} = 0$, where $\boldsymbol{\omega}$ is the vorticity vector and $\mathbf{x}$ is the position vector of each vortexline. A vortextube is formed as a bundle of vortexlines passing through a closed curve, and the strength of the vortextube is represented by the flow circulation (to be discussed later). The strength of a vortextube is constant at every cross section of the tube without any constraints; this is a consequence of the solenoidal property of vorticity vectors, i.e., $\nabla \cdot \boldsymbol{\omega} = 0$ by definition of vorticity $\boldsymbol{\omega} = \nabla \times \mathbf{u}$.

3.1. *Governing equations*

The governing equations for fluid motion are usually based on the conservation of mass and the conservation of linear momentum, i.e., Newton's second law. Assuming the body forces to be conservative, the equations can be written as,

$$\frac{D\rho}{Dt} + \rho\nabla\cdot\mathbf{u} = 0 , \tag{3.1}$$

and

$$\rho\frac{D\mathbf{u}}{Dt} = \nabla\cdot\tau_{ij} + \rho\nabla E , \tag{3.2}$$

respectively, where ρ is the fluid density, t is the time, τ_{ij} is the stress tensor, E is the body-force scalar potential (e.g., $E = -gy$), the operator D/Dt is the material derivative ($D/Dt = \partial/\partial t + \mathbf{u}\cdot\nabla$), and ∇ is the del operator.

For Newtonian fluids, the stress tensor τ_{ij} is linearly proportional to the gradient of velocity vector, and we find

$$\tau_{ij} = -p\delta_{ij}+S_{ij};\ S_{ij} = 2\mu\left(e_{ij} - \frac{1}{3}(\nabla\cdot\mathbf{u})\delta_{ij}\right);\ e_{ij} = \frac{1}{2}(\nabla u+(\nabla u)^T) , \tag{3.3}$$

where p is the pressure, defined as the isotropic part of the stress tensor, i.e., $p = 1/3$ Trace $[\tau_{ij}]$, μ is the dynamic viscosity of the fluid (the proportionality constant between the stress tensor and the velocity gradient), S_{ij} is the deviatoric stress tensor, e_{ij} is the rate-of-strain tensor, and the superscript T denotes the transpose.

If we consider the fluid to be incompressible, meaning that a fluid density is invariant as we follow a fluid parcel for an infinitesimal duration, i.e., $D\rho/Dt = 0$, then from (3.1), the divergence of the velocity vector must vanish everywhere,

$$\nabla\cdot\mathbf{u} = 0 , \tag{3.4}$$

and the stress tensor τ_{ij} is simplified as $\tau_{ij} = -p\delta_{ij} + 2\mu e_{ij}$. The conservation of linear momentum becomes

$$\rho\frac{D\mathbf{u}}{Dt} = -\nabla p + \nabla\cdot(\mu(\nabla\mathbf{u} + (\nabla\mathbf{u})^T)) + \rho\nabla E . \tag{3.5}$$

When the viscosity μ is uniform over the fluid, then using (3.4), (3.5) becomes

$$\rho\frac{D\mathbf{u}}{Dt} = -\nabla p + \mu\nabla^2\mathbf{u} + \rho\mathbf{g} , \tag{3.6}$$

which is the familiar form of the Navier-Stokes equation of linear momentum for an incompressible fluid.

The problem described by (3.4) and (3.6) is to solve for four dependent variables (i.e., $\mathbf{u} = (u, v, w)$ and p) with given boundary conditions. The basic requirements at the boundaries are that both velocity and stress remain continuous across boundaries. If $\mathbf{U}(\mathbf{x}, t)$ represents the velocity of the boundary, then continuity of velocity requires

$$\mathbf{u}(\mathbf{x}, t) = \mathbf{U}(\mathbf{x}, t) \tag{3.7}$$

on the boundary. In the absence of surface-tension, we require that the stress on each of the two surface elements on each side of the boundary be equal, i.e.,

$$(\tau_{ij} \cdot \mathbf{n})_1 + (\tau_{ij} \cdot \mathbf{n})_2 = 0 \tag{3.8}$$

on the boundary, where the unit normal vector $\mathbf{n}$ is pointing outward from each medium.

At a fluid-solid boundary, continuity of the tangential velocity must hold; this boundary condition is known as the no-slip condition. In this case, the stress in the solid is not of interest and we do not usually use stress boundary conditions. On the other hand, for deformable solid boundaries, the stress condition must be applied to determine the position of the boundary.

At a liquid-gas boundary, the density ρ and viscosity μ vary appreciably across the interface. If the inertia effects are comparable in liquid and gas, the pressure variations and frictional stresses in gas are much smaller than those in liquid. Hence, as an approximation, the stresses in the gas are considered to be negligibly small so that $(\tau_{ij}) \approx p_0 \delta_{ij}$, where p_0 is constant and δ_{ij} is the Kronecker delta function. If the unit normal vector is pointing outward from the liquid phase, then we require from (3.3) and (3.8) that, for a normal stress component at the interface,

$$p - 2\mu(\mathbf{n} \cdot e_{ij} \cdot \mathbf{n}) = p_0 \ , \tag{3.9}$$

and the tangential stress components

$$\mathbf{n} \times (e_{ij} \cdot \mathbf{n}) = 0 \ . \tag{3.10}$$

A boundary described by (3.9) and (3.10) is referred to as the free surface. Note that, contrary to a solid boundary, the condition on velocity is not useful at the free surface since we are ignoring any flow field in the gas. More detailed discussions on the boundary conditions at the liquid-gas interface will be made in Sec. 4.

3.2. *Mechanical energy equation*

At each point in a flow field, a fluid particle path is tangent to the instantaneous velocity vector and (3.2) represents the balance in force per unit volume. Hence, the rate of change in its mechanical energy, i.e., power, can be derived by taking the scalar product of the linear momentum equation (3.2) with the velocity vector, $\mathbf{u}$:

$$\mathbf{u} \cdot \left(\rho \frac{D\mathbf{u}}{Dt} \right) = \mathbf{u} \cdot (\nabla \cdot \tau_{ij} + \rho \nabla E) \,. \tag{3.11}$$

Since the conservative body force per unit mass $\nabla E (= \mathbf{g})$ is independent of time, the last term in (3.11) can be written as

$$\mathbf{u} \cdot \rho \nabla E = \rho \frac{DE}{Dt} \,. \tag{3.12}$$

Furthermore,

$$\begin{aligned} \rho \mathbf{u} \cdot \frac{D\mathbf{u}}{Dt} &= \rho u_i \left\{ \frac{\partial u_i}{\partial t} + u_j \frac{\partial u_i}{\partial x_j} \right\} \\ &= \rho \frac{\partial}{\partial t} \left(\frac{u_i u_i}{2} \right) + \rho u_j \frac{\partial}{\partial x_j} \left(\frac{u_i u_i}{2} \right) \\ &= \rho \frac{D}{Dt} \left(\frac{|\mathbf{u}|^2}{2} \right) \end{aligned} \tag{3.13}$$

and

$$\mathbf{u} \cdot (\nabla \cdot \tau_{ij}) = \nabla \cdot (\mathbf{u} \tau_{ij}) - \tau_{ij} : \nabla \mathbf{u} \,, \tag{3.14}$$

where a repeated subscript index implies summation over all values the subscript can take, and the semi-colon : denotes the inner product of two second-order tensors that yields a scalar quantity. Using (3.12), (3.13), and (3.14), (3.11) becomes, without approximation,

$$\rho \frac{D}{Dt} \left(\frac{|\mathbf{u}|^2}{2} - E \right) = \nabla \cdot (\mathbf{u} \tau_{ij}) - \tau_{ij} : \nabla \mathbf{u} \,. \tag{3.15}$$

Integrating an arbitrary fluid domain D yields

$$\frac{D}{Dt}\int_D \rho\left(\frac{|\mathbf{u}|^2}{2} - E\right) dV = \int_S (\mathbf{u}\tau_{ij}) \cdot \mathbf{n}\, ds - \int_D \tau_{ij} : \nabla \mathbf{u}\, dV\ . \qquad (3.16)$$

The term on the left-hand side represents the time rate of change in mechanical energy (sum of kinetic and potential energies) as following the lump of fluid, the first term on the right-hand side represents the rate of work done by the surface force exerting on the lump of fluid, and the last term represents the energy loss to heat. For incompressible fluids, the stress tensor is $\tau_{ij} = -p\delta_{ij} + 2\mu e_{ij}$; hence, the energy loss term can be expressed as

$$\begin{aligned}\int_D \tau_{ij} : \nabla \mathbf{u}\, d\mathbf{V} &= \int_D \left(-p\delta_{ij}\frac{\partial u_i}{\partial x_j} + 2\mu e_{ij}\frac{\partial u_i}{\partial x_j}\right) dV \\ &= \int_D 2\mu(e_{ij}e_{ij} + e_{ij}b_{ij})dV \qquad (3.17) \\ &= \int_D 2\mu|e_{ij}|^2 dV \geq 0\ .\end{aligned}$$

Note that $\nabla\mathbf{u} = \frac{1}{2}(\nabla\mathbf{u} + (\nabla\mathbf{u})^T) + \frac{1}{2}(\nabla\mathbf{u} - (\nabla\mathbf{u})^T) = e_{ij} + b_{ij}$, in which b_{ij} is the rotation tensor. In the first line, the first integrand on the right-hand side vanishes because of the incompressibility of the fluid, $\nabla \cdot \mathbf{u} = 0$; in the second line, the product of e_{ij} and b_{ij} vanishes because e_{ij} is a symmetrical tensor and b_{ij} is a skew-symmetrical tensor (see Sec. 2). The integrand in the last line, $2\mu|e_{ij}|^2$ is a positive definite and is called the energy dissipation function. Note that for an inviscid fluid ($\mu = 0$), the energy dissipation function vanishes and the mechanical energy is conserved. For a viscous fluid, mechanical energy is always dissipated unless the rate-of-strain tensor, e_{ij}, vanishes identically everywhere (no angular deformation). Even when fluid motions are rotational ($b_{ij} \neq 0$), flow conditions with no energy dissipation could occur if there is no fluid angular deformation. On the other hand, if $e_{ij} \neq 0$, mechanical energy must dissipate even when the (viscous) flow is irrotational. The irrotationality by itself does not provide any assurance about mechanical energy dissipation.

3.3. *Bernoulli equation*

Assuming a steady inviscid-fluid flow with conservative body force but allowing the flow to be rotational $b_{ij} \neq 0$ with density stratification, $\rho = \rho(\mathbf{x}, t)$, the linear-momentum equation can be written as

$$\rho(\mathbf{u} \cdot \nabla)\mathbf{u} = -\nabla p + \rho\nabla E\ . \qquad (3.18)$$

Since

$$(\mathbf{u}\cdot\nabla)\mathbf{u} = \nabla\left(\frac{|\mathbf{u}|^2}{2}\right) - \mathbf{u}\times(\nabla\times\mathbf{u}) = \nabla\left(\frac{|\mathbf{u}|^2}{2}\right) - \mathbf{u}\times\boldsymbol{\omega}\,, \tag{3.19}$$

in which $\boldsymbol{\omega}$ is the vorticity, $\boldsymbol{\omega} = \nabla\times\mathbf{u}$, (3.18) can be written as

$$\nabla\left(\frac{|\mathbf{u}|^2}{2}\right) - \mathbf{u}\times\boldsymbol{\omega} = -\frac{1}{\rho}\nabla p + \nabla E\,. \tag{3.20}$$

Since $\rho = \rho(\mathbf{x})$, so that $\frac{1}{\rho}\nabla p = \nabla(\frac{p}{\rho}) + \frac{p}{\rho^2}\nabla\rho$, hence, (3.20) can be expressed as

$$\mathbf{u}\times\boldsymbol{\omega} = \nabla\left(\frac{|\mathbf{u}|^2}{2} + \frac{p}{\rho} - E\right) + \frac{p}{\rho^2}\nabla\rho\,. \tag{3.21}$$

Taking the scalar product with the velocity vector $\mathbf{u}$ yields

$$\mathbf{u}\cdot(\mathbf{u}\times\boldsymbol{\omega}) - \frac{p}{\rho^2}\mathbf{u}\cdot\nabla\rho = \mathbf{u}\cdot\nabla\left(\frac{|\mathbf{u}|^2}{2} + \frac{p}{\rho} - E\right)\,. \tag{3.22}$$

Note that $\mathbf{u}$ is perpendicular to $(\mathbf{u}\times\boldsymbol{\omega})$ by definition; hence, the first term in (3.22) necessarily vanishes. Since $\frac{D\rho}{Dt} = 0$ for an incompressible fluid, which is $\mathbf{u}\cdot\nabla\rho = 0$ for steady flow, the second term in (3.22) also vanishes. Therefore, (3.22) is reduced to

$$\mathbf{u}\cdot\nabla\left(\frac{|\mathbf{u}|^2}{2} + \frac{p}{\rho} - E\right) = 0\,. \tag{3.23}$$

Since $E = -gy$, this indicates that $\nabla\left(\frac{|\mathbf{u}|^2}{2} + \frac{p}{\rho} + gy\right)$ is orthogonal to $\mathbf{u}$ or $\frac{|\mathbf{u}|^2}{2} + \frac{p}{\rho} + gy =$ constant along the direction of $\mathbf{u}$, i.e., along a streamline, although the value of the constant may vary on a different streamline.

Now, assuming a homogeneous fluid, i.e., $\rho =$ constant, and taking a scalar product on (3.21) with the vorticity $\boldsymbol{\omega}$ instead of $\mathbf{u}$, then (3.21) becomes

$$\boldsymbol{\omega}\cdot(\mathbf{u}\times\boldsymbol{\omega}) = \boldsymbol{\omega}\cdot\nabla\left(\frac{|\mathbf{u}|^2}{2} + \frac{p}{\rho} - E\right) = 0\,, \tag{3.24}$$

since $\boldsymbol{\omega}$ is perpendicular to $(\mathbf{u}\times\boldsymbol{\omega})$ by definition. In this case, $\nabla\left(\frac{|\mathbf{u}|^2}{2} + \frac{p}{\rho} + gy\right)$ is orthogonal to $\boldsymbol{\omega}$ or $\frac{|\mathbf{u}|^2}{2} + \frac{p}{\rho} + gy =$ constant along the direction of $\boldsymbol{\omega}$, i.e., along a vortex line.

Under the assumption of steady inviscid flow of a homogeneous fluid, we find from (3.21) that, if $\mathbf{u} \times \boldsymbol{\omega} = 0$, $\frac{|\mathbf{u}|^2}{2} + \frac{p}{\rho} + gy =$ constant everywhere. This means that if the flow is irrotational, i.e., $\boldsymbol{\omega} = 0$, or if the flow is two-dimensional, then $\frac{|\mathbf{u}|^2}{2} + \frac{p}{\rho} + gy =$ constant everywhere.

Now, if $\mathbf{u} \times \boldsymbol{\omega} \neq 0$, from (3.23) and (3.24), $\mathbf{u} \cdot \nabla \left(\frac{|\mathbf{u}|^2}{2} + \frac{p}{\rho} - E \right) = \boldsymbol{\omega} \cdot \nabla \left(\frac{|\mathbf{u}|^2}{2} + \frac{p}{\rho} - E \right) = 0$; hence, $\frac{|\mathbf{u}|^2}{2} + \frac{p}{\rho} + gy =$ constant along a streamline as well as along a vortex line. A surface of $\frac{|\mathbf{u}|^2}{2} + \frac{p}{\rho} + gy =$ constant is called the Lamb surface.

3.4. *Euler's integral*

For irrotational motion, i.e., $\boldsymbol{\omega} = \nabla \times \mathbf{u} = 0$ everywhere, we can conveniently identify the velocity field by the gradient of a scalar function, viz., velocity potential ϕ:

$$\mathbf{u} = \nabla \phi \, , \tag{3.25}$$

which identically satisfies the irrotationality,

$$\nabla \times (\nabla \phi) \equiv 0 \, , \tag{3.26}$$

and the continuity for incompressible fluids can be represented by the Laplace equation:

$$\nabla \cdot \mathbf{u} = \nabla^2 \phi = 0 \, . \tag{3.27}$$

Now, consider incompressible and homogeneous fluid ($\rho =$ constant) where the fluid can be viscous, then the equation of motion (3.6) can be written as

$$\frac{\partial \mathbf{u}}{\partial t} + \nabla \left(\frac{|\mathbf{u}|^2}{2} \right) - \mathbf{u} \times \boldsymbol{\omega} = -\frac{1}{\rho} \nabla p + \frac{\mu}{\rho} \nabla^2 \mathbf{u} + \nabla E \, . \tag{3.28}$$

In terms of a velocity potential ϕ, (3.28) can be expressed as

$$\frac{\partial}{\partial t}(\nabla \phi) + \nabla \left(\frac{|\nabla \phi|^2}{2} \right) = -\frac{1}{p} \nabla p + \frac{\mu}{\rho} \nabla^2 (\nabla \phi) + \nabla E \, . \tag{3.29}$$

Therefore, for homogeneous fluids,

$$\nabla \left(\frac{\partial \phi}{\partial t} + \frac{|\nabla \phi|^2}{2} + \frac{p}{\rho} - \frac{\mu}{\rho} \nabla^2 \phi - E \right) = 0 \, . \tag{3.30}$$

Since the fluid is incompressible, i.e., $\nabla^2\phi = 0$, the viscous term in (3.30) vanishes even if the fluid viscosity is not zero. Integrating (3.30) spatially, we obtain

$$\frac{\partial \phi}{\partial t} + \frac{1}{2}|\nabla\phi|^2 + \frac{p}{\rho} + gy = f(t) , \tag{3.31}$$

where $f(t)$ is an arbitrary function of time t. To distinguish this equation from the Bernoulli equation that is valid along a streamline as described in Subsec. 3.3, (3.31) is often referred to as Euler's integral.

Since our interest is to solve for the velocity field but not the velocity potential itself, we can redefine the velocity potential as

$$\phi' = \phi - \int^t f(t)dt , \tag{3.32}$$

without causing difficulty in retrieving $\mathbf{u} = \nabla\phi$. Euler's integral (3.31) is reduced to

$$\frac{\partial \phi'}{\partial t} + \frac{1}{2}|\nabla\phi'|^2 + \frac{p}{\rho} + gy = 0 , \tag{3.33}$$

which is valid everywhere in the fluid domain as long as the flow is irrotational and incompressible and the fluid is homogeneous but the fluid viscosity does not need to be zero.

This form is used in the theory of water waves to describe a boundary condition connecting surface pressure to water velocity at the air-water interface. If we assume pressure along the water surface to be constant (say, zero), the normal component of the dynamic boundary condition at the free surface $y = \eta(x, z, t)$ can be written as (neglecting the surface-tension effects)

$$\frac{\partial \phi}{\partial t} + \frac{1}{2}|\nabla\phi|^2 + g\eta = 0 \quad \text{on} \quad y = \eta(x, z, t) . \tag{3.34}$$

3.5. *Vorticity equation*

Consider the Navier-Stokes equation of linear momentum for an incompressible fluid of the form

$$\frac{\partial \mathbf{u}}{\partial t} + \nabla\left(\frac{|\mathbf{u}|^2}{2}\right) - \mathbf{u} \times \boldsymbol{\omega} = -\frac{1}{\rho}\nabla p + \frac{1}{\rho}\nabla \cdot (2\mu e_{ij}) + \nabla E , \tag{3.35}$$

with allowing the variations in ρ and μ, but assuming that the body force is conservative. Since vorticity is $\boldsymbol{\omega} = \nabla \times \mathbf{u}$, the equation describing the vorticity field can be derived by taking the curl of (3.35). (Note that this is similar to how the mechanical-energy equation was derived by taking the scalar product of the Navier-Stokes equation with a velocity vector $\mathbf{u}$.) To do this, the following operations should be observed:

$$\nabla \times \left(\nabla \left(\frac{|\mathbf{u}|^2}{2} \right) - \mathbf{u} \times \boldsymbol{\omega} \right) = (\mathbf{u} \cdot \nabla)\boldsymbol{\omega} - (\boldsymbol{\omega} \cdot \nabla)\mathbf{u} \, . \tag{3.36}$$

Note that the incompressibility $\nabla \cdot \mathbf{u} = 0$ and the identity $\nabla \cdot \boldsymbol{\omega} = 0$ were used to derive (3.36). And

$$\nabla \times \left(-\frac{1}{\rho} \nabla p \right) = -\nabla \left(\frac{1}{\rho} \right) \times \nabla p \, , \tag{3.37}$$

because $\nabla \times \nabla p \equiv 0$ (since p is a scalar), and

$$\nabla \times \left(\frac{1}{\rho} \nabla \cdot (2\mu e_{ij}) \right) = \nabla \frac{1}{\rho} \times \nabla \cdot (2\mu e_{ij}) + \frac{1}{\rho} \nabla \times (\nabla \cdot (2\mu e_{ij})) \, , \tag{3.38}$$

and finally,

$$\nabla \times \nabla E \equiv 0 \, , \tag{3.39}$$

because E is a scalar quantity for a conservative body force.

Using (3.36), (3.37), (3.38), (3.39), the curl of (3.35) yields the vorticity equation in the form:

$$\frac{D\boldsymbol{\omega}}{Dt} = \underset{\text{(a)}}{(\boldsymbol{\omega} \cdot \nabla)\mathbf{u}} - \underset{\text{(b)}}{\nabla \left(\frac{1}{\rho} \right) \times \nabla p} + \underset{\text{(c)}}{\nabla \left(\frac{1}{\rho} \right) \times (\nabla \cdot S_{ij})} + \underset{\text{(d)}}{\frac{1}{\rho} \nabla \times (\nabla \cdot S_{ij})} \, , \tag{3.40}$$

in which $S_{ij} = 2\mu e_{ij}$ is the deviatoric stress tensor for incompressible fluids. The left-hand side of (3.40) represents the rate of change in vorticity $\boldsymbol{\omega}$ as following a fluid parcel, which is determined by the following factors:

(a) $(\boldsymbol{\omega} \cdot \nabla)\mathbf{u}$ represents the change of angular velocity due to the change in the moment of inertia. Note that $\nabla \mathbf{u}$ represents stretch-compression and rotation effects of a fluid element. Consider a fluid element of spherical shape and initial vorticity vector to be $\boldsymbol{\omega}_o$. If a fluid element is stretched, then, as demonstrated in Fig. 4, the vorticity in the stretched direction must increase based on the conservation of the angular momentum. The converse is true for compression of a fluid element. If a fluid element is bent, the initial vorticity vector is redistributed to the direction of its bending, since vortex lines move with the fluid (to be discussed later in Subsec. 3.6). Vortex stretching and bending effects are the major driving mechanisms for three-dimensional turbulence.

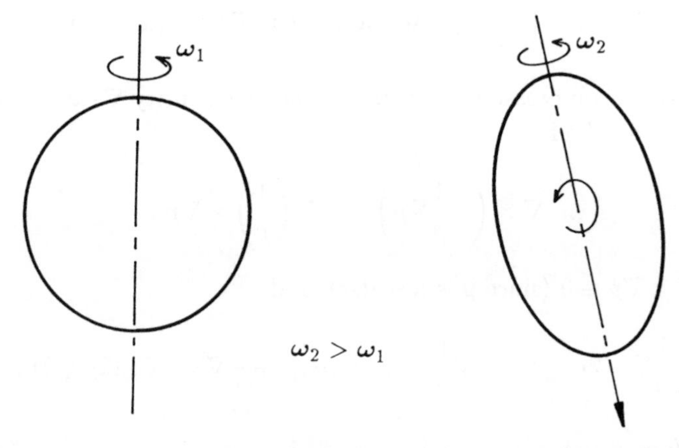

Fig. 4. Variation of the vorticity of a fluid parcel due to stretching and bending.

(b) $-\nabla\left(\frac{1}{\rho}\right) \times \nabla p \left(= \frac{1}{\rho^2}(\nabla\rho \times \nabla p)\right)$ represents the baroclinic torque, which can change the angular momentum. When the isobar and isopycnal surfaces are not parallel, i.e., $\nabla\rho \times \nabla p \neq 0$, the condition is termed baroclinic, whereas if the isobar and isopycnal surfaces are parallel, $\nabla\rho \times \nabla p = 0$, this condition is termed barotropic. Consider a fluid element of spherical shape as shown in Fig. 5. Suppose we have a homogeneous fluid, then the center of mass coincides with the location of the centroid of the fluid parcel. Since pressure force is always acting perpendicular to the surface and the lines of action must pass through the centroid, they cannot create any torque to the spherical (homogeneous) fluid element. On the other hand, when $\nabla\rho \neq 0$, the mass center is displaced from its centroid. Hence, if the resultant pressure force has a component in the direction perpendicular to the line connecting between the mass center and centroid, the fluid element is rotated by the torque created. This fluid rotation-creation mechanism is called baroclinic torque.

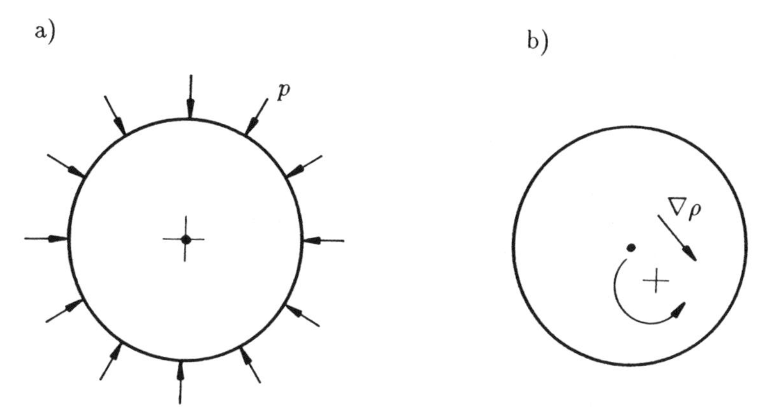

Fig. 5. Vorticity creation by baroclinic torque. The center of mass of a fluid parcel is indicated by +, and the centroid of the spherical volume where the lines of action of pressure force pass through is indicated by •. a) no mechanism to cause fluid rotation for a homogeneous fluid, b) with density variation present, fluid rotation can caused by baroclinic torque.

(c) $\nabla(\frac{1}{\rho}) \times (\nabla \cdot S_{ij})$ represents the viscous shear torque (Yeh, 1991). The term $\nabla \cdot S_{ij}$ represents the net viscous force acting on the surface of a fluid parcel (see (2.3) for the physical interpretation of the divergence operator). Even if the viscous force is uniformly acting on the fluid parcel, when $\nabla \rho \neq 0$, fluid rotation can be created due to nonuniform acceleration of fluid with different densities; this is demonstrated in Fig. 6.

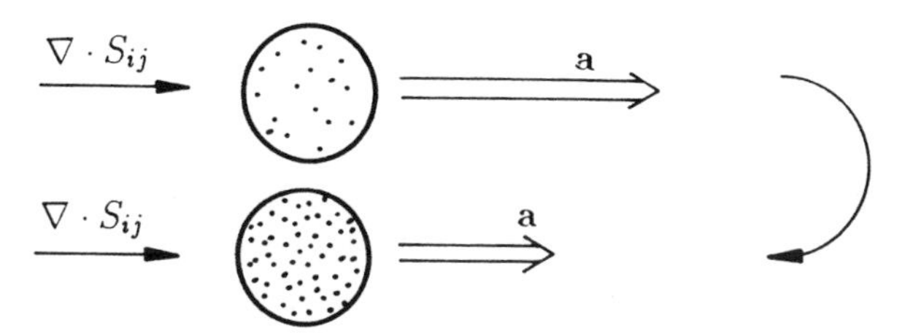

Fig. 6. Vorticity creation by viscous shear torque. Two adjacent fluid parcels with different densities are accelerated at the different rate under the same net viscous forces, consequently fluid rotation can be created.

(d) $\frac{1}{\rho}\nabla \times (\nabla \cdot S_{ij})$ represents the creation and transfer of vorticity via viscous effects. For instance, if the fluid has uniform viscosity, then, $\frac{1}{\rho}\nabla \times (\nabla \cdot S_{ij}) = \frac{1}{\rho}\nabla \times (\nabla \cdot (2\mu e_{ij})) = \frac{\mu}{\rho}\nabla^2 \boldsymbol{\omega}$. Suppose we consider a fluid domain D, then

$$\int_D \frac{\mu}{\rho} \nabla^2 \boldsymbol{\omega} \, dV = \int_S \frac{\mu}{\rho} \nabla \boldsymbol{\omega} \cdot \mathbf{n} \, ds \tag{3.41}$$

by the divergence theorem. The term $-\frac{\mu}{\rho}\nabla\boldsymbol{\omega}\cdot\mathbf{n}$ represents the flux at which vorticity $\boldsymbol{\omega}$, diffuses across the surface S. Hence, this term represents the vorticity transfer from the surroundings via viscosity, but not the creation of vorticity within the fluid volume D. When the fluid viscosity is not uniform, then $\frac{1}{\rho}\nabla \times (\nabla \cdot S_{ij})$ represents both vorticity diffusion (3.41) and the creation of fluid rotation due to the nonuniform viscous forces.

For a homogeneous fluid, the vorticity equation (3.40) reduces to

$$\frac{D\boldsymbol{\omega}}{Dt} = (\boldsymbol{\omega} \cdot \nabla)\mathbf{u} + \nu \nabla^2 \boldsymbol{\omega} \,, \tag{3.42}$$

where $\nu = \frac{\mu}{\rho}$ is the kinematic viscosity. Note that the pressure term (usually unknown) is eliminated because of $\nabla \times \nabla p = 0$. Furthermore, for two-dimensional flows, say $\mathbf{u} = (u, v, 0), \mathbf{x} = (x, y, 0)$, there is only one component of the vorticity vector $\boldsymbol{\omega} = (0, 0, \omega_3)$. Hence, the vorticity equation becomes the scalar equation of the form:

$$\frac{D\omega_3}{Dt} = \nu \left(\frac{\partial^2 \omega_3}{\partial x^2} + \frac{\partial^2 \omega_3}{\partial y^2} \right) \tag{3.43}$$

This is a diffusion equation. Furthermore, if a fluid is inviscid, then

$$\frac{D\omega_3}{Dt} = 0 \tag{3.44}$$

In other words, in a two-dimensional, inviscid, incompressible, barotropic flow with conservative body forces, the vorticity is carried along with the fluid motion without changing its magnitude.

3.6. *Integral properties of fluid rotation*

Vorticity represents the rate of fluid rotation, more specifically, twice the angular velocity. Rotation of a fluid parcel can be also measured in terms of flow circulation Γ, which is defined by

$$\Gamma = \int_C \mathbf{u} \cdot d\mathbf{x} = \int_S \boldsymbol{\omega} \cdot d\mathbf{A} \,, \tag{3.45}$$

by the Stokes theorem, where $\mathbf{x}$ is the position vector of a closed integration contour C, and $\mathbf{A}$ is the area vector of the surface S whose boundary is a single closed curve C. The rate of change in circulation Γ following fluid parcels that make up the curve C can be found to be

$$\frac{D\Gamma}{Dt} = \int_C \left(\frac{D\mathbf{u}}{Dt}\right) \cdot d\mathbf{x} \, , \tag{3.46}$$

where D/Dt is the material derivative (see, for example, Lighthill (1986) for the derivation of (3.46)). Applying Newton's second law of motion (3.2) to the fluid parcels along C, (3.46) becomes

$$\frac{D\Gamma}{Dt} = \int_C \left(\frac{1}{\rho}\nabla \cdot \tau_{ij} + \nabla E\right) \cdot d\mathbf{x} \, . \tag{3.47}$$

Under the action of a conservative body force, the integration of ∇E along a closed path vanishes

$$\int_C \nabla E \cdot d\mathbf{x} = \int_C dE = 0 \, , \tag{3.48}$$

and (3.47) becomes

$$\frac{D\Gamma}{Dt} = \int_C \left(\frac{1}{\rho}\nabla \cdot \tau_{ij}\right) \cdot d\mathbf{x} \, . \tag{3.49}$$

Assuming the fluid to be incompressible, inviscid, and homogeneous, i.e., $\tau_{ij} = -p\delta_{ij}$, then (3.49) is reduced to the well-known Kelvin theorem:

$$\frac{D\Gamma}{Dt} = -\frac{1}{\rho}\int_C \nabla p \cdot d\mathbf{x} = -\frac{1}{\rho}\int_C dp = 0 \, , \tag{3.50}$$

for the integration around the closed curve C. Equation (3.50) states that the flow circulation Γ around a closed curve C moving with the fluid remains constant. This leads to Helmholtz's theorem, stating that a vortex tube moves with the fluid; consequently, a vortex line is a material line. In other words, if a flow field is initially irrotational, flow circulation cannot be created and the flow remains irrotational in an inviscid, homogeneous fluid domain: this is often called Lagrange's theorem. Because the integration surface S is arbitrary in (3.45), no vorticity is created in the fluid domain under such conditions. If there is finite flow circulation present initially around a closed curved C, the strength of circulation remains constant. However, vorticity $\boldsymbol{\omega}$, in a position enclosed by the curve C, can vary because the area of surface S in (3.45) can vary by fluid deformation.

In the case of an inhomogeneous and inviscid fluid, flow circulation can be produced within the fluid domain whenever the fluid is displaced from a state in which the pressure gradient ∇p, and the density gradient $\nabla \rho$, are parallel. Kelvin's theorem (3.50), can be modified to be

$$\frac{D\Gamma}{Dt} = -\int_C \left(\frac{1}{\rho}\nabla p\right) \cdot d\mathbf{x} = \int_S \frac{1}{\rho^2}(\nabla \rho \times \nabla p) \cdot d\mathbf{A} \,. \tag{3.51}$$

This is often called Bjerknes' theorem (see, for example, Lamb, 1932). Equation (3.51) represents the time rate of change in circulation by "baroclinic torque" following a fluid parcel (physical interpretation of baroclinic torque was discussed for the term (b) in (3.40)).

If we further generalize (3.51) by including viscous forces with a uniform viscosity, (3.49) becomes

$$\frac{D\Gamma}{Dt} = \int_S \left[\frac{1}{\rho^2}(\nabla \rho \times \nabla \rho) + \frac{\mu}{\rho}\nabla^2 \boldsymbol{\omega} - \frac{1}{\rho^2}(\nabla \rho \times \mu \nabla^2 \mathbf{u})\right] \cdot d\mathbf{A} \,, \tag{3.52}$$

where μ is the dynamic viscosity of the fluid. The second term on the right-hand side of (3.52) represents the diffusion of vorticity from fluid parcels adjacent to the boundary of the integration surface S. Hence, this does not represent fluid-rotation creation within the fluid domain, but represents vorticity transfer due to viscous diffusion (recall the discussion associated with (3.41)). To demonstrate this, consider a two-dimensional flow for simplicity (i.e., no variation in the direction perpendicular to the surface S). If the fluid density is uniform then (3.52) can be written as,

$$\frac{D\Gamma}{Dt} = \frac{\mu}{\rho}\int_S \nabla^2 \boldsymbol{\omega} \cdot d\mathbf{A} = \frac{\mu}{\rho}\int_C (\nabla_{\|}\omega_{\perp}) \cdot \mathbf{n}\, d\mathbf{x} \,, \tag{3.53}$$

by the two-dimensional form of the divergence theorem. In (3.53), $\omega_{\perp}$ is a component of the vorticity vector which is normal to the surface S, $\nabla_{\|}$ denotes the two-dimensional gradient operator taken on the surface $\hat{S}$ which contains the integration surface S ($S \in \hat{S}$, see Fig. 7), and $\mathbf{n}$ is the normal unit vector in the two-dimensional space $\hat{S}$ pointing outward from the surface domain s. Equation (3.53) represents the time rate of change in flow circulation due to vorticity diffusion across the boundary of integration surface S that is enclosed by the line C. Hence, the flow circulation is changed due to the flux of vorticity from the surroundings. Since the closed line C is arbitrary, suppose that the line C is expanded to the boundary of the entire (homogeneous) fluid domain. It is then evident that, if the flow is initially irrotational everywhere, fluid rotation cannot be created anywhere within the uniform fluid domain, except by the vorticity diffusion flux from the boundary.

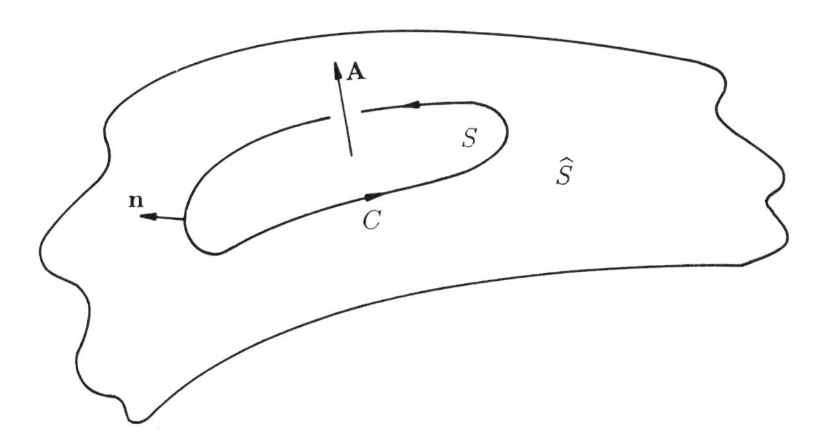

Fig. 7. A closed integration contour C on a surface $\hat{S}$; the surface S is enclosed by $C$and is defined by the area vector $\mathbf{A} : S \in \hat{S}$; $\mathbf{n}$ is the normal unit vector in the two-dimensional space S pointing outward from the surface S.

The last term in (3.52) represents torque produced by viscous shear forces due to the nonhomogeneity of the fluid. In the closed-contour integral form, the viscous term (including the vorticity diffusion) can be expressed by

$$\int_C \frac{\mu}{\rho} \nabla^2 \mathbf{u} \cdot d\mathbf{x} = \int_C \frac{1}{\rho} (\nabla \cdot S_{ij}) \cdot d\mathbf{x} , \tag{3.54}$$

where S_{ij} is the deviatoric stress tensor (3.3). The term $\nabla \cdot S_{ij}$ represents a net viscous force acting upon a fluid particle (see (2.3) for the physical interpretation of the divergence operator). Hence, even if the value of $\nabla \cdot S_{ij}$ is uniform, the variation in fluid density along C will cause fluid rotation; this rotation-creation mechanism is termed viscous-shear torque (physical interpretation of viscous-shear torque was discussed for term (c) in (3.40)).

Another important property associated with fluid rotation is found from the divergence theorem identity (see (2.6)):

$$\int_D \boldsymbol{\omega}\, dV = \int_S \mathbf{n} \times \mathbf{u}\, ds\,. \tag{3.55}$$

This indicates that if there is no tangential component of the velocity along every portion of the boundaries surrounding the fluid domain, then the net amount of vorticity within the domain is zero. This means that, when fluid rotation is created within the fluid domain (e.g., by baroclinic torque or viscous-shear torque), rotation of the opposite sign must be generated at the same time so as to maintain the net amount of vorticity vector to be nil, as long as there is no tangential motion in all the boundaries surrounding the fluid domain. Even vortex stretching and bending processes as described in term (a) in (3.40) must follow this restriction of the net amount of vorticity vector to be zero. This is not so if the boundaries are in motion so that the net tangential velocity vector is nonzero.

A vortex tube cannot end at any ordinary point in the fluid domain. This is true since $\nabla \cdot \boldsymbol{\omega} = 0$ by identity,

$$\int_D \nabla \cdot \boldsymbol{\omega}\, dV = \int_S \boldsymbol{\omega} \cdot \mathbf{n}\, ds = 0\,. \tag{3.56}$$

Suppose there is only one vortex tube in a fluid domain, and all the vortical flow motion is confined within the tube and a flow circulation around a contour of tube circumference is constant and finite. If the vortex tube is terminated within the domain D, the vortex tube must penetrate through the surface S only once. Hence, $\int_S \omega \cdot \mathbf{n}\, ds \neq 0$ contradicts (3.56), proving that vortex tubes cannot end within the fluid domain.

It is emphasized that (3.55) and (3.56) are exact mathematical results and no approximation nor assumption is involved. As long as the definition of vorticity is $\nabla \times \mathbf{u}$, the statements and interpretations made from (3.55) and (3.56) are also exact.

3.7. *Irrotational flows*

As discussed in Subsec. 3.4, if the flow is irrotational everywhere, the velocity field can be identified by the (scalar) velocity potential ϕ: $\mathbf{u} = \nabla\phi$. Conversely, the existence of the velocity potential ϕ identically satisfies the irrotationality, $\nabla \times \mathbf{u} = \nabla \times (\nabla\phi) \equiv 0$.

A solution is unique for an irrotational flow of incompressible fluid in a simply connected region. (The region is said to be simply connected if there exists a surface S which spans closed curve C within the fluid domain.) The uniqueness can be shown by considering a fluid domain D, which is bounded by a surface S. For an irrotational flow the problem can be presented by the Laplace equation $\nabla^2\phi = 0$ in D with the kinematic boundary condition $\frac{\partial\phi}{\partial n} \equiv \nabla\phi\cdot\mathbf{n} = \mathbf{U}\cdot\mathbf{n}$ on S. Now, assume both ϕ_1 and ϕ_2 to be solutions to this problem, viz., the solution is not unique. Then, $\phi_3 = \phi_1 - \phi_2$ is also a solution of $\nabla^2\phi_3 = 0$ in D with $\frac{\partial\phi_3}{\partial n} = \nabla\phi_3\cdot\mathbf{n} = \frac{\partial}{\partial n}(\phi_1 - \phi_2) = 0$ on S. The kinetic energy of ϕ_3 is

$$\frac{1}{2}\rho\int_D(\nabla\phi_3)^2 dV = \frac{1}{2}\rho\int_D\{\nabla\cdot(\phi_3\nabla\phi_3) - \phi_3\nabla^2\phi_3\}dV = \frac{1}{2}\rho\int_S\phi_3\frac{\partial\phi_3}{\partial n}ds = 0\ . \tag{3.57}$$

Since the kinetic energy is positive definite, (3.57) can vanish if, and only if, the integrand vanishes everywhere identically. Therefore,

$$\nabla\phi_3 = 0 \Rightarrow \nabla\phi_1 = \nabla\phi_2\ . \tag{3.58}$$

Hence, the velocity field of an irrotational flow is unique. This proof for uniqueness indicates that the trivial solution (i.e., $\nabla\phi = 0$ everywhere) is the only solution to the problem with homogeneous boundary conditions, i.e., $\nabla^2\phi = 0$ in D and $\nabla\phi\cdot\mathbf{n} = 0$ on S (stationary boundary). Note that this is not always true for flows described by the full Navier-Stokes equations.

Now, let us consider two velocity fields $\mathbf{u}_1$ and $\mathbf{u}_2$ in a simply connected fluid domain and assume that both velocity fields have the same distribution of vorticity,

$$\nabla\times\mathbf{u}_1 = \nabla\times\mathbf{u}_2 = \boldsymbol{\omega} \quad \text{in } D\ , \tag{3.59}$$

and satisfy the same boundary conditions: $\mathbf{u}_1\cdot\mathbf{n} = \mathbf{U}\cdot\mathbf{n}$ and $\mathbf{u}_2\cdot\mathbf{n} = \mathbf{U}\cdot\mathbf{n}$ on S. If we take the difference $\mathbf{u} = \mathbf{u}_2 - \mathbf{u}_1$, then $\nabla\times\mathbf{u} = 0$ in D and $\mathbf{u}\cdot\mathbf{n} = 0$ on S. Hence, the difference in velocities $\mathbf{u}_1$ and $\mathbf{u}_2$ satisfies the irrotational flow field in the homogeneous boundary, i.e., $\nabla^2\phi = 0$ in D and $\nabla\phi\cdot\mathbf{n} = 0$ on S. This leads to $\mathbf{u} = 0$ everywhere in D. Therefore, a velocity field with a given vorticity field and boundary conditions is unique.

Next we consider again two velocity fields, $\mathbf{u}_1$ and $\mathbf{u}_2$, where $\nabla\times\mathbf{u}_1 = \nabla\times\mathbf{u}_2 = \boldsymbol{\omega}$ in D as the previous case but $\mathbf{u}_1\cdot\mathbf{n} = 0$ and $\mathbf{u}_2\cdot\mathbf{n} = \mathbf{U}\cdot\mathbf{n}$ on S instead. Again $\mathbf{u}_2 - \mathbf{u}_1$ is an irrotational velocity field and $(\mathbf{u}_2 - \mathbf{u}_1)\cdot\mathbf{n} = \mathbf{U}\cdot\mathbf{n}$ on S. If the fluid domain is simply connected, then $(\mathbf{u}_2 - \mathbf{u}_1)$ is unique and can be represented by $\nabla\phi$. Hence, $\mathbf{u}_2 = \nabla\phi + \mathbf{u}_1$, where $\nabla\phi$ satisfies the boundary condition, $\mathbf{U}\cdot\mathbf{n} = \nabla\phi\cdot\mathbf{n}$, and $\mathbf{u}_1$ has a vorticity field $\boldsymbol{\omega}$ with the homogeneous boundary condition.

Now let's compute the kinetic energy of the flow field $\mathbf{u}_2$:

$$\begin{aligned}\frac{1}{2}\rho\int_D(\mathbf{u}_2\cdot\mathbf{u}_2)dV &= \frac{1}{2}\rho\int_D(\nabla\phi+\mathbf{u}_1)\cdot(\nabla\phi+\mathbf{u}_1)dV \\ &= \frac{1}{2}\rho\int_D(\nabla\phi)^2dV + \rho\int_D(\mathbf{u}_1\cdot\nabla\phi)dV + \frac{1}{2}\rho\int_D(\mathbf{u}_1\cdot\mathbf{u}_1)dV\,. \end{aligned} \tag{3.60}$$

But because $\nabla\cdot\mathbf{u}_1 = 0$ in D and $\mathbf{u}_1\cdot\mathbf{n} = 0$ on S,

$$\int_D(\mathbf{u}_1\cdot\nabla\phi)dV = \int_D\nabla\cdot(\phi\mathbf{u}_1)dV = \int_S\phi\mathbf{u}_1\cdot\mathbf{n}\,ds = 0\,. \tag{3.61}$$

Therefore, (3.60) becomes

$$\frac{1}{2}\rho\int_D(\mathbf{u}_2\cdot\mathbf{u}_2)dV = \frac{1}{2}\rho\int_D|\nabla\phi|^2 + |\mathbf{u}_1|^2dV\,. \tag{3.62}$$

The kinetic energy of any vortical flow is equal to the kinetic energy of the irrotational flow satisfying the boundary condition plus the kinetic energy of the vortical flow with the homogeneous boundary condition. Also, since the value of (3.62) is greater than $\frac{1}{2}\rho\int_D|\nabla\phi|^2dV$, the irrotational flow has the least kinetic energy among all the flows satisfying the given boundary conditions. This is called Kelvin's minimum energy theorem.

Irrotational-flow problems are to solve the Laplace equation for ϕ in the domain D. The Laplace equation is elliptic and independent of time: the time dependence can arise from the boundary conditions. The instantaneous velocity of the boundary uniquely determines the velocity potential, i.e., irrotational motion is independent of the flow's past history. This is because the only relevant force involved in irrotational flows is the pressure force that propagates with infinite speed in an incompressible fluid medium. All of the effects from the past flow conditions are kept in the vortical (rotational) part of the motion. Recall that fluid rotation can be generated only at a boundary for a homogeneous fluid, and once generated, vorticity is transported by advection and diffusion. Such processes are not instantaneous and depend on the conditions in the past.

The decomposition of velocity field that we have just discussed, e.g., $\mathbf{u} = \nabla\phi + \mathbf{u}_1$, is now generalized. Suppose $\mathbf{u}$ is a velocity vector field which can be decomposed into its solenoidal and irrotational parts, i.e., $\mathbf{u} = \mathbf{u}_s + \mathbf{u}_i$ such that

$$\nabla \cdot \mathbf{u}_s = 0 \text{ (solenoidal)} , \tag{3.63}$$

$$\nabla \times \mathbf{u}_i = 0 \text{ (irrotational)} , \tag{3.64}$$

and

$$\nabla \cdot \mathbf{u}_i = \nabla \cdot \mathbf{u} = \vartheta(\mathbf{x}, t) , \tag{3.65}$$

$$\nabla \times \mathbf{u}_s = \nabla \times \mathbf{u} = \boldsymbol{\omega}(\mathbf{x}, t) . \tag{3.66}$$

This decomposition is often called Helmholtz's decomposition. Note that this decomposition does not yield a unique velocity field $\mathbf{u}$ since $\hat{\mathbf{u}}_s = \mathbf{u}_s + \nabla\phi$ and $\hat{\mathbf{u}}_i = \mathbf{u}_i - \nabla\phi$ gives another solution for $\mathbf{u}$. The uniqueness is restored by the specification of boundary conditions. For example, if the values of $\mathbf{u}_s$ are chosen at the boundaries, then the boundary conditions for $\mathbf{u}_i = \mathbf{u} - \mathbf{u}_s$ are now determined to satisfy the boundary conditions of the velocity field $\mathbf{u}$. The specification of boundary conditions assumes that $\nabla\phi = 0$ at the boundary; hence $\nabla\phi = 0$, everywhere in the flow domain. Therefore, this decomposition becomes unique.

Defining the vector potential $\mathbf{B}$ and the scalar potential Φ, this decomposition can be written as

$$\mathbf{u} = \mathbf{u}_s + \mathbf{u}_i = \nabla \times \mathbf{B} + \nabla\Phi . \tag{3.67}$$

Note that

$$\nabla \cdot \mathbf{u} = \nabla \cdot (\nabla \times \mathbf{B}) + \nabla^2\Phi = \vartheta(\mathbf{x}, t) , \tag{3.68}$$

and

$$\nabla \times \mathbf{u} = \nabla \times (\nabla \times \mathbf{B}) + \nabla \times (\nabla\Phi) = \boldsymbol{\omega}(\mathbf{x}, t) . \tag{3.69}$$

Since $\nabla \times (\nabla \times \mathbf{B}) = \nabla(\nabla \cdot \mathbf{B}) - \nabla^2\mathbf{B}$, once the vorticity field $\boldsymbol{\omega}(\mathbf{x}, t)$ and the divergence $\vartheta(\mathbf{x}, t)$ are known, then the velocity field can be uniquely identified by a pair of Poisson equations:

$$\nabla^2\mathbf{B} = -\boldsymbol{\omega}(\mathbf{x}, t) , \tag{3.70}$$

by specifying $\nabla \cdot \mathbf{B} = 0$, and

$$\nabla^2\Phi = \vartheta(\mathbf{x}, t) . \tag{3.71}$$

4. Boundary Conditions at the Air-Water Interface

In order to formulate the proper boundary conditions applied to the air-water interface, the following approach is taken. First, the air-water interface is viewed as a fluid-fluid material interface of a contact discontinuity. Based on continuum mechanics, general conditions to describe the jump discontinuities in flow and fluid properties are derived. Then, the derived conditions are reduced for more specific jump conditions which are applied at a contact (material) discontinuity. The boundary conditions at the air-water interface are then formulated based on the conservation of mass, linear momentum, and mechanical energy at the jump discontinuities. Vorticity conditions at the interface are discussed in Sec. 6.

4.1. *Jump conditions*

Let $D = D(t)$ denote an arbitrary material volume which is moving with the fluid and $\mathbf{F}(\mathbf{x}, t)$ is an integrable scalar or vector function of position $\mathbf{x}$ and time t. Suppose that the function $\mathbf{F}$ and velocity $\mathbf{u}$ are discontinuous across the surface Σ within the domain D, where the surface Σ separates the domain D into the subdomains D_1 and D_2 (see Fig. 8). Then, the transport theorem can be written in the following modified form (Serrin, 1959):

$$\frac{D}{Dt}\int_D \mathbf{F}\, dV = \int_{D_1+D_2} \left(\frac{\partial \mathbf{F}}{\partial t} + \nabla \cdot (\mathbf{uF})\right) dV + \int_\Sigma [[\mathbf{F}(\mathbf{u}\cdot\mathbf{n} - \mathbf{U}\cdot\mathbf{n})]] ds\ , \quad (4.1)$$

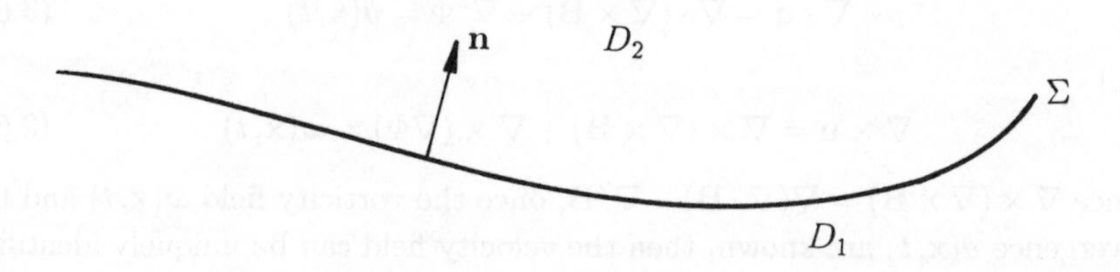

Fig. 8. A definition sketch for the jump conditions.

where the operator D/Dt denotes the material derivative, i.e., $D/Dt = (\partial/\partial t + \mathbf{u} \cdot \nabla)$, $\mathbf{n}$ is the normal unit vector, $\mathbf{u}$ is the fluid particle velocity, $\mathbf{U}$ is the velocity of the discontinuity, and the double bracket denotes the jump of the quantity across the discontinuity, e.g.,

$$[[\mathbf{F}]] = \underset{\substack{\mathbf{x}\to\xi\in\Sigma \\ \mathbf{x}\in D_2}}{Lim}\ \mathbf{F} - \underset{\substack{\mathbf{x}\to\xi\in\Sigma \\ \mathbf{x}\in D_1}}{Lim}\ \mathbf{F}\ . \tag{4.2}$$

Taking the domain D to be a thin slab containing the surface Σ, and taking the limit of the thickness of the slab to vanish, we obtain from (4.1)

$$\int_{\Sigma} [[\mathbf{F}(\mathbf{u}\cdot\mathbf{n} - \mathbf{U}\cdot\mathbf{n})]]ds = 0\ , \tag{4.3}$$

as long as the integrand is bounded. Because the portion of Σ can be arbitrarily small, it follows that the integrand itself must be zero for an arbitrary scalar or vector function $\mathbf{F}$,

$$[[\mathbf{F}(\mathbf{u}\cdot\mathbf{n} - \mathbf{U}\cdot\mathbf{n})]] = 0\ . \tag{4.4}$$

When the function $\mathbf{F}$ is discontinuous at Σ, the jump condition for the time rate of change of the quantity $\mathbf{F}$ as following a fluid parcel is described by (4.4).

If the function $\mathbf{F}$ represents fluid density ρ, the conservation of mass becomes

$$\frac{D}{Dt}\int_{D_1} \rho dV + \frac{D}{Dt}\int_{D_2} \rho dV = \frac{D}{Dt}\int_{D} \rho dV = 0\ , \tag{4.5}$$

since the fluid density ρ is integrable in D. From (4.1) and (4.4), the jump condition for the mass conservation across the surface Σ is

$$[[\rho(\mathbf{u}\cdot\mathbf{n} - \mathbf{U}\cdot\mathbf{n})]] = 0\ . \tag{4.6}$$

A material boundary (i.e., a contact discontinuity) is defined by the condition, $\mathbf{u}\cdot\mathbf{n} - \mathbf{U}\cdot\mathbf{n} = 0$. This condition is often referred to as the kinematic boundary condition, which is a special condition of (4.6).

We now expand (4.4) to more general jump conditions by adopting the concept of a weak solution for a gas-dynamics shock problem (e.g., see Kevorkian, 1990). Consider a conservation equation of the form

$$\frac{D\mathbf{F}}{Dt} + \mathbf{F}\nabla\cdot\mathbf{u} + \nabla\cdot\mathbf{G} + \mathbf{H} = 0\ , \tag{4.7}$$

where $\mathbf{F}, \mathbf{G}$, and $\mathbf{H}$ can be scalar, vector, or tensor functions of position and time, as long as they are consistent in the equation. Now, let $\psi = \psi(\mathbf{x}, t)$ be any continuously differentiable scalar function such that $\psi = 0$ on the surface S of a domain D.

Since (4.7) holds everywhere in the domain D,

$$\int_D \psi \left(\frac{D\mathbf{F}}{Dt} + \mathbf{F}\nabla \cdot \mathbf{u} + \nabla \cdot \mathbf{G} + \mathbf{H} \right) dV = 0 \,. \tag{4.8}$$

This leads to the relation

$$\int_D \nabla \cdot (\mathbf{F}\mathbf{u}\psi + \mathbf{G}\psi) dV = \int_D \left((\mathbf{F}\mathbf{u} + \mathbf{G}) \cdot \nabla\psi - \psi \left(\frac{\partial \mathbf{F}}{\partial t} + \mathbf{H} \right) \right) dV \,. \tag{4.9}$$

Now, if $\psi, \mathbf{F}$, and $\mathbf{G}$ are continuously differentiable, then using the Gauss theorem, we find

$$\int_D \nabla \cdot (\mathbf{F}\mathbf{u}\psi + \mathbf{G}\psi) dV = \int_S \psi(\mathbf{F}\mathbf{u} + \mathbf{G}) \cdot \mathbf{n} \, ds = 0 \,, \tag{4.10}$$

since $\psi = 0$ on S. Therefore,

$$\begin{aligned} &\int_D \left((\mathbf{F}\mathbf{u} + \mathbf{G}) \cdot \nabla\psi - \psi \left(\frac{\partial \mathbf{F}}{\partial t} + \mathbf{H} \right) \right) dV \\ &\quad = \int_D \left(\mathbf{F}\frac{D\psi}{Dt} + \mathbf{G} \cdot \nabla\psi - \psi\mathbf{H} \right) dV - \frac{\partial}{\partial t} \int_D \mathbf{F}\psi dV = 0 \,. \end{aligned} \tag{4.11}$$

Note that the integrand in (4.11) does not involve any spatial derivatives of $\mathbf{F}, \mathbf{G}$ or $\mathbf{H}$; the only derivatives that occur are for ψ. We now use this feature to define a weak solution of (4.7) as one that satisfies (4.11) in D for any smooth function ψ that vanishes on ∂D. If the functions $\mathbf{F}, \mathbf{G}$, and $\mathbf{H}$ are continuous in D, then (4.10) is valid; hence, (4.11) is equivalent to (4.7). Even if the functions $\mathbf{F}, \mathbf{G}$, and $\mathbf{H}$ are discontinuous and are not differentiable, as long as they are integrable, then the integrations over D in (4.11) remain well defined.

Suppose that there is a surface Σ in the domain D where the discontinuities in the functions $\mathbf{F}, \mathbf{G}$, and $\mathbf{H}$ occur, and the domain D is separated by Σ into two parts, D_1 and D_2 (see Fig. 8). Since in each subdomain D_1 and D_2, the functions $\mathbf{F}, \mathbf{G}$, and $\mathbf{H}$ are continuous, using (4.9), (4.10), and (4.11), the Gauss theorem gives

$$\begin{aligned} &\int_{D_1} \left(\mathbf{F}\frac{D\psi}{Dt} + \mathbf{G} \cdot \nabla\psi - \psi\mathbf{H} \right) dV - \frac{\partial}{\partial t} \int_{D_1} \mathbf{F}\psi dV \\ &\quad = \int_{S_1} \psi(\mathbf{F}\mathbf{u} + \mathbf{G}) \cdot \mathbf{n} \, ds = \int_\Sigma \psi(\mathbf{F}(\mathbf{u} \cdot \mathbf{n} - \mathbf{U} \cdot \mathbf{n}) + \mathbf{G} \cdot \mathbf{n}) ds \,, \end{aligned} \tag{4.12}$$

and

$$\int_{D_2}\left(\mathbf{F}\frac{D\psi}{Dt}+\mathbf{G}\cdot\nabla\psi-\psi\mathbf{H}\right)dV-\frac{\partial}{\partial t}\int_{D_2}\mathbf{F}\psi dV$$
$$=\int_{S_2}\psi(\mathbf{Fu}+\mathbf{G})\cdot\mathbf{n}\,ds=-\int_{\Sigma}\psi(\mathbf{F}(\mathbf{u}\cdot\mathbf{n}-\mathbf{U}\cdot\mathbf{n})+\mathbf{G}\cdot\mathbf{n})ds\ . \tag{4.13}$$

Note that the integrals over Σ on which $\psi \neq 0$ do not vanish: $\psi = 0$ on the surface S of a domain D but not on Σ. Note that $(\mathbf{u}\cdot\mathbf{n}-\mathbf{U}\cdot\mathbf{n})$ is the relative normal flow velocity on Σ because the discontinuity surface Σ does not need to be a material surface, and the negative sign in (4.13) indicates that the unit normal vector $\mathbf{n}$ on Σ points outward from the subdomain D_1 (inward to the subdomain D_2) as shown in Fig. 8. Taking the domain D to be a thin slab containing the surface Σ, and taking the limit of the thickness of the slab to vanish,

$$\int_{D_1+D_2}\left(\mathbf{F}\frac{D\psi}{Dt}+\mathbf{G}\cdot\nabla\psi-\psi\mathbf{H}\right)dV-\frac{\partial}{\partial t}\int_{D_1+D_2}\mathbf{F}\psi dV=0\ , \tag{4.14}$$

because the integrands are well defined. Adding (4.12) and (4.13) for D_1 and D_2 yields

$$\int_{\Sigma}\psi[[\mathbf{F}(\mathbf{u}\cdot\mathbf{n}-\mathbf{U}\cdot\mathbf{n})+\mathbf{G}\cdot\mathbf{n}]]ds=0\ . \tag{4.15}$$

Since (4.15) is true for arbitrary ψ on Σ, the integrand with the double square bracket (see (4.2)) must vanish, and we obtain the jump condition for the conservation equation of the form (4.7):

$$[[\mathbf{F}(\mathbf{u}\cdot\mathbf{n}-\mathbf{U}\cdot\mathbf{n})+\mathbf{G}\cdot\mathbf{n}]]=0\ . \tag{4.16}$$

Note that this equation is valid even if Σ is not a material surface. For a contact discontinuity (i.e., Σ is a material surface), the kinematic condition must satisfy, i.e., $(\mathbf{u}\cdot\mathbf{n}-\mathbf{U}\cdot\mathbf{n})=0$ on Σ. Hence, we find

$$[[\mathbf{G}\cdot\mathbf{n}]]=0\ . \tag{4.17}$$

This result (4.17) is valid as long as $\mathbf{F}, \mathbf{G}$, and $\mathbf{H}$ are smooth functions of their arguments in the individual subdomains D_1 and D_2.

4.2. *Conservation of linear momentum and mechanical energy at the interface*

Consider the equation of linear momentum of the form

$$\rho \frac{D\mathbf{u}}{Dt} = \nabla \cdot \tau_{ij} + \rho \nabla E \,. \tag{4.18}$$

This can be written in the form:

$$\frac{D(\rho \mathbf{u})}{Dt} + \rho \mathbf{u}(\nabla \cdot \mathbf{u}) = \nabla \cdot \tau_{ij} + \rho \nabla E \,, \tag{4.19}$$

where τ_{ij} is the stress tensor and E is the body-force scalar potential, i.e., for gravity, $E = -gy$. Then, (4.16) immediately provides the jump condition for the conservation of linear momentum by taking $\mathbf{F} = \rho \mathbf{u}, \mathbf{G} = -\tau_{ij}, \mathbf{H} = -\rho \nabla E$,

$$[[\rho \mathbf{u}(\mathbf{u} \cdot \mathbf{n} - \mathbf{U} \cdot \mathbf{n})]] - [[\tau_{ij} \cdot \mathbf{n}]] = 0 \,, \tag{4.20}$$

or

$$\rho(\mathbf{u} \cdot \mathbf{n} - \mathbf{U} \cdot \mathbf{n})[[\mathbf{u}]] - [[\tau_{ij} \cdot \mathbf{n}]] = 0 \,, \tag{4.21}$$

because of mass conservation (4.6). Now if the discontinuity surface Σ is a contact discontinuity, then the kinematic condition requires that $(\mathbf{u} \cdot \mathbf{n} - \mathbf{U} \cdot \mathbf{n})$ vanish on Σ. Hence the jump condition for the conservation of linear momentum across a contact discontinuity becomes the well-known dynamic condition at a material surface, i.e., the stress acting on the surface must be continuous:

$$[[\tau_{ij} \cdot \mathbf{n}]] = 0 \,. \tag{4.22}$$

Now consider the conservation of mechanical energy of the form (3.15), i.e.,

$$\rho \frac{D}{Dt}\left(\frac{|\mathbf{u}|^2}{2} - E\right) = \nabla \cdot (\mathbf{u}\tau_{ij}) - \tau_{ij} : \nabla \mathbf{u} \,, \tag{4.23}$$

or, by using the conservation of mass (3.1) and $E = -gy$,

$$\frac{D}{Dt}\left(\rho \frac{|\mathbf{u}|^2}{2} + \rho g y\right) + \left(\rho \frac{|\mathbf{u}|^2}{2} + \rho g y\right)(\nabla \cdot \mathbf{u}) = \nabla \cdot (\tau_{ij} u_i) - \tau_{ij} \frac{\partial u_i}{\partial x_j} \,. \tag{4.24}$$

By taking $\mathbf{F} = \rho \frac{|\mathbf{u}|^2}{2} + \rho g y, \mathbf{G} = -\tau_{ij} u_i, \mathbf{H} = \tau_{ij} \frac{\partial u_i}{\partial x_j}$, (4.16) becomes

$$\left[\left[\rho\left(\frac{|\mathbf{u}|^2}{2} + gy\right)(\mathbf{u} \cdot \mathbf{n} - \mathbf{U} \cdot \mathbf{n})\right]\right] - [[\tau_{ij} u_i n_j]] = 0 \,, \tag{4.25}$$

or

$$\rho(\mathbf{u}\cdot\mathbf{n}-\mathbf{U}\cdot\mathbf{n})\left[\left[\frac{|\mathbf{u}|^2}{2}+gy\right]\right]-[[\tau_{ij}u_in_j]]=0\ . \qquad (4.26)$$

At a contact discontinuity,

$$[[\tau_{ij}u_in_j]]=0\ . \qquad (4.27)$$

This means that the rate of work done by the one fluid at the interface must be identical to the rate of work received by the other fluid. The mechanical energy cannot be stored at the boundary. Since from the dynamic boundary condition, $[[\tau_{ij}n_j]]=0$, this energy boundary condition implies that

$$[[\mathbf{u}]]=0, \qquad (4.28)$$

i.e., the fluid velocity must be continuous across the material boundary. In other words, it is necessary to satisfy the no-slip boundary condition at the material boundary and this arises as a consequence of the jump condition of the mechanical-energy equation.

4.3. *Surface-tension effects*

When liquid is in contact with another substance, there is a free interfacial energy present between them. This appears to contradict (4.27) which states that no energy can be stored at the interface. It is because this interfacial energy cannot be described based on the continuum hypothesis but arises from the molecular-level dynamics, i.e., the interfacial energy is associated with the difference between the inward attraction of the molecules in the interior of each phase and those at the surface of contact. Let us consider a soap film that is stretched linearly with a distance dx; hence, the work done in stretching the film is γLdx, where γ is the surface-tension (force per unit length) and L is the width of the film. This work could be written as γdA where $dA=Ldx$. Therefore, instead of the quantity, force per unit length, the surface-tension γ can be interpreted as an energy per unit area, and the surface-tension is often called surface free energy. The term "surface-tension" implies that a liquid surface is extended by stretching the molecules on the surface, while the term "surface free energy" implies energy is required to form more surface by bringing molecules from the interior of the liquid to the surface (Adamson, 1990).

Consider a small liquid-gas interface identified by two orthogonal elements ξ and ζ on the surface. Since the surface is small enough, the curvatures

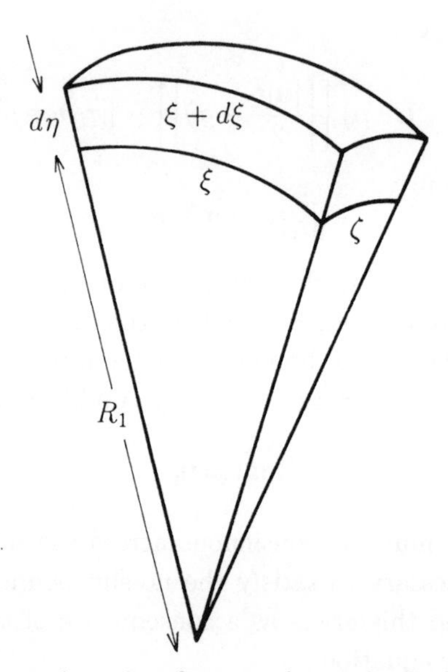

Fig. 9. A change in an interface area due to the displacement $d\eta$.

of the surface can be identified by the two radii of curvature R_1 and R_2, which are in two normal plane sections at right angles to one another, just as the planes q and q' described in Sec. 2. Hence, ξ and ζ are regarded as elements of the circumference of circles with radii R_1 and R_2. Suppose the surface is displaced by a small amount of $d\eta$ outward, the change in area is $\delta A = (\xi + d\xi)(\zeta + d\zeta) - \xi\zeta = \xi d\zeta + \zeta d\xi$ in the differential sense. Now, the work done in changing the area by surface-tension force is

$$\text{Work}_\gamma = \gamma \delta A = \gamma(\xi d\zeta + \zeta d\xi) , \tag{4.29}$$

while the work done by the pressure difference Δp between two media is

$$\text{Work}_p = \Delta p \xi \zeta d\eta . \tag{4.30}$$

From the sketch shown in Fig. 9, we can readily find the relation by similar triangles,

$$\frac{\xi + d\xi}{R_1 + d\eta} = \frac{\xi}{R_1} \Rightarrow d\xi = \frac{\xi d\eta}{R_1} , \tag{4.31}$$

and, similarly,

$$\frac{\zeta + d\zeta}{R_2 + d\eta} = \frac{\zeta}{R_2} \Rightarrow d\zeta = \frac{\zeta d\eta}{R_2} . \tag{4.32}$$

If the condition is at equilibrium, (4.29) and (4.30) must balance. Substituting (4.31) and (4.32) into (4.29) and equating (4.29) and (4.30) yields

$$\Delta p = \gamma \left(\frac{1}{R_1} + \frac{1}{R_2} \right) . \tag{4.33}$$

This relationship is called Laplace's formula. Note that for a plane surface, Δp vanishes since the two radii become infinity, i.e., no pressure discontinuity across a plane surface. As discussed in Sec. 2, the quantity $\frac{1}{2}(\frac{1}{R_1} + \frac{1}{R_2}) = -\frac{1}{2}(a_1 + a_2)$ is called the mean curvature where a_1 and a_2 are the principal curvatures. The mean curvature, i.e., the average of principal curvatures, is independent of the directions of the orthogonal pair (see (2.26)), and can be evaluated by taking a trace of the curvature tensor. From (2.12) and (2.16), the curvature tensor κ_{ij} can be expressed as

$$\kappa_{ij} = -\nabla \left(\frac{\nabla F}{n} \right) = -\frac{(\delta_{ik} - n_i n_k) \cdot \dfrac{\partial^2 F}{\partial x_k \partial x_l} \cdot (\delta_{lj} - n_l n_j)}{n} , \tag{4.34}$$

where $n = |\nabla F| = \sqrt{\left(\frac{\partial F}{\partial x}\right)^2 + \left(\frac{\partial F}{\partial y}\right)^2 + \left(\frac{\partial F}{\partial z}\right)^2}$ and $F(\mathbf{x}, t) = 0$ is the function representing the location of the interface. The geometric quantity $\left(\frac{1}{R_1} + \frac{1}{R_2}\right)$ is now given by

$$\left(\frac{1}{R_1} + \frac{1}{R_2} \right) = -\mathbf{Trace}[\kappa_{ij}] = \nabla \cdot \left(\frac{\nabla F}{n} \right) . \tag{4.35}$$

For a water-wave problem, the free surface can be described by

$$F(x, y, z, t) = y - \eta(x, z, t) = 0 . \tag{4.36}$$

Substituting (4.36) into (4.35) yields

$$\frac{1}{R_1} + \frac{1}{R_2} = -\frac{\eta_{xx}(1 + \eta_z^2) + \eta_{zz}(1 + \eta_x^2) - 2\eta_{xz}\eta_x\eta_z}{(1 + \eta_x^2 + \eta_z^2)^{\frac{3}{2}}} . \tag{4.37}$$

where subscripts represent partial differentiation. For a two-dimensional problem in the x-y plane, surface-tension effect can be expressed as

$$\frac{1}{R} = -\frac{\eta_{xx}}{(1 + \eta_x^2)^{\frac{3}{2}}} . \tag{4.38}$$

Assuming the equilibrium condition (4.33) to be valid, the dynamic boundary condition (4.22) is modified to

$$[[\tau_{ij} \cdot \mathbf{n}]] = \gamma \left(\frac{1}{R_1} + \frac{1}{R_2} \right) \mathbf{n} \, . \tag{4.39}$$

Taking the time rate of work (power) per unit area done by surface-tension effect from (4.29), the mechanical-energy jump condition (4.27) can be modified as

$$[[\tau_{ij} u_i n_j]] = \gamma \frac{1}{A} \frac{dA}{dt} = \gamma \frac{d}{dt} \ln A \, . \tag{4.40}$$

When the surface-tension is constant at the liquid-gas interface, only the normal component of stress is affected by the surface tension. Nonetheless, such a condition is rather ideal and the air-water interface in most natural environments exhibits variations in surface-tension. The surface-tension variations can be caused by, for example, the introduction of surface-active substances (surfactant) with concentrations that vary from point to point, the variation of surface temperature since the surface-tension of the liquid-gas interface is a function of temperature, or the presence of a varying electric charge on the liquid surface. The surface-tension variation from one point to another creates the appearance of tangential stresses on the liquid surface. Note that the surface-tension effects tend to reduce the area of the surface, and the liquid surface is moved in a direction from lower to higher surface tension. The magnitude of tangential stress is determined by the surface-tension gradient, i.e., the tangential force per unit area exerted on the surface is $\nabla\gamma$. This surface-tension-gradient driven motion is known as capillary convection or the Marangoni effect. Our dynamic boundary condition (4.39) can be now written as

$$[[\tau_{ij} \cdot \mathbf{n}]] = \gamma \left(\frac{1}{R_1} + \frac{1}{R_2} \right) \mathbf{n} - \nabla_\lambda \gamma \, , \tag{4.41}$$

where ∇_λ is the gradient operator in the directions tangent to the surface. For Newtonian incompressible fluids, (4.41) can be written as (see (3.3))

$$[[-p\delta_{ij} \cdot \mathbf{n} + \mu(\nabla \mathbf{u} + (\nabla \mathbf{u})^T) \cdot \mathbf{n}]] = \gamma \left(\frac{1}{R_1} + \frac{1}{R_2} \right) \mathbf{n} - \nabla_\lambda \gamma \, , \tag{4.42}$$

hence, the tangential stress components are associated with the viscous effects, and the capillary convection (the Marangoni effect) is impossible for inviscid fluids.

5. Free-Surface Conditions of Water-Wave Problems

If the flow is initially irrotational, it will remain irrotational. This is true for an incompressible, homogeneous, and inviscid fluid, according to Lagrange's theorem, interpreted from Kelvin's theorem (3.50). This is because there is no mechanism to create any torque within homogeneous fluids — based on the conservation of angular momentum, no rotational motion can be created without the action of torque. For an inviscid fluid, the only source to create fluid rotation is at the boundaries by baroclinic torque (3.51), but the created fluid rotation cannot escape from the boundary and remains there in the form of a vortex sheet. This is because, according to the kinematic boundary condition, (inviscid) fluid particles at the boundary cannot be detached from the boundary, and any rotational (vortical) motion must be carried with the fluid according to Helmholtz's theorem (also interpreted from Kelvin's theorem (3.50)). For viscous fluids, the creation of fluid rotation is also limited at the boundaries as long as the fluids are homogeneous. Once fluid rotation is created at the boundary, the rotational motion can, however, escape from the boundary and be transported to the fluid domain by viscous diffusion. Rotational fluid motions injected into the flow domain from the boundary can be advected with the fluid motions (according to Helmholtz's theorem) and are also continually diffused by viscosity.

Consider water-wave motions. As stated, fluid rotations can be created only at boundaries, and initial transport of rotational motion from the boundary to the fluid interior is a gradual process by viscous diffusion. The oscillatory wave motion creates the fluid rotation of opposite directions one after the other. Hence, away from the boundary, previously created fluid rotation is canceled out with the newly created rotation in the opposite direction, and rotational motion is confined to the region adjacent to the boundary, i.e., the boundary layer, and fluid motion away from the boundary remains irrotational.

Once it is decided to neglect the vortical (rotational) flow near the boundaries and to treat the flow as an irrotational-flow problem, the velocity field can be identified by the (scalar) velocity potential ϕ ($\mathbf{u} = \nabla\phi$, then the irrotationality of the flow is automatically satisfied, $\nabla \times \mathbf{u} = \nabla \times (\nabla\phi) = 0$). The four dependent variables, (u, v, w) and p, can be replaced by a single function ϕ. Once the velocity potential is determined, the velocity field is retrieved by taking the gradient of ϕ and the pressure field can be evaluated by solving the Euler integral (3.33). By expressing a flow by the velocity potential, nonlinear vector equations of motion (the Navier-Stokes equations) can be replaced by a much simpler linear scalar equation, i.e., Laplace's equation, $\nabla \cdot \mathbf{u} = \nabla^2 \phi = 0$.

5.1. *Water-wave problem formulation*

Let us first examine the traditional formulation for a water-wave problem. Assuming Newtonian, incompressible, homogeneous fluids, irrotational flows with no surface-tension effects and constant pressure at the free surface $y = \eta(x, z, t)$, the problem can be formulated in terms of velocity potential:

$$\nabla^2 \phi = 0 \tag{5.1}$$

in a fluid domain D. The kinematic boundary condition on the (material) surface of the air-water interface is

$$\mathbf{u} \cdot \mathbf{n} - \mathbf{U} \cdot \mathbf{n} = 0 \ , \tag{5.2}$$

where $\mathbf{u}$ is the water particle velocity and $\mathbf{U}$ is the velocity of the boundary (i.e., water surface). If we define the location of the surface by (4.36), i.e., $F(x, y, z, t) = y - \eta(x, z, t) = 0$, then the unit vector normal to the surface is found by (2.11). Since the value of the function F is always constant at the surface, $dF/dt = 0$ at the surface, or $\frac{\partial F}{\partial t} + \mathbf{U} \cdot \nabla F = 0$. Therefore, the velocity of the boundary which is normal to the boundary can be found to be

$$\mathbf{U} \cdot \mathbf{n} = \frac{-\dfrac{\partial F}{\partial t}}{|\nabla F|} \ . \tag{5.3}$$

Substituting (5.3), (2.11) and (4.36) into (5.2) yields the kinematic boundary condition for the free surface

$$\frac{\partial \eta}{\partial t} = -\frac{\partial \phi}{\partial x}\frac{\partial \eta}{\partial x} + \frac{\partial \phi}{\partial y} - \frac{\partial \phi}{\partial z}\frac{\partial \eta}{\partial z} \quad \text{on } y = \eta(x, z, t) \ . \tag{5.4}$$

Now, the kinematic boundary condition contains two unknowns: the velocity potential ϕ and the location of the free surface η. Therefore, to close this problem, another independent boundary condition must be imposed at the free surface, viz., the dynamic boundary condition. Assuming that the air pressure exerted on the free surface is constant,

$$p = \text{constant} \quad \text{on } y = \eta(x, z, t) , \tag{5.5}$$

and using the Euler integral yields the dynamic boundary condition in the form (3.34):

$$\frac{\partial \phi}{\partial t} + \frac{1}{2}|\nabla \phi|^2 + g\eta = 0 \quad \text{on } y = \eta(x, z, t) . \tag{5.6}$$

Note that the constant value of pressure was taken to be zero without loss of generality. Furthermore, at rigid stationary boundaries that confine the fluid domain, we require

$$\nabla \phi \cdot \mathbf{n} = 0 . \tag{5.7}$$

Equations (5.1), (5.4), (5.6), and (5.7) are the basis for the traditional water-wave formulation and a majority of water-wave problems are analyzed with this formulation. Even with all of the simplifying assumptions made, solutions to this problem remain intractable. Sources of difficulties evidently arise from the free-surface boundary conditions (5.4) and (5.6), which are nonlinear and the boundary location on which the conditions are applied is not known *a priori*. To circumvent these difficulties, we often resort to a standard mathematical approach — the representation of functions (such as ϕ and η) by asymptotic series. Methods to solve the above problem are referred to elsewhere, e.g., Wehausen and Laitone (1960).

At the lowest approximation, (5.4) and (5.6) are reduced to

$$\frac{\partial \eta}{\partial t} = \frac{\partial \phi}{\partial y} \quad \text{on } y = 0 , \tag{5.8}$$

and

$$\frac{\partial \phi}{\partial t} + g\eta = 0 \quad \text{on } y = 0 , \tag{5.9}$$

For waves propagating in water of uniform depth h, one of the solutions to the linearized problem is found to be

$$\phi(x, y, t) = \frac{a\sigma}{k} \frac{\cosh k(y + h)}{\sinh kh} \sin(kx - \sigma t) , \tag{5.10}$$

where a is the wave amplitude ($\eta = a\cos(kx - \sigma t)$), σ is the wave angular frequency ($= 2\pi/T$), k is the wavenumber ($= 2\pi/L$), T is the wave period, and L is the wavelength. This solution satisfies the irrotationality of the flow, and represents the oscillatory features in the x direction and the exponential decaying behavior in the negative y direction.

5.2. *Energy considerations*

Now, recall that the rate of change in mechanical energy as following a lump of fluid can be expressed by (3.16), and for incompressible fluids, $\tau_{ij} = -p\delta_{ij} + 2\mu e_{ij}$. Hence, (3.16) can be written as

$$\frac{D}{Dt}\int_D \rho\left(\frac{u^2}{2} + gy\right) dV = \int_S \{-p\delta_{ij} + 2\mu e_{ij}\} u_i n_j ds - \int_D 2\mu |e_{ij}|^2 dV \ . \qquad (5.11)$$

The last term represents the energy dissipation rate. Since $e_{ij} = \frac{1}{2}(\nabla\mathbf{u} + (\nabla\mathbf{u})^T) = \frac{\partial^2\phi}{\partial x_i \partial x_j}$, the energy dissipation rate $\dot{\varepsilon}$ averaged over one wave period and one wavelength can be found by substituting (5.10) into

$$\begin{aligned}\dot{\varepsilon} &= \frac{1}{LT}\int_0^T \int_D 2\mu |e_{ij}|^2 dV dt \\ &= 2\mu \frac{1}{LT}\int_0^T \int_0^L \int_{-h}^0 \left(\left(\frac{\partial^2\phi}{\partial x^2}\right)^2 + 2\left(\frac{\partial^2\phi}{\partial x \partial y}\right)^2 + \left(\frac{\partial^2\phi}{\partial y^2}\right)^2\right) dy\, dx\, dt\ ,\end{aligned} \qquad (5.12)$$

which is

$$\dot{\varepsilon} = 2\mu \frac{ka^2\sigma^2}{\tanh kh} \neq 0\ . \qquad (5.13)$$

This evidently indicates that the solution to the linearized wave-motion problem for an irrotational flow dissipates its energy at the rate of (5.13). The irrotational fluid motion described by (5.10) involves the angular deformation of fluid parcels; hence, the rate-of-strain tensor e_{ij} is finite even though the rotational tensor $b_{ij} = \frac{1}{2}(\nabla\mathbf{u} - (\nabla\mathbf{u})^T)$ vanishes everywhere in the fluid domain. (Note that, according to (2.1), $\boldsymbol{\omega} = \varepsilon_{ijk} b_{jk}$; hence, if $\boldsymbol{\omega} = 0$, then $b_{ij} = 0$.)

Now for irrotational, incompressible (but not necessarily inviscid) flows, the Euler integral is applicable

$$\rho\frac{\partial\phi}{\partial t} + \frac{1}{2}\rho|\nabla\phi|^2 + p + \rho g y = 0\ . \qquad (5.14)$$

Hence, the mechanical energy for irrotational flows can be expressed as

$$ME = \frac{1}{2}\rho|\nabla\phi|^2 + \rho g y = -\rho\frac{\partial\phi}{\partial t} - p \,. \tag{5.15}$$

The integrated mechanical energy contained in the domain D is

$$\varepsilon(t) = \int_D \left(\frac{1}{2}\rho|\nabla\phi|^2 + \rho g y\right) dV = -\int_D \left(\rho\frac{\partial\phi}{\partial t} + p\right) dV \,, \tag{5.16}$$

and the time rate of change in mechanical energy in the fluid domain shown in Fig. 10 is

$$\frac{D}{Dt}\int_D \varepsilon dV = \dot{\varepsilon} = \int_D \rho\nabla\phi\cdot\nabla\left(\frac{\partial\phi}{\partial t}\right) dV + \int_S \left(\frac{1}{2}\rho|\nabla\phi|^2 + \rho g y\right) \mathbf{U}\cdot\mathbf{n}\, ds \,, \tag{5.17}$$

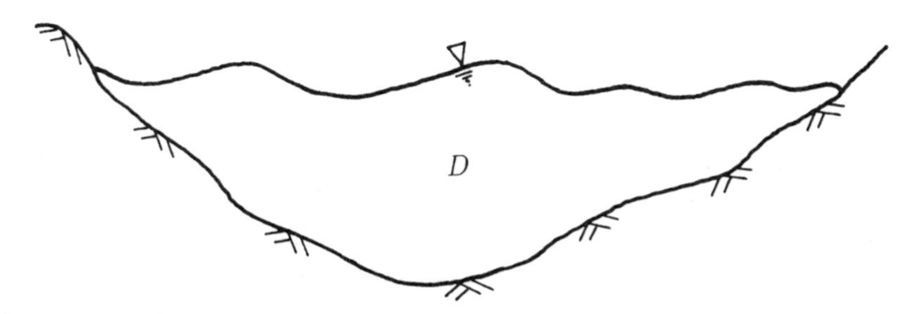

Fig. 10. A sketch of fluid domain for the energy consideration.

using the transport theorem (2.8). In (5.17), $\mathbf{U}$ is the velocity of the boundary and $\mathbf{n}$ is the unit vector pointing normal to the boundary. Using (5.16), the time-rate change in mechanical energy can be expressed as

$$\begin{aligned} \dot{\varepsilon} &= \int_D \rho\nabla\cdot\left(\frac{\partial\phi}{\partial t}\nabla\phi\right) dV - \int_S \left(\rho\frac{\partial\phi}{\partial t} + p\right) \mathbf{U}\cdot\mathbf{n}\, ds \\ &= \int_S \left\{\rho\frac{\partial\phi}{\partial t}(\nabla\phi\cdot\mathbf{n} - \mathbf{U}\cdot\mathbf{n}) - p\mathbf{U}\cdot\mathbf{n}\right\} ds \,. \end{aligned} \tag{5.18}$$

The first integrand vanishes due to the kinematic condition at material boundaries, $\nabla\phi\cdot\mathbf{n} - \mathbf{U}\cdot\mathbf{n} = 0$. The second integrand also vanishes because, as shown in Fig. 10, the bottom part of the boundary is a stationary rigid body boundary, i.e., $\mathbf{U} \cdot \mathbf{n} = 0$, and at the rest of the boundary, i.e., at the free surface, the pressure is assumed to be constant; hence, the integration of $p\mathbf{U} \cdot \mathbf{n}$ over the free surface must vanish due to the conservation of fluid volume. Therefore,

$$\dot{\varepsilon} = 0 \, . \tag{5.19}$$

Evidently, (5.19) contradicts (5.13). This inconsistency is not due to the approximation involved in (5.10). Suppose, instead of (5.10), we used a more accurate approximation by including the higher-order terms, then the correction in (5.13) due to the higher-order terms would be at most $\mathrm{O(ak)}^4$ that cannot make the energy dissipation described in (5.13) vanish.

To find the cause of this contradiction, let us consider the Navier-Stokes equation of linear momentum for an irrotational (but not necessarily inviscid) flow of the form:

$$\begin{aligned} \frac{D}{Dt}(\nabla\phi) &= -\frac{1}{\rho}\nabla p + \frac{\mu}{\rho}\nabla^2(\nabla\phi) + \mathbf{g} \\ &= -\frac{1}{\rho}\nabla p + \mathbf{g} \, . \end{aligned} \tag{5.20}$$

Even though the fluid viscosity is present, the viscous term vanishes for an incompressible fluid, because $\nabla^2\phi = 0$. As far as the linear momentum is concerned, it appears that there is no viscous effect when the flow is irrotational. Consider, for example, a fluid parcel that has initially a cubic shape as shown in Fig. 11. Under pure angular deformation of a fluid parcel without rotation, the viscous force acting on one of the fluid-parcel surfaces is canceled out with the one acting on the opposite surface (see Fig. 11). Note that the rate-of-strain tensor (which measures the angular deformation) is symmetric and can be reduced to its diagonal form with the principal axes. The linear deformation in one principal axis has identical magnitude in the direction opposite to it; the force associated with such deformation does not have a net effect on the linear momentum.

Now let us consider the equation of mechanical energy,

$$\rho\frac{D}{Dt}\left(\frac{|\mathbf{u}|^2}{2} + \rho g y\right) = \nabla \cdot (\mathbf{u}\tau_{ij}) - \tau_{ij} : \nabla\mathbf{u} \, , \tag{5.21}$$

where $\tau_{ij} = -p\delta_{ij} + 2\mu e_{ij}$; $e_{ij} = \frac{1}{2}(\nabla\mathbf{u} + (\nabla\mathbf{u})^T)$, for incompressible fluids. Basically, (5.21) states that the rate of change in mechanical energy following

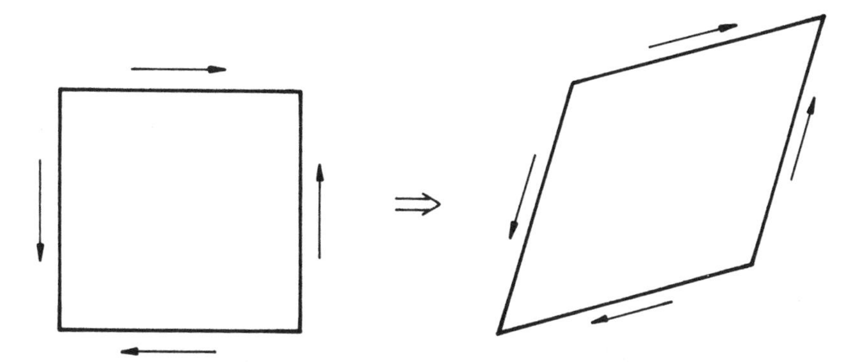

Fig. 11. A sketch to describe pure angular deformation of a fluid parcel.

a fluid parcel is equal to the time rate of work done on the fluid parcel by the surroundings, $\nabla \cdot (\mathbf{u}\tau_{ij})$, minus the dissipation of mechanical energy to internal energy, $\tau_{ij} : \nabla \mathbf{u}$. For irrotational flows (but not necessarily inviscid fluids), the energy dissipation term can be written as

$$\tau_{ij} : \nabla \mathbf{u} = 2\mu |e_{ij}|^2 = 2\mu \left| \frac{\partial^2 \phi}{\partial x_i \partial x_j} \right|^2 , \tag{5.22}$$

and the work done by the surface force can be expressed as

$$\nabla \cdot (\mathbf{u}\tau_{ij}) = -\nabla \cdot (p\nabla\phi) + 2\mu \left| \frac{\partial^2 \phi}{\partial x_i \partial x_j} \right|^2 . \tag{5.23}$$

Substituting (5.22) and (5.23) into (5.21) and integrating over the fluid domain (Fig. 10) gives the time rate of change in mechanical energy of the specified fluid domain, which is expressed as

$$\begin{aligned} \int_D \{\nabla \cdot (\mathbf{u}\tau_{ij}) - \tau_{ij} : \nabla \mathbf{u}\} dV &= -\int_D \nabla \cdot (p\nabla\phi) dV \\ &= -\int_S p\nabla\phi \cdot \mathbf{n}\, ds = 0 , \end{aligned} \tag{5.24}$$

since $\nabla\phi \cdot \mathbf{n} = 0$ on a stationary solid boundary, and p is constant at the free surface (again, the value of $p\nabla\phi \cdot n$ integrated over the entire free surface must vanish because the fluid volume is conserved).

This demonstrates evidently that the rate of energy dissipation is exactly balanced with a portion of the rate of work done by the surroundings, more specifically, the work done via viscous forces. When the pressure is constant at the free surface, the work done by the surroundings does not vanish because

$$\begin{aligned}\int_D \nabla \cdot (\mathbf{u}\tau_{ij}) dV &= 2\mu \int_S \{\nabla \cdot (\nabla\phi \circ \nabla\phi)\} \cdot \mathbf{n}\, ds \\ &= 2\mu \int_{S_f} \{\nabla \cdot (\nabla\phi \circ \nabla\phi)\} \cdot \mathbf{n}\, ds \\ &\neq 0\,, \end{aligned} \tag{5.25}$$

where S_f denotes the free-surface boundary. Hence, in order for the solution (5.10) to be valid, a specific amount of energy must be pumped into the fluid domain at the free surface via viscous force. This is true for any irrotational-flow solutions based on the traditional water-wave formulations: (5.1), (5.4), and (5.6). Basically, the condition $p =$ constant (or more conveniently, $p = 0$, as specified in (5.6)) at the free surface, is not sufficient for the stress-free dynamic boundary condition unless we assume the fluid to be inviscid. In fact, if there were indeed no stresses at the boundary, there would be no mechanism to feed any energy into the fluid domain. It is emphasized that the traditional water-wave formulation does not explicitly require the fluid to be inviscid, although the imposed dynamic boundary condition of the form, $p = 0$, at the free surface (instead of the true vanishing stress condition) implicitly assumes that viscous stress at the surface is negligible.

Suppose there is no artificial force to pump the energy on the surface and the free surface is truly a stress-free plane. Then, our irrotational flow solution breaks down, the flow near the surface becomes vortical (to be discussed in Subsec. 5.3), and the wave motion must attenuate as it propagates. Potential flows cannot dissipate energy unless the same amount of energy is supplied by the surroundings. This complies with Kelvin's minimum energy theorem (see (3.62)), which states that irrotational flow has the least mechanical energy among all flows satisfying given boundary conditions.

To see this more explicitly, recall that the free-surface dynamic boundary condition without surface-tension effects is written as

$$\tau_{ij} \cdot \mathbf{n} = 0 \quad \text{on } y = \eta(x,t)\,. \tag{5.26}$$

For our linearized two-dimensional water wave problem, (5.26) becomes

$$\tau_{yy} = -p + 2\mu \frac{\partial v}{\partial y}; \quad \tau_{yx} = \mu\left(\frac{\partial u}{\partial y} + \frac{\partial v}{\partial x}\right), \quad \text{on } y = 0\,. \tag{5.27}$$

If we use our linear solution (5.10) for the water-wave problem, we obtain:

$$\tau_{yy} = -p + 2\mu \frac{a\sigma k}{\tanh kh} \sin(kx - \sigma t) \neq 0 \; ; \quad \tau_{yx} = 2\mu \, a\sigma k \cos(kx - \sigma t) \neq 0 \; . \tag{5.28}$$

Then, the rate at which these forces do work on the fluid is

$$\tau_{yy} v + \tau_{yx} u = -pa\sigma \sin(kx - \sigma t) + 2\mu \frac{a^2 \sigma^2 k}{\tanh kh} \; . \tag{5.29}$$

The time average value of (5.29) is $\frac{2\mu a^2 \sigma^2 k}{\tanh kh}$, which is identical to (5.13). Note that for a constant value of p at the free surface, the pressure force does not contribute to the mean rate of work on the free surface.

5.3. *Free-surface boundary layer*

At a free surface, both normal and tangential components of stresses are zero unless there are surface-tension effects. However, these free-surface conditions are, in general, incompatible with irrotational (viscous) motions; specifically, the condition of zero tangential stress at the free-surface boundary cannot be satisfied by irrotational motions. Hence, in general, there is a jump in vorticity at the boundary, and a boundary layer is formed along the surface. To illustrate this, let us consider the free surface expressed in "local" cylindrical coordinates (r, θ, z) as shown in Fig. 12. The vorticity in the z direction can be written as

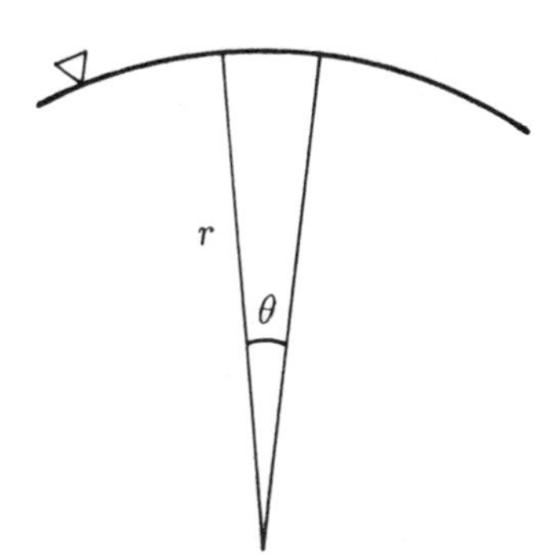

Fig. 12. A sketch of the free-surface boundary expressed with local cylindrical coordinates.

$$\omega_z = \frac{1}{r} \frac{\partial v_r}{\partial \theta} - \frac{1}{r} \frac{\partial}{\partial r} (r v_\theta) \; , \tag{5.30}$$

where v_r and v_θ are fluid velocities in the radial (r) and azimuthal (θ) directions, respectively. Equation (5.30) can be further arranged as

$$\omega_z = -\left(\frac{1}{r}\frac{\partial v_r}{\partial \theta} + r\frac{\partial}{\partial r}\left(\frac{v_\theta}{r}\right)\right) + \frac{2}{r}\frac{\partial v_r}{\partial \theta} - \frac{2v_\theta}{r} \,. \tag{5.31}$$

Note that the term in the parenthesis is twice the off-diagonal term of the rate-of-strain tensor (e_{ij}) that must vanish at the stress-free boundary, if the fluid viscosity is finite. Because the term in the parenthesis must vanish at the free surface, vorticity can be expressed by

$$\omega_z = \frac{2}{r}\frac{\partial v_r}{\partial \theta} - \frac{2v_\theta}{r} \,, \tag{5.32}$$

at the free-surface boundary.

Now, consider an irrotational flow in which a velocity field is computed in terms of velocity potential. The irrotationality of the flow field requires $\boldsymbol{\omega} = \nabla \times \mathbf{u} = \nabla \times (\nabla\phi) \equiv 0$ everywhere, including the free surface. For irrotational flows, (5.31) becomes (Batchelor, 1967)

$$\omega_z = 0 = -\left(\frac{1}{r}\frac{\partial^2 \phi}{\partial r \partial \theta} + r\frac{\partial}{\partial r}\left(\frac{1}{r^2}\frac{\partial \phi}{\partial \theta}\right)\right) + \frac{2}{r}\frac{\partial^2 \phi}{\partial \theta \partial r} - \frac{2}{r^2}\frac{\partial \phi}{\partial \theta} \,. \tag{5.33}$$

In order to satisfy both the vanishing stress condition and the irrotationality at the free surface, the following condition is required:

$$\frac{1}{r}\frac{\partial^2 \phi}{\partial r \partial \theta} + r\frac{\partial}{\partial r}\left(\frac{1}{r^2}\frac{\partial \phi}{\partial \theta}\right) = \frac{2}{r}\frac{\partial^2 \phi}{\partial \theta \partial r} - \frac{2}{r^2}\frac{\partial \phi}{\partial \theta} = 0 \,. \tag{5.34}$$

When the free surface is stationary (or after making the free surface stationary by taking a proper moving coordinate system), there is no radial velocity component at the surface where the locus is fitted by local cylindrical coordinates, i.e., $\partial\phi/\partial r = 0$ at all points on the surface (see Fig. 12). Then, (5.34) is satisfied if the surface curvature is zero ($r \to \infty$, i.e., a plane surface) or if there is no tangential component of the fluid velocity at the surface. Otherwise, there must be the vorticity jump with the amount

$$\Delta\omega_z = \frac{1}{r}\frac{\partial^2 \phi}{\partial r \partial \theta} + r\frac{\partial}{\partial r}\left(\frac{1}{r^2}\frac{\partial \phi}{\partial \theta}\right) \,, \tag{5.35}$$

at the free surface. This evidently shows that the irrotational flow is generally incompatible with the free-surface dynamic condition of viscous fluids, consequently, a boundary layer is formed along the free surface.

A similar observation at the stress-free surface can be made for general rotational flows. Again, when the free surface is stationary (by choosing a proper coordinate system if necessary), the term $\frac{\partial v_r}{\partial \theta} = 0$ and (5.32) becomes

$$\omega_z = -\frac{2v_\theta}{r} \quad \text{or} \quad 2\kappa v_\theta \ , \tag{5.36}$$

where κ is curvature ($\kappa = -1/r$). This result was discussed by Longuet-Higgins (1953). Equation (5.36) indicates that at a plane free surface ($\kappa = 0$), the vorticity is always nil, $\boldsymbol{\omega} = 0$. On the other hand, when the surface curvature is non-zero, the flow is always vortical with the magnitude of (5.36) (or (5.32) for a non-stationary surface) as long as the tangential velocity component is finite at the surface. Note that according to (5.36), vorticity at the surface is independent of fluid viscosity: the vorticity does not vanish in the limit of small viscosity. The following physical interpretation for (5.36) was provided by Longuet-Higgins (1992). Because the stress must vanish at the (curved) free surface, $\tau_{ij} \cdot \mathbf{n} = 0$, there is no relative deformation of the fluid parcel at the free surface, hence the fluid-parcel motion along the surface must behave like a rigid-body motion without deformation. Since the rigid-body rotation, Ω, is the ratio of the tangential orbital velocity to its radius, v_θ/r, and the vorticity is twice the rate of rotation, $\omega = 2\Omega$, (5.36) results.

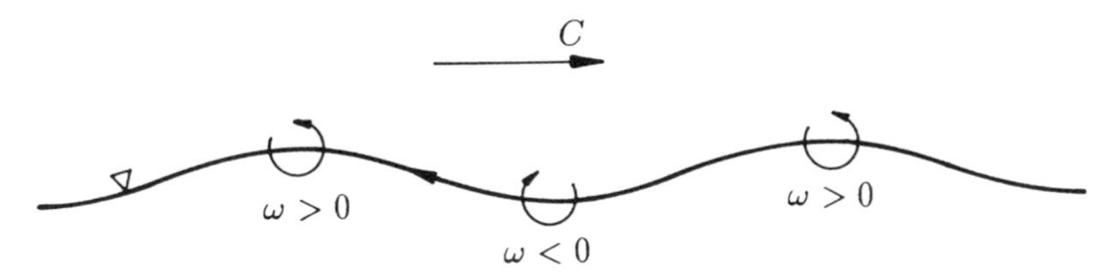

Fig. 13. Vorticity distribution along the free surface of the wave propagating from left to right.

In the coordinates moving with the wave phase velocity c, a uniform linear wave train can be written by $\eta = a\cos kx$, where a is the wave amplitude. Since the surface curvature is $\kappa \cong \eta_{xx}$ as the lowest order approximation of (4.38) and $v_\theta \cong -c$ in the moving coordinates, we find $\omega = 2\kappa v_\theta = -2c\eta_{xx} = ak^2 c\cos kx$.

Thus, ω is positive at the wave crests and negative in the troughs as shown in Fig. 13.

The same conclusions presented in this section can be drawn by using more general curvilinear coordinates (instead of "local" cylindrical coordinates) to describe the free surface (Longuet-Higgins, 1953; Batchelor, 1967). It is emphasized that the vorticity at the surface described in (5.36) is based on the assumption of vanishing stress at the interface, limited to two-dimensional stationary flows, and does not provide information of the rate of vorticity production at the interface, which will be discussed in Sec. 6.

5.4. *Wave attenuation*

As shown in Subsec. 5.2, unless the (air) flow above the air-water interface feeds the energy into the water domain at the rate $\dot{\varepsilon} = 2\mu\, a^2\sigma^2 k/\tanh kh$ through the surface, the irrotational-flow theory becomes invalid. On the other hand, if we assume that there is no stress being exerted on the surface, i.e., $\tau_{ij} \cdot \mathbf{n} = 0$, it can be viewed that, as the lowest-order approximation, the irrotational wave motion loses mechanical energy at the rate of $\dot{\varepsilon} = 2\mu\, a^2\sigma^2 k/\tanh kh$ via viscous effects in the fluid interior. It is emphasized that, in a real fluid environment, wave-energy dissipation takes place not only within the fluid interior, but also in the boundary layers formed along the rigid boundaries, e.g., the side walls and bottom of a wave tank. Because the effects of free surface are the focus here, the topics of energy dissipation associated with rigid-body boundary layers are simply referred to elsewhere (e.g., Mei and Liu, 1973) and are excluded from our discussion. Note that, for deep-water waves, a majority of energy dissipation takes place in the interior of the fluid owing to the fact that the rate-of-strain tensor does not vanish in irrotational wave motions (5.13). Since the value of energy density of the wave motion is $\varepsilon = \frac{1}{2}\rho g a^2$ (see, e.g., Dean & Dalrymple, 1991) and the energy dissipation rate for deep-water waves ($kh \to \infty$) is $\dot{\varepsilon} = 2\mu\, a^2\sigma^2 k$, we can write

$$\frac{\dot{\varepsilon}}{\varepsilon} = -4\frac{\mu k \sigma^2}{\rho g} = -4\nu k^2 \,, \tag{5.37}$$

using the dispersion relation for deep-water waves, $\sigma^2 = gk$. Since the mechanical energy is proportional to the square of wave amplitude ($\varepsilon \propto a^2$), the wave amplitude decays in time according to

$$a(t) = a_0 e^{-2\nu k^2 t} = a_0 e^{-\beta t}, \tag{5.38}$$

where β is the damping coefficient with the e-folding time $\beta^{-1} = (2\nu k^2)^{-1}$. The time varying wave amplitude $a(t)$ can be incorporated into the potential flow solution by writing,

$$\eta(x,t) = \Re\{a_0 e^{ikx} e^{\alpha t}\} \quad \text{and} \quad \phi(x,t) = -\Im\left\{\frac{a_0\alpha}{k} e^{ky} e^{ikx} e^{\alpha t}\right\} , \tag{5.39}$$

where $\alpha = -\beta - i\sigma$, $\Re\{\bullet\}$ and $\Im\{\bullet\}$ denote the real and imaginary parts respectively. Note that this modified solution satisfies the Laplace equation $\nabla^2\phi = 0$ and the boundary condition at $y \to -\infty$. At the upper free-surface boundary, the kinematic boundary condition requires (5.8). Substituting (5.39) into (5.8) yields

$$-a_0\beta e^{-\beta t}\cos(kx - \sigma t) + a_0\sigma e^{-\beta t}\sin(kx - \sigma t) = a_0\sigma e^{-\beta t}\sin(kx - \sigma t) , \tag{5.40}$$

which is satisfied approximately only if $|\beta| \ll |\sigma|$, or $2\nu k^2 \ll \sigma$ or $\sigma/\nu k^2 \equiv \mathbf{R} \gg 1$. Hence, viscous energy dissipation leads to simple attenuation in wave amplitude only if the wave Reynolds number $\mathbf{R}$ is large.

The wave attenuation rate predicted by (5.38) appears to be artificial because it is based on the no-stress free-surface condition which causes a very gradual wave attenuation. Even if we consider the presence of the air above the water surface, the dynamic influence of the air motion on the water motion should be minute, because the density of the air ($\rho_a \approx 1.25$ kg/m^3) is much smaller than that of water ($\rho_w \approx 1,000$ kg/m^3). However, this minute dynamic effect of the air could alter significantly the very slow process of wave attenuation, which is also caused by the meager effect of the free surface.

5.5. *Two-layer irrotational-flow systems*

If the free surface is viewed as a fluid-fluid (air-water) interface and assuming incompressible irrotational flows in both fluid domains, the problem is formulated to solve the Laplace equations for velocity potentials in each fluid domain with the kinematic conditions at the interface:

$$\frac{\partial\eta}{\partial t} + \frac{\partial\phi_1}{\partial x}\frac{\partial\eta}{\partial x} = \frac{\partial\phi_1}{\partial y}; \ \frac{\partial\eta}{\partial t} + \frac{\partial\phi_2}{\partial x}\frac{\partial\eta}{\partial x} = \frac{\partial\phi_2}{\partial y} \quad \text{at } y = \eta(x,t) . \tag{5.41}$$

For inviscid fluids, neglecting the surface-tension effect, the dynamic condition is

$$p_1 = p_2 \quad \text{at} \quad y = \eta(x,t) , \tag{5.42}$$

where pressure p can be expressed by the Euler integral of the form

$$p_1 = -\left(\rho_1 \frac{\partial \phi_1}{\partial t} + \frac{1}{2}\rho_1 (\nabla \phi_1)^2 + \rho_1 g \eta\right) \quad \text{and}$$

$$p_2 = -\left(\rho_2 \frac{\partial \phi_2}{\partial t} + \frac{1}{2}\rho_2 (\nabla \phi_2)^2 + \rho_2 g \eta\right) . \tag{5.43}$$

Based on this formulation, Holyer (1979) computed the highest wave characteristics using the Padé approximants. Compared with the results of Cokelet (1977) for the highest wave under the free-surface condition, i.e., $p = 0$ at $y = \eta$ instead of (5.42), Holyer found that a wave with air above water surface can be 2% higher than that based on the free-surface condition. The maximum value of integrated wave properties such as the mean wave momentum and the mean energy are not influenced significantly by the presence of air above the water surface. Holyer's results clearly indicate that as far as the potential-flow dynamics are concerned, the presence of air on the water surface has very little influence on the dynamics of the water motions, i.e., the free-surface condition yields very good approximations for the irrotational water-wave problems.

While detailed wave characteristics and behaviors of a periodic (quasi-steady) solution are not affected by the existence of air, the presence of air appears to play a significant role in the determination of its unsteady behavior. It is well known among surfers that wave-breaking characteristics are different depending on the wind direction. On a day with offshore wind blowing, breakers tend to be a plunging type — most adventurous surfers prefer this type of breaker. While on a day with onshore wind blowing, the breakers tend to be a spilling type. The effects of wind on wave breaking are examined experimentally in a laboratory environment by Douglass (1990), and his results confirm the surfers notion as described.

The quasi-steady phenomenon such as a bore, a hydraulic jump, or a spilling breaker generates strong vortical flows at the surface roller — the surface roller is a quasi-steady recirculating flow region at the front face of a bore. Nonetheless, the integrated characteristics of those flows can be modeled accurately without considering the presence of air. For example, upstream and downstream conditions of a hydraulic jump can be predicted adequately without taking into account the air above the water surface; a hydraulic jump is usually solved with the shallow-water wave equations based on the assumptions of a hydrostatic pressure field and uniform flow conditions (e.g., see Henderson, 1966). A spilling breaker can be modeled adequately by treating the white cap as a gravity current of reduced density riding on the potential flow; the density current consists of a water-air mixture that continually entraps the water from the irrotational flow underneath (Longuet-Higgins & Turner, 1974). On the other hand, at least some of the physical interpretations associated with vortical fluid motions at the water surface cannot be made correctly without the consideration of the presence of air. The creation of rotational motion at the front face of bore cannot be explained properly without the consideration of air presence (Yeh, 1991). From the basic view point of the angular-momentum conservation, a fluid parcel cannot initiate its rotation unless there is torque exerted on it. As discussed in Subsec. 3.6, torque can occur only when the fluid is inhomogeneous in density and/or viscosity. Based on this argument, the presence of the air is essential for density and viscosity gradients and, hence, for the creation of fluid rotation.

As we discussed in Subsec. 5.2, to maintain a periodic motion of irrotational incompressible (but viscous) flow in deep water, the energy rate, $\dot{\varepsilon} = 2\mu a^2 \sigma^2 k$, must be fed into the flow domain through the water surface. Suppose we assume the free-surface conditions, i.e., no stress on the water surface. Then, no work can be done at the surface and no energy can be input into the flow domain. A uniform periodic irrotational motion is no longer possible and the energy imbalance results in the attenuation of wave motion with the rate of an e-folding time $\beta^{-1} = (2\nu k^2)^{-1}$ (see (5.38)). In other words, energy attenuation of deep-water waves, which is a slowly-varying time-dependent phenomenon (for a high Reynolds number), is controlled by the condition imposed at the water-surface boundary. Hence, it is anticipated that the effects of air appear to play an important role in wave attenuation.

The presence of air above the water surface was considered in the analysis of Dore (1978). Taking the coordinates at the mean air-water interface, the x axis pointing horizontally, the y axis pointing upwards, the velocity fields in the air and water were expressed with the Helmholtz decomposition (see (3.67) for a general form):

$$u = \frac{\partial \phi}{\partial x} + \frac{\partial \psi}{\partial y} \text{ and } v = \frac{\partial \phi}{\partial y} - \frac{\partial \psi}{\partial x}, \tag{5.44}$$

where ψ is the stream function, the irrotational part of the flow satisfies the Laplace equation $\nabla^2 \phi = 0$, and the vortical part satisfies the diffusion equation, $\frac{\partial \psi}{\partial t} = \nu \nabla^2 \psi$, for the linearized problem. Using the conditions $u \to 0$ and $v \to 0$ as $y \to \pm\infty$, the vortical flow part in each fluid domain (i.e., air and water) is solved for the interfacial boundary layer for progressive waves. The irrotational flow part is solved in each fluid domain by utilizing the linearized kinematic boundary condition and (viscous) tangential dynamic boundary condition. The computed velocity fields in the air and water domain lead to the total energy density and total energy dissipation function, just like we computed earlier for the water domain alone, i.e., (5.37). From that procedure, Dore found that the wave attenuation rate (the damping coefficient):

$$\beta = \frac{\sqrt{2}\sigma}{\mathbf{R}_w^{1/2}}\sqrt{\frac{\rho_a \mu_a}{\rho_w \mu_w}} + 2\nu_w k^2 \ , \tag{5.45}$$

where σ is the angular frequency of the wave motions, k is the wavenumber, ν_w is the kinematic viscosity of the water, $\nu_w = \mu_w/\rho_w$, $\mathbf{R}_w$ is the wave Reynolds number, $\mathbf{R}_w \equiv \sigma/\nu_w k^2$, and the subscripts a and w denote the fluid properties of the air and water respectively. Note that the second term on the right-hand side is the attenuation rate for the "free-surface" (stress-free) boundary at the interface; taking a limit $\rho_a \to 0, \mu_a \to 0$, (5.45) reduces to (5.38).

Using $\rho_a = 1.25$ kg/m^3 and $\mu_a = 1.76 \times 10^{-5} N \cdot s/m^2$ for the air, and $\rho_w = 1{,}000$ kg/m^3, $\mu_w = 1.31 \times 10^{-3} N \cdot s/m^2$ for the water, (5.45) can be written as

$$\frac{\beta}{\sigma} = 0.0058 \times \mathbf{R}_w^{-1/2} + 2\mathbf{R}_w^{-1} \ . \tag{5.46}$$

The influence of air on wave attenuation becomes important in comparison with the condition of no fluid above the water surface when $\mathbf{R}_w > 119{,}000$, which is equivalent to $k < 7.39$/m (i.e., wave length > 0.85 m). Hence, the effect of air on wave attenuation cannot be neglected for most of the gravity waves; on the other hand, the air effect on the attenuation rate appears to be insignificant for short waves such as gravity-capillary or capillary waves.

5.6. *Surfactant effects*

If surface-tension γ at the air-water interface is constant, the net work done by the surface-tension effect on the water during one wave oscillation is zero. This can be readily shown by the following. Consider a wave motion approximated by (5.10). At the lowest-order approximation, the force per unit area associated with surface-tension is $\gamma\frac{\partial^2 \eta}{\partial x^2}$ according to (4.39) and (4.38). The rate at which the surface-tension force works on the water, i.e., energy rate (power), is $\gamma\frac{\partial^2 \eta}{\partial x^2} v$, in which v is the vertical velocity component. Since v and $\gamma\frac{\partial^2\eta}{\partial x^2}$ are out-of-phase by $\pi/2$, the integration of the product over one wave period vanishes. Hence, there is no net work done on the water by the surface-tension effect. This is not the case when the surface tension is not constant over the air-water interface. As discussed in Subsec. 4.3, the surface-tension can be varied by the presence of surface-active substances (surfactant) with concentrations that vary from point to point and/or by the surface temperature variations. The variation in surface-tension creates the appearance of tangential stresses on the surface.

In a natural environment, the surface-tension at the water surface is often contaminated by a single molecular layer of surfactant, i.e., a monolayer. A monolayer is a film considered to be only one molecule thick, and the thickness is small enough ($\approx$ 20Å) that gravitational effects are negligible. Once the air-water interface is covered with a monolayer, either by absorption from a solution or by spreading, the interface property becomes elastic so that it tends to resist the periodic surface expansion and compression associated with wave motions. Consequently, wave motions with the contaminated surface dissipate energy by viscous friction. The effect of resistance to tangential motion at the interface on wave motion depends on the surface elastic modulus (there is an analogy between a monolayer covered surface and a stretched elastic membrane).

The evaluation of locally varying surface stress is complicated and in most cases, only a formal treatment is possible based on the principles of surface rheology (see Adamsons, 1990; Edwards, Brenner, and Wasan, 1991). One of them is the surface shear viscosity μ_s, which is the two-dimensional counter part of the viscosity. For example, if two parallel line elements along the x direction in the surface are in a motion with a finite relative velocity in the x direction, the equilibrium force per unit length (not per area) of the element is

$$\Delta\gamma = \mu_s \frac{du}{dz} \ . \tag{5.47}$$

Another rheological property is the surface dilational modulus $\hat{E}$, which is the counterpart of bulk modulus of elasticity:

$$\Delta\gamma = \hat{E}\frac{1}{A}\frac{dA}{dt} \ , \tag{5.48}$$

where A is the geometric area of the surface. Note that the equilibrium quantity corresponding to $\hat{E}$ is the modulus of surface elasticity E defined by

$$E = \frac{d\gamma}{d\ln A} \, . \tag{5.49}$$

The effects of surface-tension variation can be included in the dynamic boundary condition as shown in (4.42). Neglecting the stresses in the air domain, (4.42) becomes the condition for the water domain,

$$\left[-p\delta_{ij} \cdot \mathbf{n} + \mu(\nabla \mathbf{u} + (\nabla \mathbf{u})^T) \cdot \mathbf{n}\right] = -\gamma \left(\frac{1}{R_1} + \frac{1}{R_2}\right) \mathbf{n} + \nabla_\lambda \gamma \, , \tag{5.50}$$

where ∇_λ is the gradient operator in the directions tangent to the interface. To solve the problem involved with surface-tension variations, the gradient of surface-tension must be expressed by other parameters, usually related to the horizontal displacement ξ of the surface due to the fluid motion on the surface, together with the aforementioned rheological properties of the film such as surface dilational modulus and surface shear viscosity. The surface does change its area and shape when the surface stress is applied. Therefore, both dilational and shear effects play a role in determination of the surface-tension gradient.

Most of the analyses on surfactant effects are based on the effects of surface dilation only and the effects of its shape change (surface shear viscosity) are not important, hence usually neglected. Using the equilibrium condition (5.49) for small deformations, the change in surface-tension can be related to the relative change in area by

$$\nabla_\lambda \gamma = \frac{d\gamma}{d\ln A} \nabla_\lambda (\ln A) = \frac{E}{A} \nabla_\lambda A \, . \tag{5.51}$$

If we consider a first-order approximation of two-dimensional flows in the x-y plane for simplicity, (5.51) reduces to, using (5.49),

$$\nabla_\lambda \gamma = \frac{d\gamma}{dx} = \frac{d\gamma}{d\ln A} \frac{1}{A} \frac{dA}{dx} = E \frac{\partial^2 \xi}{\partial x^2} \, , \tag{5.52}$$

where ξ is the horizontal displacement of the surface element due to the fluid motion (Lucassen-Reynders and Lucassen, 1969). Note that, in (5.52), the relative change in the surface element area A is related to the horizontal displacement ξ of the surface by $\Delta A/A = \partial\xi/\partial x$. Substituting (5.52) and (4.38) into (5.50) yields

$$\mu \left(\frac{\partial u}{\partial y} + \frac{\partial v}{\partial x}\right) = E \frac{\partial^2 \xi}{\partial x^2} \, , \tag{5.53}$$

and

$$-p + 2\mu\frac{\partial v}{\partial y} = \gamma\frac{\partial^2 \eta}{\partial x^2} , \tag{5.54}$$

in the tangential and normal components respectively.

Expressing the velocity field with the Helmholtz decomposition of the form (5.44), and taking the "normal mode" assumption:

$$\phi(x, y, t) = \hat{\phi}e^{ky}e^{i(kx+\sigma t)} \text{ and } \psi(x, y, t) = \hat{\psi}e^{my}e^{i(kx+\sigma t)} \tag{5.55}$$

the diffusion equation,

$$\frac{\partial \psi}{\partial t} = \nu\nabla^2\psi ,$$

yield

$$m^2 = k^2 + (i\,\sigma/\nu) . \tag{5.56}$$

Substituting (5.44), (5.55), (5.56), and the expression of pressure by the Euler integral (3.33) into (5.53) and (5.54) yield a pair of algebraic equations (Lucassen, 1968):

$$\hat{\phi}(i\rho gk - i\rho\sigma^2 - 2\mu\sigma k^2 + i\gamma k^3) + \hat{\psi}(\rho gk + 2i\mu\sigma mk + \gamma k^3) = 0 , \tag{5.57}$$

and

$$\hat{\phi}(2i\mu\sigma k^2 + Ek^3) + \hat{\psi}(\mu\sigma(m^2 + k^2) - iEmk^2) = 0 . \tag{5.58}$$

For a given value of wave frequency σ, two complex roots for the wavenumber can be obtained by setting the determinant of (5.57) and (5.58) to vanish in order to keep solutions for $\hat{\phi}$ and $\hat{\psi}$ non-trivial (Lucassen, 1968). The imaginary part of the wavenumber represents the wave attenuation rate. One of the roots represents the surface wave and the other represents the longitudinal wave (or the Marangoni wave). The longitudinal wave is the wave associated with the 'elastic' monolayer of surfactant, which is caused by compression and expansion of the molecular film situated at the air-water interface. The longitudinal waves are attenuated very quickly, and their wavelength and damping coefficient depend strongly on the value of the modulus of surface elasticity E but depend little on the surface-tension. When the wavelengths of the surface wave and the longitudinal wave coincide, the surface wave shows a resonance-like effect. Lucassen-Reynders and Lucassen (1969) showed that, for this condition, the surface expansion and compression which normally take place at the locations of wave troughs and crests, respectively, are shifted by a half wavelength by the

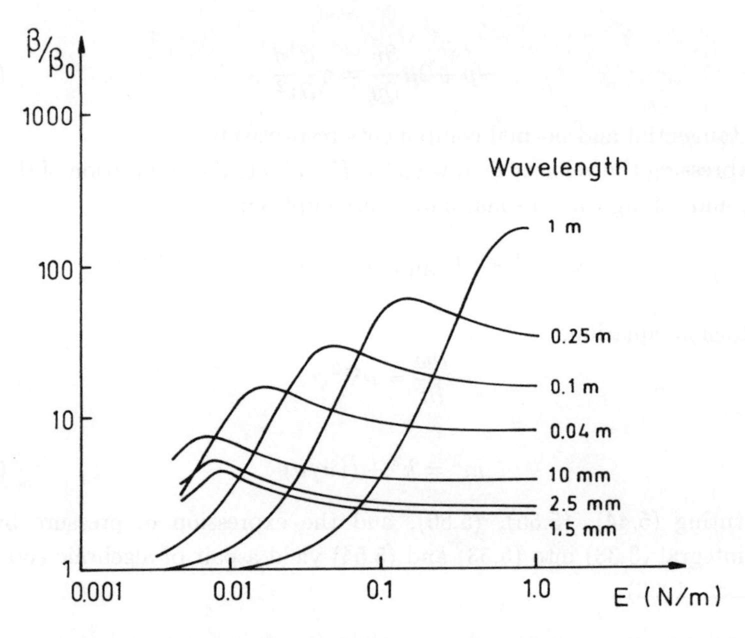

Fig. 14. Ratio of the damping coefficient β with a monolayer surface to that β_0 with a clean surface as a function of surface dilational modulus. (after Lucassen, 1982)

longitudinal waves of the surface layer. Due to this effect, the velocity shear at the surface becomes greater than that under the condition of a rigid surface, and the damping coefficient takes its maximum value which is approximately twice the limited value expected for an infinite value of E (i.e., at a rigid surface).

Lucassen (1968) numerically demonstrated that the real part of a wavenumber is influenced very little by the value of the modulus of surface elasticity E, while the imaginary part (i.e., the damping coefficient) strongly depends on E. As shown in Fig. 14, depending on the wave length (or wavenumber) and the value of E, the maximum value of the damping coefficient can attain the magnitude 10 to more than 100 times greater than the damping coefficient of the wave with a clean (stress-free) surface. If there is no physical limitation in the value of E, Lucassen (1982) found that, as shown in Fig. 14, the longer the wavelength (i.e., the greater the wave period), the larger the maximum relative increase in the value of the damping coefficient (i.e., the ratio of the damping coefficient with the monolayer surface to that with a clean surface is greater for the longer wave). This indicates that the potential effect of surfactant on wave damping must increase with increasing wavelength. Nonetheless, in reality, the value of E in the natural environment is limited by the type of substance available. Monolayer surfactants in the natural environment have the values of E typically in the range of 10^{-3} to $5 \times 10^{-2} N/m$ (Alpers & Hühnerfuss, 1989). For this range of E, based on Fig. 14, the maximum relative increase in damping coefficient occurs for a wavelength of less than 10 cm. In this wavelength regime, the maximum damping coefficient is typically more than 10 times, but less than 100 times, the damping rate with a clean surface.

6. Vorticity Conditions at the Free Surface

In a homogeneous fluid domain, fluid rotation cannot be created within its own fluid domain except at the boundaries as discussed in Subsecs. 3.5 and 3.6. If the water surface is viewed as the interface of a two-layered fluid system, both a rigid boundary and a fluid-fluid interface can be the sources of fluid rotation. Nonetheless, there are fundamental differences in characteristics between these two sources.

If a rigid boundary itself is not rotating, the (nonzero) vorticity component normal to the boundary does not exist; this is a consequence of the no-slip condition. Since vorticity is solenoidal, $\nabla \cdot \boldsymbol{\omega} = 0$, (3.56) indicates that a vortex tube cannot end at a nonrotating rigid boundary. On the other hand, if some rotation was imposed on the rigid boundary, then vortex tubes could end at the boundary; even so, vorticity which is normal to the boundary must have a uniform value everywhere on the rotating rigid boundary. These constrained behaviors at the rigid boundary are not applicable to the conditions at a fluid-fluid interface. Vortex tubes can extend freely from one side to the opposite side of the fluid across the interface, and the normal component of vorticity to the interface no longer needs to be uniform along the surface of the interface. It is noted that these properties of vortex tubes are not always applicable to vortex lines. Along a vortex tube, its strength, i.e., flow circulation, must take a constant value due to the solenoidal property of vorticity. However, no such condition is imposed to a vortex line; its strength, i.e., vorticity, can be varied along the vortex line. This is because by definition, a vortex line is an integral curve that is tangent to vorticity vectors, and a vortex line represents the direction but not the magnitude of fluid rotation. Hence, for an isolated case, a vortex line can end at a stationary rigid boundary where its vorticity component normal to the boundary vanishes. An example of the attachment of a vortex line to the boundary associated with a whirlwind was discussed by Lighthill (1963); in his Figure II.3, the vortex tube spreads out to infinity at the boundary and one of the vortex lines ends at the stagnation point on the boundary.

Considering the fact that vortex tubes can be extended from the water to air domain, it is evident that the presence of air can play important roles in the determination of vorticity conditions at the air-water interface. Hence, in this section, the jump conditions for a material boundary, which were developed in Subsec. 4.1, are extended to the vorticity conditions.

6.1. *Vorticity jump condition*

Keeping fluid-compressibility effects, the general form of vorticity equation can be written as

$$\frac{D\boldsymbol{\omega}}{Dt} + \boldsymbol{\omega}(\nabla \cdot \mathbf{u}) = (\boldsymbol{\omega} \cdot \nabla)\mathbf{u} + \nabla \times \left(\frac{1}{\rho}\nabla \cdot \tau_{ij}\right)$$
$$= \nabla \cdot (\mathbf{u} \circ \boldsymbol{\omega}) + \nabla \times \left(\frac{1}{\rho}\nabla \cdot \tau_{ij}\right), \qquad (6.1)$$

(see Subsec. 3.5). Here, we used the identity that the divergence of vorticity vanishes identically, i.e., $\nabla \cdot \boldsymbol{\omega} = \nabla \cdot (\nabla \times \mathbf{u}) \equiv 0$. In (6.1), $\circ$ denotes the dyadic multiplication, i.e., $\mathbf{u} \circ \boldsymbol{\omega} = u_i \omega_j$. To find the jump condition corresponding to the vorticity equation, we need to expand the compatibility equation (e.g., 4.16) to incorporate the last term of (6.1). Setting $\mathbf{A} = (\frac{1}{\rho}\nabla \cdot \tau_{ij})$, consider the divergence theorem of the form (2.6):

$$\int_D \nabla \times \mathbf{A}\, dV = \int_S \mathbf{n} \times \mathbf{A}\, ds\,. \qquad (6.2)$$

Following the derivation in Subsec. 4.1 together with the use of (6.2), the corresponding jump condition for the conservation equation of the form:

$$\frac{D\mathbf{F}}{Dt} + \mathbf{F}\nabla \cdot \mathbf{u} + \nabla \cdot \mathbf{G} + \nabla \times \mathbf{A} + \mathbf{H} = 0 \qquad (6.3)$$

is readily found to be

$$[[\mathbf{F}(\mathbf{u} \cdot \mathbf{n} - \mathbf{U} \cdot \mathbf{n})]] + [[\mathbf{G} \cdot \mathbf{n}]] + [[\mathbf{n} \times \mathbf{A}]] = 0\,. \qquad (6.4)$$

It is understood that the double bracket denotes the jump as expressed in (4.2). Again, at a contact discontinuity, $\mathbf{u}\cdot\mathbf{n}-\mathbf{U}\cdot\mathbf{n}=0$ and (6.4) become

$$[[\mathbf{G}\cdot\mathbf{n}]]+[[\mathbf{n}\times\mathbf{A}]]=0\ . \tag{6.5}$$

Comparing (6.1) and (6.3), and substituting $\mathbf{F}=\boldsymbol{\omega}$, $\mathbf{G}=-\mathbf{u}\circ\boldsymbol{\omega},\mathbf{A}=-\left(\frac{1}{\rho}\nabla\cdot\tau_{ij}\right)$, $\mathbf{H}=0$, (6.4) immediately yields the jump condition across a discontinuity:

$$[[\boldsymbol{\omega}(\mathbf{u}\cdot\mathbf{n}-\mathbf{U}\cdot\mathbf{n}]]-[[(\mathbf{u}\circ\boldsymbol{\omega})\cdot\mathbf{n}]]-\left[\left[\mathbf{n}\times\left(\frac{1}{\rho}\nabla\cdot\tau_{ij}\right)\right]\right]=0\ . \tag{6.6}$$

At the contact discontinuity, $\mathbf{u}\cdot\mathbf{n}-\mathbf{U}\cdot\mathbf{n}=0$, we find

$$[[\mathbf{u}(\boldsymbol{\omega}\cdot\mathbf{n}]]+\left[\left[\mathbf{n}\times\left(\frac{1}{\rho}\nabla\cdot\tau_{ij}\right)\right]\right]=0\ , \tag{6.7}$$

by using $(\mathbf{u}\circ\boldsymbol{\omega})\cdot\mathbf{n}=u_i\omega_j n_j=\mathbf{u}(\omega\cdot\mathbf{n})$. Since both the normal and tangential components of fluid velocity $\mathbf{u}$ are continuous across the contact discontinuity (as demonstrated in (4.28), this must be so in order to satisfy the jump condition of the mechanical energy), the normal component of vorticity is also continuous, i.e.,

$$[[\mathbf{u}(\omega\cdot\mathbf{n})]]=\mathbf{u}\,[[\omega\cdot\mathbf{n}]]=0\ . \tag{6.8}$$

Consequently, the first and second terms of (6.6) vanish independently and the jump conditions for the conservation of vorticity become,

$$\left[\left[\mathbf{n}\times\left(\frac{1}{\rho}\nabla\cdot\tau_{ij}\right)\right]\right]=0\ . \tag{6.9}$$

This demonstrates that the tangential components of the net surface force acting on a fluid parcel per unit mass is continuous across the contact discontinuity (see (2.3) for the physical interpretation of the divergence of stress tensor). Since the body force per unit mass, $\mathbf{g}$, is constant across the discontinuity, we can write (6.9) as

$$\left[\left[\mathbf{n}\times\left(\frac{1}{\rho}\nabla\cdot\tau_{ij}+\mathbf{g}\right)\right]\right]=0\ . \tag{6.10}$$

Based on the conservation of linear momentum (3.2), (6.10) is equivalent to

$$\left[\left[\mathbf{n}\times\frac{D\mathbf{u}}{Dt}\right]\right]=0\ , \tag{6.11}$$

i.e., fluid acceleration tangential to the surface must be continuous. In spite of the fact that fluid velocities must be continuous across a contact discontinuity (4.28), the no-slip boundary condition by itself does not require the fluid acceleration to be continuous, since a fluid parcel is not a rigid body but is capable of deforming itself. For example, let us consider a two-dimensional flow $\mathbf{u} = (u(y), v, 0)$ at a horizontal interface in the x-z plane which is allowed to displace in the y direction ($v \neq 0$). Because of the no-slip condition, it is evident that $\partial u/\partial t = 0$ and $u\partial u/\partial x = 0$. Then, the vorticity jump condition (6.11) requires that $v\partial u/\partial y$ must vanish, which is not obvious.

At a contact discontinuity separating the fluid domain 1 from domain 2, (6.9) can be written as

$$\left\{\mathbf{n} \times \left(\frac{1}{\rho}\nabla \cdot \tau_{ij}\right)\right\}_1 = \left\{\mathbf{n} \times \left(\frac{1}{\rho}\nabla \cdot \tau_{ij}\right)\right\}_2 . \tag{6.12}$$

Neglecting surface-tension effects, the dynamic boundary condition that is obtained from the conservation of linear momentum is that the stress acting on the surface is continuous:

$$[[\tau_{ij} \cdot \mathbf{n}]] = 0 , \tag{6.13}$$

and a form similar to (6.9) can be obtained by taking a cross product with a unit normal vector $\mathbf{n}$:

$$[[\mathbf{n} \times (\tau_{ij} \cdot \mathbf{n})]] = 0 . \tag{6.14}$$

Note that (6.9) is evidently different from the dynamic boundary condition. Contrary to the dynamic condition of (6.14) that represents the tangential component of the stress acting on the surface, (6.9) or (6.11) indicates that the fluid acceleration tangential to a contact discontinuity must be continuous, which is a kinematic statement because vorticity is a kinematic quantity by definition. Another point made is that the term $\nabla \cdot \tau_{ij}$ is a field quantity associated with a (three-dimensional) fluid parcel volume (i.e., $\nabla \cdot \tau_{ij} \equiv \underset{\mathbf{V}\to 0}{Lim} \frac{1}{V} \int_S \tau_{ij} \cdot \mathbf{n}\, ds$), whereas the dynamic conditions like $\tau_{ij} \cdot \mathbf{n}$ is a quantity defined on a (two-dimensional) plane.

For Newtonian fluids, the stress tensor τ_{ij} is expressed by (3.3). For a fluid-fluid interface of incompressible fluids, (6.9) can be written as

$$\left[\left[\frac{1}{\rho}(\mathbf{n} \times \nabla)p\right]\right] + \left[\left[\frac{\mu}{\rho}\mathbf{n} \times (\nabla \times \boldsymbol{\omega})\right]\right] = 0 . \tag{6.15}$$

But since

$$\begin{aligned}\mathbf{n}\times(\nabla\times\boldsymbol{\omega}) &= -\mathbf{n}\cdot\nabla\boldsymbol{\omega}+\nabla(\mathbf{n}\cdot\boldsymbol{\omega})-\boldsymbol{\omega}\cdot\nabla\mathbf{n}\\ &= -\mathbf{n}\cdot\nabla(\mathbf{n}\times(\boldsymbol{\omega}\times\mathbf{n}))-(\mathbf{n}\times(\boldsymbol{\omega}\times\mathbf{n}))\cdot\nabla\mathbf{n}+\mathbf{n}\times(\nabla\times(\mathbf{n}(\boldsymbol{\omega}\cdot\mathbf{n})))\end{aligned} \tag{6.16}$$

(6.15) can be expressed by

$$\left[\left[\frac{1}{\rho}(\mathbf{n}\times\nabla)p\right]\right]-\left[\left[\frac{\mu}{\rho}\{\mathbf{n}\cdot\nabla\boldsymbol{\xi}-\boldsymbol{\xi}\cdot\kappa_{ij}-\mathbf{n}\times(\nabla\times\boldsymbol{\zeta})\}\right]\right]=0\,, \tag{6.17}$$

or, since $\mathbf{n}\times(\nabla\times(\mathbf{n}(\boldsymbol{\omega}\cdot\mathbf{n})))=\mathbf{n}\times(\nabla\times\boldsymbol{\zeta})=\nabla_\lambda(\mathbf{n}\cdot\boldsymbol{\omega})$,

$$\left[\left[\frac{1}{\rho}(\mathbf{n}\times\nabla)p\right]\right]-\left[\left[\frac{\mu}{\rho}\{\mathbf{n}\cdot\nabla\boldsymbol{\xi}-\boldsymbol{\xi}\cdot\kappa_{ij}-\nabla_\lambda(\mathbf{n}\cdot\boldsymbol{\omega})\}\right]\right]=0 \tag{6.18}$$

where $\boldsymbol{\zeta}$ is the normal component of vorticity, $\boldsymbol{\zeta}=\mathbf{n}(\boldsymbol{\omega}\cdot\mathbf{n})$, $\boldsymbol{\xi}$ is the tangential component of vorticity, $\boldsymbol{\xi}=\mathbf{n}\times(\boldsymbol{\omega}\times\mathbf{n})$, and κ_{ij} is the two-dimensional curvature tensor on the discontinuity surface having no component normal to the surface because $\mathbf{n}$ is a unit vector, $\kappa_{ij}=-\nabla_\lambda\mathbf{n}$ or (2.16), ∇_λ is the gradient operator along the surface of the discontinuity. (Note that both $\boldsymbol{\zeta}$ and $\boldsymbol{\xi}$ are vector quantities whose resultant is $\boldsymbol{\omega}$, and κ_{ij} is symmetric, according to (2.16), hence, $\nabla\times\mathbf{n}=0$.) Equation (6.18) represents the vorticity jump condition for incompressible Newtonian fluids at a contact discontinuity.

At a fluid-fluid interface, the first term of (6.18), $[[\frac{1}{\rho}(\mathbf{n}\times\nabla)p]]$, represents the production of fluid rotation due to the difference in fluid acceleration caused by the tangential pressure gradient; it is equivalent to fluid rotation caused by 'baroclinic torque' (see (3.51) or the term (b) in (3.40)) that is normally nonzero at an interface of density discontinuity. The term, $[[\frac{\mu}{\rho}\mathbf{n}\cdot\nabla\boldsymbol{\xi}]]$ in (6.18) represents the diffusion of vorticity, i.e., the amount of vorticity transferred to the interface by molecular diffusion (see (3.41) or (3.53)). The term, $[[\frac{\mu}{\rho}\boldsymbol{\xi}\cdot\kappa_{ij}]]$, represents the effect of bending vorticity along the curved interface. It is noted that the vector quantity, $\boldsymbol{\xi}\cdot\kappa_{ij}$, points in the direction tangential to the surface because κ_{ij} is the two-dimensional curvature tensor with zero normal component. As discussed later, the last term, $[[\frac{\mu}{\rho}\nabla_\lambda(\mathbf{n}\cdot\boldsymbol{\omega})]](=[[\frac{\mu}{\rho}\mathbf{n}\times(\nabla\times\boldsymbol{\zeta})]])$, represents the production of vorticity that is normal to the interface.

At a fluid-fluid interface, the vorticity component normal to the interface is not uniform hence, the tangential component of the term $\nabla(\mathbf{n}\cdot\boldsymbol{\omega})$ does not vanish. On the other hand, the normal component of the term $\nabla(\mathbf{n}\cdot\boldsymbol{\omega})$ is canceled out with a part of the term, $-\mathbf{n}\cdot\nabla\boldsymbol{\omega}$, in (6.16), whereby the form (6.18) results.

Just like the vorticity equation (e.g., (3.42) and (3.43)), this compatibility jump condition is grossly different between the full three-dimensional expression and the two-dimensional expression (in the vertical plane). If we consider the two-dimensional problem, then (6.18) is reduced to

$$\left[\left[\frac{1}{\rho}(\mathbf{n}\times\nabla)p\right]\right]-\left[\left[\frac{\mu}{\rho}\mathbf{n}\cdot\nabla\xi\right]\right]=0\,. \tag{6.19}$$

Hence, for a two-dimensional flow problem, there is no mechanism at a fluid-fluid interface for vorticity bending nor the production of vorticity normal to the interface. These missing flow behaviors appear to be very important in a real (three-dimensional) flow environment. Although two-dimensional models are often a good approximation for irrotational flows (e.g., wave motions), similar two-dimensional models for vortical flow at and near the interface may not be a proper approximation.

6.2. *Boundary condition at a rigid body boundary*

The term, $(\frac{1}{\rho}\nabla\cdot\tau_{ij})_1$, in (6.12) represents the net surface force acting on a unit mass, hence, if region 1 is a solid then, by Newton's second law, it represents the acceleration of the solid, $\mathbf{a}_B$, minus the body-force (i.e., gravity $\mathbf{g}$) effect. Equation (6.12) then becomes

$$\begin{aligned}\mathbf{n}\times\mathbf{a}_B-\mathbf{n}\times\mathbf{g}&=\left\{\mathbf{n}\times(\frac{1}{\rho}\nabla\cdot\tau_{ij})\right\}_{\text{fluid}}\\&=-\frac{1}{\rho}(\mathbf{n}\times\nabla)p+\frac{\mu}{\rho}\{\mathbf{n}\cdot\nabla\boldsymbol{\xi}-\boldsymbol{\xi}\cdot\kappa_{ij}-\nabla_\lambda(\mathbf{n}\cdot\boldsymbol{\omega})\}\,.\end{aligned} \tag{6.20}$$

On a rigid boundary surface, the tangential component of the term $\nabla_\lambda(\mathbf{n}\cdot\boldsymbol{\omega})$ vanishes because $\mathbf{n}\cdot\boldsymbol{\omega}$ is constant on a rigid surface. Then, (6.20) can be written as

$$\frac{\mu}{\rho}\mathbf{n}\cdot\nabla\boldsymbol{\xi}=\mathbf{n}\times\left(\mathbf{a}_B+\frac{1}{\rho}\nabla p-\mathbf{g}\right)+\frac{\mu}{\rho}\boldsymbol{\xi}\cdot\kappa_{ij}\,. \tag{6.21}$$

An identical form to (6.21) can be derived by taking the tangential component of the conservation equation of linear momentum considering the fact that $\nabla_\lambda(\mathbf{n}\cdot\boldsymbol{\omega})=0$ at a stationary rigid boundary (e.g., see Wu, Wu and Wu, 1988). It is emphasized that (6.21) is an appropriate boundary condition because the rigid-body acceleration $\mathbf{a}_B$ is given *a priori*. This is not the case, on the other hand, for the condition at a fluid-fluid interface. The fluid acceleration at the interface is usually not given but it is the value to be determined. By totally neglecting the effects of fluid in region 1 in (6.12), a similar form of (6.21) could be derived at the (vanishing stress) free surface by replacing $\mathbf{a}_B$ by $D\mathbf{u}/Dt$ of the fluid 2. Then, the resulting equation no longer serves as a boundary condition: it simply represents the tangential components of the field equation based on the conservation of linear momentum. Furthermore, although a flow at a fluid-fluid interface could be expressed by equating the tangential component of the conservation of linear momentum in each fluid at the interface, which would lead to the same form of (6.18), such a derivation still requires the condition (6.11): (6.11) is the result based on the discontinuity jump condition for the vorticity equation.

When the rigid boundary is flat and horizontal in the x-z plane, (6.21) reduces to

$$-\frac{\mu}{\rho}\frac{\partial \boldsymbol{\omega}}{\partial y} = -\frac{1}{\rho}(\mathbf{n} \times \nabla)p - \mathbf{n} \times \mathbf{a}_B \ , \tag{6.22}$$

at the boundary, $y = 0$, which was derived by Morton (1984). Morton's result (6.22) describes, for example, the vorticity creation associated with Stokes' first and second problems, i.e., flow induced by a wall impulsively accelerating in its own plane and by a tangentially oscillating wall respectively.

If the rigid body boundary is stationary on the x-z plane $\left(\frac{D\mathbf{u}}{Dt} \equiv \mathbf{a}_B = 0\right)$, then (6.22) reduces to

$$-\frac{\mu}{\rho}\frac{\partial \omega_x}{\partial y} = \frac{1}{\rho}\frac{\partial p}{\partial z} \ , \tag{6.23}$$

which shows that the x component of vorticity ω_x is created at the boundary, $y = 0$, by the pressure gradient in the z direction. Equation (6.23) is the vorticity generation condition at a flat rigid plate at $y = 0$, derived by Lighthill (1963).

The fundamental form of (6.23) clearly demonstrates that vorticity created at a solid boundary is always tangential to the boundary surface unless the boundary itself is rotating. Because the momentum transfer by advection is absent at a solid boundary, momentum transfer by molecular diffusion must exactly balance the pressure gradient; this is a consequence of the conservation of linear momentum at a no-slip stationary boundary. Created vorticity diffuses out from the boundary into the fluid domain. Once the vorticity gets into the fluid domain, it can manifest itself by the effects of stretching, bending and further diffusion. It is emphasized that in a fluid of uniform density and viscosity, a net amount of vorticity in the fluid domain can be altered only by molecular diffusion of vorticity at no-slip solid boundaries; fluid rotation cannot be produced nor diminished within the fluid domain.

6.3. *Free-surface boundary condition*

If we assume that the magnitude of the stress tensor is uniform everywhere in one fluid domain, i.e., $\nabla \cdot \tau_{ij} \equiv 0$ (note that it is not equivalent to assume the "free-surface" condition at the air-water interface where the stress vanishes on the interface, $\tau_{ij} \cdot \mathbf{n} = 0$, without surface-tension effects), then for the other fluid, e.g., water, (6.9) becomes

$$\mathbf{n} \times \left(\frac{1}{\rho} \nabla \cdot \tau_{ij} \right) = 0 \quad \text{on } y = \eta(x,z,t) , \tag{6.24}$$

where η is the location of the free surface. Using (6.18), (6.24) becomes

$$(\mathbf{n} \times \nabla) p - \mu \mathbf{n} \cdot \nabla \boldsymbol{\xi} + \mu \boldsymbol{\xi} \cdot \kappa_{ij} + \mu \nabla_\lambda (\mathbf{n} \cdot \boldsymbol{\omega}) = 0 . \tag{6.25}$$

Integrating (6.25), the magnitude of the normal component of vorticity is found to be

$$\mathbf{n} \cdot \boldsymbol{\omega} = |\boldsymbol{\zeta}| \ \mathrm{sgn}\{\mathbf{n} \cdot \boldsymbol{\omega}\} = \int \left(\frac{-1}{\mu} (\mathbf{n} \times \nabla) p + \mathbf{n} \cdot \nabla \boldsymbol{\xi} - \boldsymbol{\xi} \cdot \kappa_{ij} \right) \cdot d\mathbf{r} , \tag{6.26}$$

where $\mathbf{r}$ is the position vector of the points on the interface. The normal component of vorticity is generated by the effects of three mechanisms shown in the integrand of (6.26), namely,

1. baroclinic torque,
2. viscous diffusion of vorticity to the interface, and
3. vorticity bending due to the curvature of the surface, respectively.

This represents the mechanisms for the manifestation and persistence of the vorticity normal to the interface; such flow phenomena are reported from the laboratory observations and numerical predictions (e.g., Bernal & Kwon, 1989; Kachman, Koshimoto & Bernal, 1991; Weigand & Gharib, 1992; Ohring & Lugt, 1992). For example, Fig. 15 demonstrates free-surface deformations caused by the interactions of a vortex ring with the clean air-water interface at inclined incidence (70 degrees). The free-surface images were taken using the

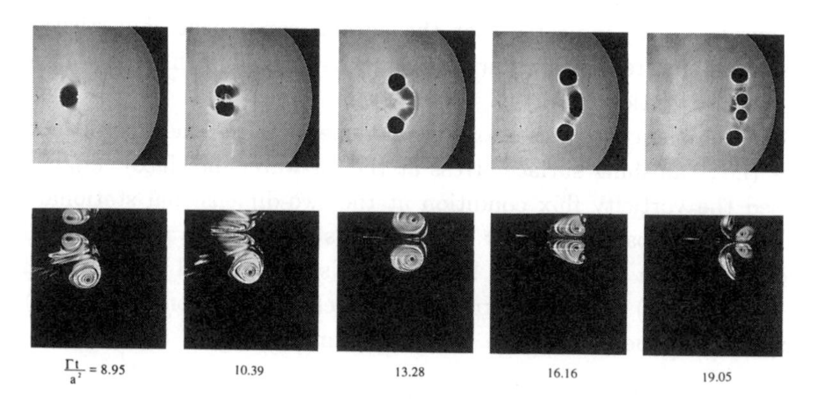

Fig. 15. Interaction of a vortex ring with a clean free surface. Top photographs show the vortex pattern on the surface by the shadowgraph images, and on the bottom row, the side view photographs are shown using the laser-induced fluorescence technique. (after Kachman, Koshimoto, and Bernal, 1991)

shadowgraph technique, and the side-view flow images were visualized using the Laser-Induced Fluorescent technique. As soon as the vortex ring impacted the water surface, a pair of vertical vortices appeared, followed by another pair of vortices which corresponds to the impingement of the bottom part of the original vortex ring. The couple of vertical-vortex pair formations persisted for a long time and behaved like those in a two-dimensional flow.

Now, if we assume the interface to be barotropic, the first term of the integrand in (6.26) is nil. Because of the immediate formation of the vortex pair at the surface in the experimental results, it is unlikely that the gradual process of vorticity diffusion is important, hence, the second term of the integrand is neglected. (The experimental results in Fig. 15 by Kachman, Koshimoto & Bernal (1991) clearly indicate that the normal vortex motions were formed in a very short time after the vortex ring reached the water surface.) Then, (6.26) indicates that the production of the normal component of vorticity at the interface is mainly due to the vorticity bending effects. Considering Helmholtz's theorem, which was interpreted from (3.50), a vortex tube is advected with the fluid, if the viscous-diffusion effect is neglected. Hence, when a vortex tube is ascending parallel to the interface which is initially a horizontal plane, the vorticity bending would take place immediately when the interface is disturbed from the horizontal plane. This is because the interface is a fluid (material) surface and the vortex tube is also transported with the fluid material. In other words, (6.26) indicates that the normal component of vorticity must coexist with the free-surface curvature that is aligned with to the tangential component of vorticity.

Based on the conservation of linear momentum and the free-surface conditions (i.e., vanishing surface stress at the air-water interface), Lugt (1987) proposed the vorticity flux condition at the two-dimensional stationary free surface; the approach is similar to Lighthill's (1963) derivation of (6.23) for the condition at a rigid no-slip boundary. Using the local polar coordinates (r, θ) (see Fig. 12) with the corresponding velocity components (u_r, u_θ), the vorticity flux at the free surface, $r = R$, is found to be

$$\frac{\mu}{\rho}\frac{\partial \xi}{\partial r} = \frac{\mu}{\rho}\left(\frac{\partial^2 u_\theta}{\partial r^2} - \frac{1}{R}\frac{\partial^2 u_r}{\partial \theta \partial r}\right) = \frac{1}{R}\frac{\partial}{\partial \theta}\left(\frac{1}{2}u_\theta^2 + \frac{p}{\rho} + gR\sin\theta\right), \tag{6.27}$$

where ξ is the vorticity component normal to the r-θ plane and the combined term in the last bracket represents the total head. At first glance, (6.27) appears to be different from (6.25). Note that (6.27) was derived from the momentum equation in the θ direction by imposing the conditions of vanishing tangential stress and $u_r = 0$ on the surface $r = R$. Zero tangential stress in Lugt's steady-flow and circular-surface model implies that the azimuthal force must vanish along the surface, hence, fluid acceleration in the θ direction must be in exact balance with the body (gravity) force, i.e.,

$$\frac{\partial}{\partial \theta}\left(\frac{1}{2}u_\theta^2 + gR\sin\theta\right) = 0\,. \tag{6.28}$$

Therefore, (6.27) becomes

$$\frac{\mu}{\rho}\frac{\partial \xi}{\partial r} = \frac{1}{R\rho}\frac{\partial p}{\partial \theta}, \tag{6.29}$$

which is equivalent to (6.25) for two-dimensional flows. (Note that, for two-dimensional flows, the last two terms in (6.25) vanish identically, and (6.29) is the same form as (6.23), the result found by Lighthill (1963) for a rigid plane boundary.)

It must be emphasized that the results of (6.25) and (6.26) are based on the assumption that the divergence of the stress tensor vanishes identically in a one fluid domain. In the case of the air-water interface, this assumption seems to be similar to the "free-surface" condition, in which stresses vanish at the interface. The free-surface condition can be represented by the dynamic condition and expressed as

$$\tau_{ij} \cdot \mathbf{n} = 0 \ , \tag{6.30}$$

at the interface. This traditional no-stress condition is an approximation based on the substantial density difference between two-fluid media, and as discussed in Subsec. 5.5, it does provide a good approximation for a wide variety of water flows. Because the air density is negligibly small compared with that of water, the air-flow conditions do not affect dynamically the flow conditions of water. This is not the case for the vorticity jump condition described in (6.12).

Vorticity is a kinematic quantity, hence, the smallness of the air density does not provide any justification for neglecting air motions. In fact, (6.12) evidently demonstrates that a small magnitude of stress tensor τ_{ij} does not mean a small magnitude of $\mathbf{n} \times (\frac{1}{\rho}\nabla \cdot \tau_{ij})$ because the density also becomes small for air. In terms of vorticity, it appears that the air-water interface must be considered a two-fluid interface. Because the vorticity field is solenoidal by definition, the vorticity component normal to the interface must extend into the air domain. Even from this very simple view point, the effects of air on the surface vorticity should be carefully examined. For this reason, the full conditions expressed by (6.18) must be used in order to quantitatively analyze vorticity behavior at the air-water interface.

Suppose the effect of baroclinic torque is neglected, i.e., there is no pressure gradient along the interface. Also, the vorticity diffusion effect is neglected because its process is slow. Then, (6.18) can be approximated for an immediate response of the interface as

$$\left[\left[\frac{\mu}{\rho}\boldsymbol{\xi}\right]\right] \cdot \kappa_{ij} \approx -\left[\left[\frac{\mu}{\rho}\right]\right] \nabla_\lambda(\mathbf{n} \cdot \boldsymbol{\omega}) \ , \tag{6.31}$$

which indicates that, for fixed values of $\kappa_{ij}, \nabla_\lambda(\mathbf{n} \cdot \boldsymbol{\omega})$, and $\boldsymbol{\xi}$ in one of the fluid domains (either in air or water), the ratio of the magnitudes in vorticity of two fluids (air and water) is inversely proportional to the ratio of the magnitudes of the kinematic viscosity, $\nu = \mu/\rho$, of the two fluids. Since the kinematic viscosity is greater in air than in water, the magnitude of vorticity in water must be greater than that in air under the premised conditions. This observation suggests that the vorticity transfer process across the air-water interface is more efficient from the air to water domain than the process from the water to air domain, i.e., at the interface, slight changes in air vorticity cause substantial vorticity changes in water. The same conclusion can be drawn by considering the vorticity diffusion and the baroclinic-torque effects. Because air density is much smaller than water density, a small magnitude of air-pressure gradient significantly affects the vorticity field in water at the interface. In fact, this supports the result by Yeh (1991) that vortical motions in the surface roller (i.e., recirculating mean flow) of a bore (a quasi-steady broken wave) are created by baroclinic torque acting along the front face of the bore; this baroclinic torque is caused by the dynamic-pressure gradient along the air-water interface, i.e., a very small air pressure variation creates strong vortical motions at the bore front. It is, however, emphasized that the efficient vorticity transfer from air to water does not mean the dynamics of the water are immediately influenced by air; a very strong vorticity may be formed as a very thin vortex sheet at the interface, which may not directly alter the dynamics of water flow. Nonetheless, once vorticity is formed in the fluid domain, it can be diffused, advected, stretched, and bent, to form turbulent motions, which can eventually dominate the flow characteristics.

Turbulent motions in a homogeneous liquid, i.e., water, are originated from the creation of vorticity at the boundaries. Consider the fact that an irrotational flow is described with the Laplace equation which is linear. Hence, flow instability does not occur within the irrotational fluid domain, although the boundary itself can be unstable. However, once vorticity is introduced into the fluid domain through the boundaries, no matter how small it is, the flow is no longer expressed by the linear Laplace equation, but the fully nonlinear Navier-Stokes equations of motion must be applied. More specifically, the term $-\mathbf{u} \times \boldsymbol{\omega}$ in (3.28) causes the flow instability which leads the flow to turbulence. The onset of flow turbulence is in general dependent upon the magnitude of Reynolds number. The requirement of a length scale in the Reynolds number clearly indicates the importance of the boundaries; in a homogeneous fluid, other than the viscous length scale, the only proper length scale is related to the positions of the boundaries, which is the origin of vorticity. Although immediate effects of the created vorticity at the interface on the flow might be minute, once fluid rotations are introduced, flow turbulence can be triggered depending on the magnitude of the Reynolds number. From this view point, vorticity-boundary conditions should be considered to be important.

Note that surface-tension effects do not appear explicitly in (6.18). This does not mean that surface-tension plays no role in vorticity-jump conditions at a fluid-fluid interface. If the value of the surface-tension is constant, the dynamic-boundary condition (6.13) becomes

$$(\tau_{2,ij} - \tau_{1,ij}) \cdot \mathbf{n} = \gamma\left(\frac{1}{R_1} + \frac{1}{R_2}\right)\mathbf{n} , \tag{6.32}$$

where $\tau_{1,ij}$ is the stress tensor evaluated in fluid 1 at the interface, $\tau_{2,ij}$ is the stress tensor evaluated in fluid 2 at the interface, R_1 and R_2 are the principal radii of curvature of the interface, γ is the surface-tension between the two fluid media, and $\mathbf{n}$ is the unit normal vector pointing outward from fluid 1. Note that as long as γ is constant, the surface-tension force only affects the normal component of stress but not the tangential components. (As discussed in Subsecs. 4.3 and 5.6, when the interface is contaminated with surfactant, the magnitude of γ is no longer constant and surface-tension can create the tangential force.) It is emphasized that the surface-tension force acts in the plane of the interface; therefore, it cannot form a torque by itself and is not capable of directly causing fluid rotation. However, the (constant) surface-tension can create a jump in the normal component of stress across the interface; the variation in stress is what produces imbalance forces (torque) and triggers fluid rotation. In other words, the surface-tension implicitly affects the combined term, $[[\frac{1}{\rho}(\mathbf{n} \times \nabla)p - \frac{\mu}{\rho}\mathbf{n} \cdot \nabla\boldsymbol{\xi}]]$ in (6.18). Note that the second term in the bracket contains the normal viscous stress component.

7. Summary

The material discussed in this article might contain some subtlety. This is because the traditional formulation of water-wave problems does not represent the exact real-fluid conditions and some of the assumptions involved may be inappropriate for some situations: the primary assumptions of the formulation are 1) irrotational flow, 2) incompressible fluid, and 3) a constant pressure at the free surface. To clarify this and to avoid a misunderstanding of discussions, it was necessary to review fluid mechanics to some extent, including the characteristics associated with vorticity. For example, it is important to emphasize that fluid rotation cannot be created unless torque is present, just as no acceleration is caused if there is no net froce acting on the material. It is often misled by a statement such that vorticity is produced by a shear flow: the fact is that a shear flow is one of the forms of vortical flows and has nothing to do with the "creation" of fluid rotation.

The free-surface boundary conditions are analyzed in a more rigorous way than the surface of constant pressure. Yet, avoiding the molecular-scale phase transition between gas(air) and liquid(water), we treat the free surface as the air-water two-fluid material interface of fluid-property discontinuity. Based on this model, the jump conditions at the interface are derived. The derived jump condition for conservation of mass yields the kinematic boundary condition, the condition for linear momentum yields the dynamics boundary condition, and the condition for mechanical energy yields the no-slip boundary condition at the interface. It is emphasized that the no-slip condition arises as a consequence of energy conservation at the jump rather than the "assumed" condition. Molecular-scale effects at the interface must be included and incorporated in the term of surface tension. When the surface tension is uniform at the surface, only the normal component of the dynamic condition is affected. On the other hand, when the surface tension varies along the interface, the tangential component of the dynamic condition is also affected; consequently, the condition of inhomogeneous surface tension along the free surface only exists for viscous fluids; it is impossible to balance the tangential force created by the inhomogeniety of surface tension in an inviscid fluid.

It is demonstrated that irrotational water-wave motions exist only if certain energy is supplied from the surroundings through the free surface and the supplied energy rate is exactly balanced with the energy dissipation rate caused by angular deformation of fluid parcels associated with the wave motions. Unless the exact energy rate is supplied by the surroundings, the flow adjacent to the free surface becomes vortical and forms the boundary layer. For a special case of vanishing stresses at the free surface, vorticity at the free surface coincides with that of a rigid body rotation along the free surface, i.e., twice the ratio of tangential velocity to its radius of the surface curvature. Furthermore, since there is no energy input from the surroundings through the free surface, it is plausible to assume that wave-energy would dissipate and wave motion would be attenuated. In fact, it was shown that the existence of air above the surface becomes important for the determination of wave attenuation rates. The formation of surface film also affects wave attenuation. The existence of surfactant (surface film) makes the surface incompressible so that the surface is tangentially immobilized; consequently, the resulting attenuation rate becomes comparable to that of a solid boundary. The attenuation rate could become even greater than that of a solid boundary when the surface wave has the same wave length as the longitudinal wave (the Marangoni wave).

The jump condition for the vorticity equation at a fluid discontinuity was derived. At the air-water (two-fluid) material interface, the resulting jump condition requires that the tangential components of the net surface force per unit mass exerting on fluid parcels must be continuous across the interface. Using Newton's second law, this requirement is equivalent to the condition that the tangential components of fluid-parcel acceleration must be continuous across the interface. Because a fluid parcel is allowed to deform, this result is not trivial and cannot be deduced by the no-slip and kinematic conditions alone. For an incompressible fluid, the derived jump condition consists of the four terms: 1) baroclinic torque, 2) vorticity diffusion, 3) vorticity bending, and 4) production of surface normal vorticity. For a two-dimensional flow in the vertical plane, vorticity bending and surface normal vorticity cannot exist, hence the jump condition is simplified drastically, but it is not a good representation of the real air-water interface. Because vorticity is a kinematic quantity, the derived jump condition clearly demonstrates that the presence of the air cannot be ignored in the vorticity balance at the interface. Furthermore, the surface-tension effect by itself cannot create fluid rotation at the interface because the surface tension exerts the interface tangentially, i.e., no torque can be produced. Nonetheless, the surface-tension effect is important for the determination of vorticity at the interface, because it does alter the stresses acting on the interface; the stresses are the ones that can lead to the formation of torque and produce fluid rotation. In other words, fluid-rotation can be produced indirectly by the surface-tension effect.

There is no doubt that the analyses presented here are not complete but should provide some physical interpretations of this very important subject.

References

Adamson, A. W. (1990). *Physical Chemistry of Surfaces.* John Wiley & Sons, Inc. 777.

Alpers, W. and Hühnerfuss, H. (1989). The damping of ocean waves by surface films: a new look at an old problem. *J. Geophy. Res.* **94**:6251–6265.

Batchelor, G. K. (1967). *An Introduction to Fluid Mechanics.* Cambridge University Press, Cambridge. 615.

Bernal, L. P. and Kwon, J. T. (1989). Vortex ring dynamics at a free surface. *Phys. Fluids A* **1**:449–451.

Cokelet, E. D. (1977). Steep gravity waves in water of arbitrary uniform depth. *Phil. Trans. Roy. Soc. Lond. A* **286**:183–230.

Dean, R. G. and Dalrymple, R. A. (1991). *Water Wave Mechanics for Engineers and Scientists.* World Scientific. 353.

Dore, B. D. (1978). Some effects of the air-water interface on gravity waves. *Geophys. Astrophys. Fluid Dynamics* **10**:215–230.

Douglass, S. L. (1990). Influence of wind on breaking waves, *J. Wtrwy., Port, Coast. and Oc. Eng. ASCE* **116**:651–663.

Edwards, D. A., Brenner, H. and Wasan, D. T. (1991). *Interfacial Transport Processes and Rheology.* Butterworth-Heinemann. 558.

Gibbs, J. W. and Wilson, E. B. (1929). *Vector Analysis.* Yale University Press. 436.

Henderson, F. M. (1966). *Open Channel Flow.* Macmillan. 522.

Holyer, J. Y. (1979). Large amplitude progressive interfacial waves. *J. Fluid Mech.* **93**:433–448.

Kachman, N. J., Koshimoto, E. and Bernal, L. P. (1991). Vortex ring interaction with a contaminated surface at inclined incidence. *Dynamics of Bubbles and Vortices Near a Free Surface*, AMD Vol. 119, ed. Sahin, I., Tryggvason, G., and Schreyer, H. L. American Society of Mechanical Engineering. 45–58.

Kevorkian, J. (1990). *Partial Differential Equations.* Wadsworth. 547.

Lamb, H. (1932). *Hydrodynamics.* Cambridge University Press. 738.

Lighthill, M. J. (1963). Boundary layer theory. *Laminar Boundary Layers*, ed. L. Rosenhead. Clarendon Press.

Lighthill, M. J. (1986). *An Informal Introduction to Theoretical Fluid Mechanics.* Clarendon Press. 260.

Liu, P. L.-F., Synolakis, C. E., and Yeh, H. (1991). Report on the international workshop on long-wave run-up. *J. Fluid Mech.* **229**:675-688.

Longuet-Higgins, M. S. (1953). Mass transport in water waves. *Phil. Trans. Roy. Soc. Lond. A.* **245**:535–581.

Longuet-Higgins, M. S. (1992). Capillary roller and bores. *Breaking Waves*, ed. M. L. Banner and R. H. J. Grimshaw. Springer-Verlag. 21–37.

Longuet-Higgins, M. S. and Cokelet, E. D. (1976). The deformation of steep surface waves on water. I. A numerical method of computation. *Proc. Roy. Soc. Lond. A* **350**:1–26.

Longuet-Higgins, M. S. and Turner, J. S. (1974). An 'entraining plume' model of a spilling breaker. *J. Fluid Mech.* **63**:1–20.

Lucassen, J. (1968). Longitudinal capillary waves. Part 1–Theory. *Trans. Faraday Soc.* **64**:2221–2235.

Lucassen, J. (1982). Effect of surface-active material on the damping of gravity waves: a reappraisal. *J. Colloid Interface Science* **85**:52–58.

Lucassen-Reynders, E. H. and Lucassen, J. (1969). Properties of capillary waves. *Adv. Colloid Interface Sci.* **2**:347–395.

Lugt, H. J. (1987). Local flow properties at a viscous free surface. *Phys. Fluids* **30**:3647–3652.

Mei, C. C. and Liu, P. L.-F. (1973). The damping of surface gravity waves in a bounded liquid. *J. Fluid Mech.* **59**:239–256.

Morton, B. R. (1984). The generation and decay of vorticity. *Geophysics Astrophysics Fluid Dynamics* **28**:277–308.

Nadaoka, K., Masuda, M., and Suzuki, T. (1991). Development and applications of a numerical method for the analysis of wave-eddies coexisting field based on BEM and Discrete-Vortex Method. *Proc. Jap. Soc. Civil Engr.* **434**:67–76.

Ohring, S. and Lugt, H. J. (1992). Three-dimensional vortex-ring/free-surface interaction in a viscous fluid. *Free-Surface Vorticity Workshop*, San Diego.

Schwartz, M., Green, S. and Rutledge, W. A. (1960). *Vector Analysis with Applications to Geometry and Physics.* Happer & Brothers. 556.

Serrin, J. (1959). Mathematical principles of classical fluid mechanics. *Handbuch der Physik*, Vol. 8/1, ed. S. Flügge. Springer-Verlag. 125–263.

Wehausen, J. V. and Laitone, E. V. (1960). Surface waves. *Handbuch der Physik*, Vol. 9, ed. S. Flügge. Springer-Verlag. 446–778.

Weigand, A. and Gharib, M. (1992). Interaction of turbulent vortex ring with a free surface. *Free-Surface Vorticity Workshop*, San Diego.

Wu, J. Z., Wu, J. M. and Wu, C. J. (1988). A viscous compressible flow theory on the interaction between moving bodies and flow field in the (ω, ϑ) framework. *Fluid Dynamics Research* **3**:203–208.

Xü, Hongbo, (1992). Numerical study of fully nonlinear water waves in three dimensions. Ph. D. thesis. MIT, Massachusettes. 211.

Yeh, H. (1991). Vorticity-generation mechanisms in bores. *Proc. Roy. Soc. Lond. A*, **432**:215–231.

SEA FLOOR DYNAMICS

MOSTAFA A. FODA

The general interaction problem between waves in coastal waters and the seabed is considered. In particular, the role of the seabed in dissipating wave energy is examined, and various dissipation mechanisms are reviewed. A portion of the wave energy is consumed in inducing and maintaining sediment motion on the sea floor. Modes and characteristics of the resulting granular flow in waves are discussed in rather general terms. Furthermore, recent advances in studying nonlinear energy transfer into the seabed and the different modes of wave-induced sediment fluidization processes in cohesive as well as cohesionless seabeds are presented. Finally, we discuss some selected issues concerning the interaction of gravity marine structures with the seabed.

1. Introduction

In coastal waters, the benthic boundary layer (BBL) is usually considered as the zone from approximately one meter above to one meter below the sediment-water interface. Fluid and sediment motions within the BBL have significant effects on the geo-mechanical and geological imprints in the seabed, as well as the biological and geo-chemical activities near the seafloor.

The modes of flow-induced sediment motion varies from the slow creep and roll of surface particles on a flat or rippled seafloor, to the faster sediment sheet flow, to the sometimes massive bulk movement in the form of mud flows and submarine slides. These sediment movements are constantly reshaping the seabed morphology in never-ending cycles of sediment depositions and erosions. Furthermore, the resuspension of bottom sediments during stormy wave periods, especially in a muddy estuarine environment, will have important water quality implications. Pollutants buried into the seabed can be remobilized with sediment resuspension, and hence may be recycled back into the food chain through the biologically-rich shallow waters. Sediment resuspension in shallow water will also increase the water column turbidity, and therefore limit light availability to photosynthesizers. This will have obvious influences on the activities and vitality of the living benthic community. On the other hand, "warm holes" into the seafloor and sea-kelp growth and their associated surface roughness modifications are counter examples of biological activities that may have significant mechanical implications concerning the resulting exchanges of mass and momentum across the water-sediment interface. Chemical activities can further add to the complexities of the BBL processes. It is well known that the way a cohesive-sediment bed, e.g., marine mud, interacts with water waves is fundamentally different from that of a cohesionless or sandy bed. Electrochemical forces, which are negligible in a sand bed, would in fact dominate over gravity in holding together the fragile skeleton of an upper-stratum cohesive sediment bed. This makes the constitutive behavior (stress-strain relation) of cohesive mud very different from that of sand. This, in turn, makes the two types of beds respond differently to the imposed water wave loading, with wide variations in associated seabed processes such as sediment transport modes and intensities, soil fluidization and failure mechanisms, wave-damping characteristics, and others. However, regardless of the particular type of sediment, the one distinct attribute which distinguishes the mechanics of the BBL is the strong coupling between water flow and granular sediment motion. As stated above, the resulting water-sediment interaction across the BBL is important for the coastal processes of circulation, transport and exchange of material, energy and momentum into and out of the water column. Furthermore, external interferences due to, for instance, various engineering practices in the coastal and offshore regions will add new elements to the general interaction problem. For example, the introduction of a structure on the seafloor will undoubtedly influence the ensuing water-sediment interaction near the structure. On the other hand, the stability and overall performance of such a structure will be inextricably tied to what goes on in the surrounding BBL.

Although the seafloor has not received nearly as much research attention as has the "other" sea boundary, the air-sea interface, the past two decades have witnessed a real surge in research interest in the dynamics of the seafloor and BBL processes in general. In this review paper, the focus will be on some selected topics to highlight current research work in this area. The emphasis, however, will be on the general problem of seabed interaction with water waves.

Even though the dynamics of water waves is an established and a rather mature field of study, dating back to the work of Stokes (1847), the seabed has only been a recent addition to the considered dynamical problem. In the majority of studies on water waves, a seabed is either absent for deep-water studies, or is considered as a fixed, noninteracting, lower bounding surface for waves in intermediate or shallow waters. As the wave enters water depths less than about one half the wave length, the wave will start to "feel" the sea bottom. Within the context of an inviscid wave theory, the presence of a rigid, impervious seabed amounts to the introduction of a no-flux boundary condition across the bed surface. This gives rise to the two important effects of wave shoaling and wave refraction by bottom bathymetry. These effects, carried to shallow enough water depth, will lead to wave instability and breaking in the surf zone. A second boundary condition of no-slip will be imposed on the solid bed surface by further accounting for the viscosity effect. This results in the creation of a frictional boundary layer at the sea bottom, and hence the introduction of an energy sink for the water wave. The seabed can be a much more effective absorber of wave energy if we further allow the bed to become nonrigid, and account for effects such as seafloor undulation, rippling and erodibility, as well as the compliance, permeability, and consolidation of the seabed. The general problem becomes that of an *energy transfer* into the seabed, rather than of *energy dissipation* in the bottom boundary layer. The realization of the significance of the seabed as an important energy sink may have been initiated, though inadvertently, by Hasseleman *et al.* (1973) while working on the extensive JONSWAP wave data. The observed swell decay above the generally stiff sandy bed in the North Sea did not fit, in a number of important ways, predictions from a simple bottom friction model. Later studies have furnished more evidence of the active role of the seabed in extracting energy from water waves in coastal waters, and perhaps throughout the continental shelf region. Such transfer of energy would take place at different intensities over both sandy and muddy seabeds. The transferred energy will be consumed in a variety of seabed processes. Sediment transport is one such process, as a load of surface sediment is put into motion by acquiring some of that energy. Besides sediment transport, the transferred energy will contribute to raising the strain levels inside the seabed, sometimes triggering massive soil failures, such as soil liquefaction, mudflows, slumping and turbidity currents, typically under storm conditions. The process of energy transfer into the seabed becomes greatly exaggerated when the seabed becomes very soft. Indeed, an exceptionally high rate of energy transfer from water waves into the seabed can sometimes take place in coastal mudbank regions, where the upper stratum seabed deposit is made of extremely soft jel-like cohesive sediment (Wells *et al.* 1985; Tubman and Suhayda, 1976). The extraordinary ability of the gel-like bottom mud to absorb the incoming wave energy often results in the waves never reaching the shorelines behind these open-coast mud deposits (e.g., MacPherson and Kurup, 1981). Associated with this enhanced energy transfer there will clearly be a corresponding increase in other seabed processes, such as the substantially high rate of sediment suspension which results in the coffee-colored waters so characteristic of these muddy shorelines.

The main objectives of this review paper is, first, to highlight some of the recent advances in seafloor dynamics, with particular attention given to three specific applications: water-wave damping by the seabed, wave-induced granular flow on the seafloor, and wave-induced fluidization processes inside the seabed. Secondly, we wish to emphasize the interdependence among these basic seabed effects, and attempt to review relevant work within a unifying framework of a coupled seawave-seabed system. Finally, we will discuss some of the issues concerning the dynamic behavior at the interface between an offshore structure and the seabed, which is of obvious engineering importance.

2. Nondissipative Waves on a Compliant Bed

We adopt a two-dimensional cartesian coordinate system $(x,\ y)$ with x in the horizontal direction, and y pointing vertically upward. Let a plane gravity wave propagate in the positive x direction in a water layer of depth h, placed above a nonrigid bed which occupies the lower half-space $y < 0$ (Fig. 1). In order to present the physics of the interaction problem in a sharper focus, we proceed first with the simplest mathematical formulation for an inviscid incompressible water layer placed above an incompressible elastic bed. Then, we add more complexity later by introducing the main effects due to the system's viscosity and compressibility. Furthermore, the analysis will be limited at the beginning to the linear dynamics of the interaction problem, assuming small-amplitude water waves and linear constitutive behaviors for both the water and the seabed. Later on, the analysis will be extended to cover important nonlinear effects, due to finite-amplitude wave effects as well as for nonlinear constitutive behaviors.

The general equations of motion for both the water layer and the solid bed are given by

$$\rho \frac{Du}{Dt} = \frac{\partial \tau_{xx}}{\partial x} + \frac{\partial \tau_{xy}}{\partial y} \tag{1}$$

$$\rho \frac{Dv}{Dt} = \frac{\partial \tau_{xy}}{\partial x} + \frac{\partial \tau_{yy}}{\partial y} - \rho g \tag{2}$$

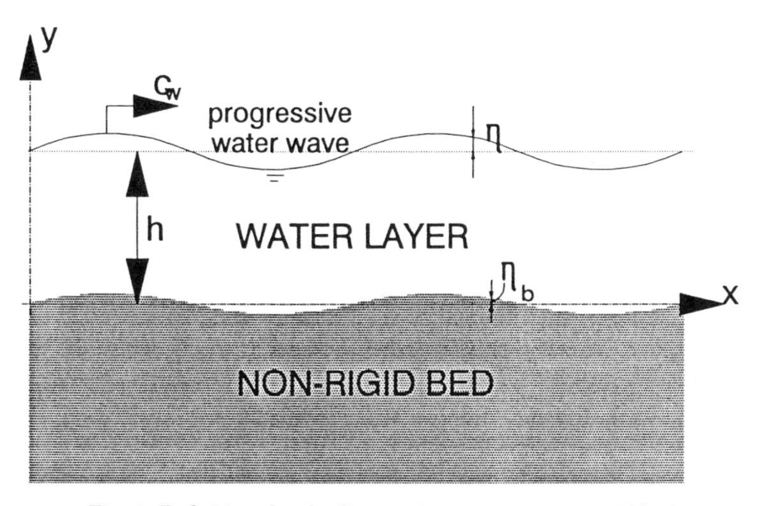

Fig. 1. Definition sketch. Progressive wave on a nonrigid bed.

$$\frac{D\rho}{Dt} + \rho\left(\frac{\partial u}{\partial x} + \frac{\partial v}{\partial y}\right) = 0 \tag{3}$$

where

$$\frac{D}{Dt} = \frac{\partial}{\partial t} + u\frac{\partial}{\partial x} + v\frac{\partial}{\partial y} \tag{4}$$

and where, in either medium, u and v are the eulerian-velocity components in the x and y directions, respectively; τ_{xx}, τ_{yy}, and τ_{xy} are the components of the plane stress-tensor in the x, y coordinate system, ρ is the bulk density of the medium, g is the gravitational acceleration, and t is time. Equations (1) and (2) are statements of conservation of momentum in the x and y directions respectively, and (3) is the continuity or conservation of mass equation. Next, we proceed to formulate the dynamical interaction problem for a harmonic small-amplitude wave with water-surface elevation $\eta(x,t)$ given by

$$\eta = Ae^{i(kx-\omega t)} \tag{5}$$

where η is measured from the still water level $y = h$, A is the wave amplitude, k is the wavenumber, and ω is the wave frequency. For small amplitude waves, the velocities are small and their products can be neglected, so that the nonlinear convective acceleration terms in the left-hand sides of (1) and (2) may be dropped. Furthermore, we neglect compressibility, or assume that

$\rho =$ constant in (3). Under these assumptions the governing equations (1), (2) and (3) are reduced to

$$\rho \frac{\partial u}{\partial t} = \frac{\partial \tau_{xx}}{\partial x} + \frac{\partial \tau_{xy}}{\partial y} \tag{6}$$

$$\rho \frac{\partial v}{\partial t} = \frac{\partial \tau_{xy}}{\partial x} + \frac{\partial \tau_{yy}}{\partial y} - \rho g \tag{7}$$

$$\frac{\partial u}{\partial x} + \frac{\partial v}{\partial y} = 0 \, . \tag{8}$$

In a simplified inviscid sea-elastic seabed system, the equations can be further simplified as follows:

(i) *Inviscid water layer.* There is only an isotropic fluid pressure p acting at any point inside the water layer, so that the stress-tensor components there would simply be given by

$$\tau_{xx} = \tau_{yy} = -p, \qquad \tau_{xy} = 0 \, . \tag{9}$$

The water flow can then be taken as irrotational, where a velocity potential ϕ can be introduced such that the horizontal and vertical velocities, u and v, are given by

$$u = \frac{\partial \phi}{\partial x}, \qquad v = \frac{\partial \phi}{\partial y} \tag{10}$$

and the pressure p is evaluated by substituting (9) and (10) in (6) and (7), and integrating once in space to get the linearized Bernoulli's equation

$$p = -\rho \frac{\partial \phi}{\partial t} - \rho g y \tag{11}$$

where ρ is the water density. From continuity (8), it follows that ϕ must satisfy Laplace's equation

$$\nabla^2 \phi = 0 \, . \tag{12}$$

(ii) *Elastic bed.* It is customary to express the solid displacements X and Y in the x and y directions, in terms of another pair of functions: a potential ϕ and a shear function ψ as

$$X = \frac{\partial \phi}{\partial x} + \frac{\partial \psi}{\partial y} \qquad Y = \frac{\partial \phi}{\partial y} - \frac{\partial \psi}{\partial x} \, . \tag{13}$$

Substituting these expressions into the continuity equation (8), we see that ϕ is still governed by Laplace's equation (incompressible material)

$$\nabla^2\phi = 0 \ . \tag{14}$$

The governing equation for ψ is obtained by introducing Hooke's law to provide the constitutive stress-strain relations for the incompressible elastic bed

$$\tau_{xx} = 2G\frac{\partial X}{\partial x} + \rho_s\frac{\partial^2\phi}{\partial t^2} + \rho_s g y \tag{15}$$

$$\tau_{yy} = 2G\frac{\partial Y}{\partial y} + \rho_s\frac{\partial^2\phi}{\partial t^2} + \rho_s g y \tag{16}$$

$$\tau_{xy} = G\left(\frac{\partial X}{\partial y} + \frac{\partial Y}{\partial x}\right) \tag{17}$$

where G is the shear modulus of the elastic solid, and ρ_s is the density of the solid bed. Substituting (15), (16) and (17) into the momentum equations (6) and (7), and then taking the curl of these equations in order to eliminate ϕ, we obtain the following equation for the shear function ψ

$$c_s^2\nabla^2\psi = \frac{\partial^2\psi}{\partial t^2} \qquad c_s = (G/\rho_s)^{1/2} \tag{18}$$

which is the conventional Kelvin's wave equation with c_s being the speed of the elastic shear-wave in the bed.

(iii) *Boundary conditions.* The formulation of the interaction problem is completed by defining the appropriate boundary conditions for the wave motion. At the water free-surface, $y = h$, we impose the linear kinematic and dynamic boundary conditions

$$\frac{\partial\eta}{\partial t} = \frac{\partial\phi}{\partial y} \tag{19}$$

$$\text{at} : y = h$$

$$\frac{\partial\phi}{\partial t} = -g\eta \ . \tag{20}$$

The kinematic condition (19) states that the motion of the free surface is determined by the water-particle velocity at the surface. The linearized dynamic condition (20) states, using Bernoulli's equation (11), that the pressure at the free surface is maintained at a constant value, or that the excess pressure (above hydrostatic) is zero. Notice also that in order to be consistent with the adopted small-amplitude assumption, the conditions (19) and (20) are applied at the still-water level $y = h$, which approximates the exact location of the free surface at $y = h + \eta$. Next, we turn to the undulating bed surface with its elevation η_b given by

$$\eta_b = A_b e^{i(kx-\omega t)} \tag{21}$$

where η_b is measured from the undisturbed bed-surface level $y = 0$, and A_b is the amplitude of the harmonic bed surface undulation. The appropriate boundary conditions at this surface would be to require the continuity of normal stress, shear stress and the continuity of normal velocity at the surface. Because we are ignoring water viscosity effects in the present model, a discontinuity in tangential velocity as well as a zero shear-stress boundary condition will have to be assumed at this surface. A better representation of these two boundary conditions will have to await the later incorporation of the viscosity effects in the analysis. The kinematic condition on η_b along with the condition of continuity of normal velocity at the water-bed interface implies from (10) and (13) that

$$\frac{\partial \eta_b}{\partial t} = \left(\frac{\partial \phi}{\partial y}\right)^+ = \frac{\partial}{\partial t}\left(\frac{\partial \phi}{\partial y} - \frac{\partial \psi}{\partial x}\right)^- . \tag{22}$$

The continuity of normal stress, from (11) and (16), requires

$$\left(\rho\frac{\partial \phi}{\partial t} + \rho g \eta_b\right)^+ = \left(2G\frac{\partial Y}{\partial y} + \rho_s\frac{\partial \phi}{\partial t} + \rho_s g \eta_b\right)^- \tag{23}$$

while the condition of zero shear requires from (17) that

$$\left(\frac{\partial X}{\partial y} + \frac{\partial Y}{\partial x}\right)^- = 0 \tag{24}$$

where, in (22), (23) and (24), the superscript + means just above the water-bed interface or at $y = 0^+$, and the superscript − means just below the interface or at $y = 0^-$.

Finally, we require that the solution decays to zero at infinite depth, i.e.,

$$\phi, : \psi \to 0 \qquad \text{as } y \to -\infty . \tag{25}$$

(iv) *Harmonic solution.* The harmonic solution to Laplace's equation (12) for $y > 0$ is given by

$$\phi = (ae^{ky} + be^{-ky})e^{i(kx-\omega t)} \qquad y > 0 \tag{26}$$

where from (19), (20) and (22) we have

$$a = b - iA_b \frac{\omega}{k} \tag{27}$$

$$b = -iA_b \frac{\omega}{2k} \frac{(kg/\omega^2 - 1)(\tanh kh + 1)}{1 - (kg/\omega^2)\tanh kh} \ . \tag{28}$$

Next, we turn to the solution in the elastic bed where the solution of Laplace's equation (14) for the displacement potential ϕ in $y < 0$, satisfying the boundary condition (25) at infinity, yields

$$\phi = ce^{ky}e^{i(kx-\omega t)} \qquad y < 0 \ . \tag{29}$$

On the other hand, the solution of Kelvin wave equation (18) for ψ in the lower half plane is given by

$$\psi = de^{k_1 y}e^{i(kx-\omega t)} \tag{30}$$

where,

$$k_1 = \sqrt{k^2 - s^2}, \qquad s^2 = \frac{\rho_s \omega^2}{G} \tag{31}$$

and where s is the elastic shear wave-number. From conditions (22) and (24), we obtain expressions for the coefficients c and d

$$c = id + \frac{A_b}{k} \tag{32}$$

$$d = 2i\frac{k}{s^2}A_b \ . \tag{33}$$

Substituting the above expressions in the remaining boundary condition (23) of continuity of normal stress at the water-bed interface, and after some manipulation, we obtain the dispersion relation of this gravity hydro-elastic wave (see, e.g., MacPherson, 1980)

$$\frac{\dfrac{\rho}{\rho_s}\left(1 - \left(\dfrac{gk}{\omega^2}\right)^2\right)\tanh kh}{\left(\dfrac{gk}{\omega^2}\right)\tanh kh - 1} + \frac{gk}{\omega^2} - \left(\frac{2k^2}{s^2} - 1\right)^2 + \left(\frac{2k^2}{s^2}\right)^2\left(1 - \frac{s^2}{k^2}\right)^{1/2} = 0 \tag{34}$$

which relates the wave frequency ω and wavenumber k. Since we are considering a conservative system, only the real roots from (34), corresponding to nondissipative waves, will be acceptable here. Solving (34) for the eigenvalue $k(\omega)$ will complete the solution for this basic interaction problem. One important property of the solution is the ratio A_b/A of the undulation amplitudes at the seabed surface vs. the free water surface. From (20) we see that this ratio is given by

$$\frac{A_b}{A} = \cosh kh - \frac{gk}{\omega^2} \sinh kh \ . \tag{35}$$

As for the roots of (34), several important limits can be derived from the general dispersion relation, and it is instructive to briefly discuss these various limits as a way of highlighting the wide range of the obtained solution.

2.1. *Waves on a stiff bed*

We first look at the case of water-wave propagation over a bed of a large elastic stiffness, $G \to \infty$, so that from the definition of s in (31) we have $k/s \to \infty$. It can be shown that at this limit the dispersion relation (34) may be reduced to

$$\frac{\frac{\rho}{\rho_s}\left(1 - \left(\frac{gk}{\omega^2}\right)^2\right)\tanh kh}{\left(\frac{gk}{\omega^2}\right)\tanh kh - 1} + \frac{gk}{\omega^2} - \frac{3}{2} = -\frac{2k^2}{s^2} + O(s^2/k^2) \ . \tag{36}$$

Notice that the right-hand side of (36) is proportional to $(k/s)^2 \gg 1$, so that in order to balance the equation it is clear that the denominator in the first term of (36) should approach zero, i.e.,

$$\omega^2 \to gk \tanh(kh) \qquad \text{as} : G \to \infty \tag{37}$$

which is the classical dispersion relation for surface water waves on a rigid bed (e.g., Lamb, 1932). Comparing (36) with (37), we notice that for a nonrigid stiff bed, the phase speed ω/k of the wave according to (36) is always slower than that from (37). Therefore, it is seen that a small seabed compliance in coastal waters will cause a slight *slow down* in the propagation speed of water waves, as compared to rigid-bed calculation. The above solution represents a "surface-wave" mode, associated with the water-layer free surface, and it is only a slight modification to the shallow water wave solution on a rigid bed.

An "internal-wave" mode associated with the bed-water interface represents a second possible free-wave solution. In classical elasticity literature, this type of free wave belongs to the class of "Stonely" interfacial wave mode (Graff, 1975). The Stonely wave root is derived from (34) by noting that the Stonely wavenumber is of the same order of magnitude as the elastic-shear wavenumber s, i.e., we have $k/s \sim O(1)$, which in turn implies for a stiff bed that k is much smaller than the characteristic gravity wave length scale ω^2/g, or that $gk/\omega^2 \ll 1.0$. At this limit, (34) is reduced to

$$\frac{\rho}{\rho_s} \tanh kh = -\left(\frac{2k^2}{s^2} - 1\right)^2 + \left(\frac{2k^2}{s^2}\right)^2 \left(1 - \frac{s^2}{k^2}\right)^{1/2} \tag{38}$$

which is the dispersion relation for a Stonely wave on the interface between the elastic half plane and the overlying water layer of thickness h. If we further assume that $kh \ll 1$ so that the left-hand side of (38) goes to zero, we get the dispersion relation for Rayleigh surface-wave in an elastic half-plane

$$16r^3 - 24r^2 + 8r - 1 = 0, \qquad r = (k/s)^2 . \tag{39}$$

Notice that the system now has a set of two distinctly different types of free wave modes. This fact will not alter by changing the constitutive behavior of the seabed, from elastic solid to inviscid fluid, to other complicated rheologies; only the description of the wave modes will get more complex. As is the case for any linear system, the general dynamic response of the seawave-seabed system will consist of a superposition of harmonics from these two types of wave modes.

Notice that in this limit of a stiff bed, the phase speed of the surface water wave is dramatically slower than the corresponding elastic bed wave speed. This leads to the obvious conclusion that at this limit, there is little dynamic coupling between the water waves and the seabed.

One noted exception to such uncoupled behavior is the rare occurrence of the phenomenon of microseism. A microseism is a very-low magnitude seismic wave which is excited in the seabed by an overlying water wave. Longuet-Higgins (1950) proposed a generation mechanism where an elastic compressional wave in the seabed is excited by the spatially-uniform pressure oscillation, which exists at the second-order level in a *standing* water wave.

Later on, it will be shown that the nonlinear coupling of water waves with elastic waves, particularly in cases of very soft beds, will give rise to important dynamical effects concerning energy transfer into the seabed.

2.2. *Waves on an inviscid liquid substrate*

The other extreme is a bed of diminishing elastic stiffness, $G \to 0$ (or $k/s \to 0$), which may correspond to an extremely soft muddy bed, so that (34) becomes

$$\frac{\dfrac{\rho}{\rho_s}\left(1-\left(\dfrac{gk}{\omega^2}\right)^2\right)\tanh kh}{\left(\dfrac{gk}{\omega^2}\right)\tanh kh - 1} + \frac{gk}{\omega^2} - 1 = 0\,. \tag{40}$$

One clear root of (40) is the deep-water surface wave relation

$$\omega^2 = gk \tag{41}$$

which is the same dispersion relation for waves in deep water of a *constant* density. It is a well-known result that a general density stratification will not change the inviscid dispersion relation of surface waves in deep water (Lamb, 1932). Another root of (40) represents a much slower internal fluid-mud wave on the interface between the two liquid layers (e.g., Lamb, 1932, p. 372)

$$\omega^2 = \frac{g\prime k \tanh(kh)}{\dfrac{\rho}{\rho_s} + \tanh(kh)}, \qquad g\prime = g\frac{\rho_s - \rho}{\rho}\,. \tag{42}$$

One interesting practical application of the above system was explored by Ting and Raichlen (1986) and Ting (1989) who investigated the possible resonant excitation of an internal fluid-mud wave inside a navigation channel by an incident surface wave. Clearly, resonance would occur if the frequency of the incident wave matches one of the natural frequencies of internal waves in the channel. Such a resonance would enhance, among other things, the fluid velocity near the bottom, leading to undesirable sediment erosion on the walls of the channel.

3. Internal Dissipation

So far, we developed solutions for a conservative water-bed system, since we did not allow for energy dissipation to take place either in the water layer or inside the seabed. Viscous damping of water waves may take place in both. Most of the energy dissipation in the water layer would typically take place in the water boundary layer just above the seabed. We postpone discussion on boundary-layer dissipation for now and focus here on wave dissipation inside the seabed. Internal viscous damping inside a seabed may be modeled by assuming that the bed behaves as (i) a very viscous liquid (Gade, 1958; Dalrymple and Liu, 1978), (ii) a viscoelastic solid (MacPherson, 1980; Hsiao and Shemdin, 1980), or (iii) a poroelastic solid (Yamamoto *et al.*, 1978; Madsen, 1978; Mei and Foda, 1981a).

3.1. *Waves on a viscous or a viscoelastic bed*

Gade (1958) and Dalrymple and Liu (1978) developed linear solutions for a dissipative water wave by assuming that the seabed may be modeled as a very viscous liquid layer. The conversion of the above-developed solution for an elastic seabed to the case of a viscous seabed can be done in a very straightforward manner. This is simply because, within the context of harmonic motion in an incompressible medium, the stress-strain relation for a viscous fluid is very similar in appearance to that of an elastic solid. In a Newtonian fluid, the *deviatoric* stress $\boldsymbol{\tau}\prime$ (i.e., the stress tensor in excess of the isotropic pressure) is directly proportional to *the rate-of-strain* $\partial\mathbf{e}/\partial t$ where $\mathbf{e}$ represents the material's strain tensor. On the other hand, deviatoric stress $\boldsymbol{\tau}\prime$ in an elastic solid is, instead, proportional to the strain $\mathbf{e}$ (not the rate of strain), i.e., we have

$$\boldsymbol{\tau}\prime = G\mathbf{e} \qquad \text{for an elastic solid} \tag{43}$$

and

$$\boldsymbol{\tau}\prime = \mu\frac{\partial\mathbf{e}}{\partial t} = -i\omega\mu\mathbf{e} \qquad \text{for a viscous fluid} \tag{44}$$

where μ is the fluid viscosity. By comparing (43) with (44) we see that we only need to replace G in the elastic–solid relation by $-i\omega\mu$ to recover the viscous-fluid relation. Therefore, replacing G by $-i\omega\mu$ in the already obtained solution will extend the solution to the case of a viscous bed. In much the same way, we may cover the case of a general viscoelastic bed by replacing the shear modulus G by the complex viscoelastic shear modulus G_e

$$G_e = G - i\omega\mu\ . \tag{45}$$

Substituting (45) into the dispersion relation (34), we observe that s is now a generalized complex parameter defined in terms of G_e as

$$s = \sqrt{\rho_s\omega^2/G_e}\ . \tag{46}$$

The solution to (34) will clearly yield a complex eigen-wavenumber k for a general viscoelastic bed. The real part of k, $\mathrm{Re}(k) = k_r$, will combine with the real wave-frequency to give the wave's phase speed $c_w = \omega/k_r$, and the imaginary part of k, $\mathrm{Im}(k) = k_i$, will give the spatial damping rate of the dissipative wave. Therefore, from (5) we see that the harmonic wave-surface elevation will take on the form

$$\eta = Ae^{-k_i x} e^{ik_r(x - c_w t)} . \tag{47}$$

Clearly, wave energy dissipation is intimately related to the imaginary part k_i and there are various ways of calculating the rate of wave energy dissipation. However, we postpone discussion of such calculations until we review other models of dissipative seabeds.

3.2. *Waves on a poroelastic bed*

Another mode of seabed dissipation is due to seabed permeability. Propagating waves will force an oscillating seepage flow through the pores of the bed, and the associated friction on the walls of the internal pores will give rise to another energy sink for waves in the coastal waters. The flow produced by waves in permeable beds has been studied by many researchers including Putnam (1949), Ried and Kajiura (1957), Sleath (1970), Liu (1974), Massel (1976), Yamamoto *et al.* (1978), Madsen (1978), Packwood and Perrigrine (1980), and others. The simplest model of a permeable bed is that of a *rigid* porous bed. The resulting seepage flow obeys Darcy's law

$$nU = -\frac{K}{\mu}\frac{\partial p}{\partial x} \tag{48}$$

$$nV = -\frac{K}{\mu}\frac{\partial p}{\partial y} \tag{49}$$

where U and V are the horizontal and vertical seepage velocity components, respectively, p is the pore-pressure, K is the bed's specific permeability coefficient, and n is the porosity (pore volume per unit bulk volume) of the bed. Notice that Darcy's law, as stated above, is based on two main assumptions: (i) the seepage flow is *laminar*, and (ii) the flow has negligible inertia. Therefore, a necessary condition for the validity of Darcy's law is the requirement that Reynold's number of the seepage flow, based on the average pore size, should be small (e.g., Bear, 1972; Scheidegger, 1974). Implementing the continuity condition (8) on the seepage flow (U, V), we see from (48) and (49) that the pore-pressure p satisfies Laplace's equation

$$\nabla^2 p = 0 \,. \tag{50}$$

It is straightforward to show that for a harmonic wave with a free-surface elevation η given by (5), the solution to (50) is

$$p = p_o e^{ky} e^{i(kx-\omega t)} \tag{51}$$

where

$$p_o = \frac{\rho g A}{\cosh(kh)} \tag{52}$$

and where p_o is the inviscid-wave solution for bottom pressure in a water layer of depth h, assuming a rigid impervious bottom. A simple dimensional-analysis argument (e.g., Mei and Foda, 1981a) can be used to confirm that the slow seepage flow in and out of the porous bed will have negligible correction effect on the bottom wave pressure p_o as given by (52).

Next, we may relax the rigid-bed assumption and adopt instead the poro-elastic model of Biot (1941, 1956) which assumes that the solid skeleton of the porous medium deforms elastically, obeying Hooke's Law, while employing a generalized Darcy's law to describe the seepage-flow balance. Instead of (48) and (49), we have for the balance of pore-water momentum in the x and y directions respectively

$$\rho \frac{\partial U}{\partial t} = -\frac{\partial p}{\partial x} - \frac{n\mu}{K}(U-u) \tag{53}$$

$$\rho \frac{\partial V}{\partial t} = -\frac{\partial p}{\partial y} - \frac{n\mu}{K}(V-v) \tag{54}$$

where, now, fluid inertia is included in the general momentum balance equations. Notice that in this deformable porous medium the seepage flow is defined as the velocity of the pore-water *relative to* the velocity of the solid skeleton, i.e., $U-u$, $V-v$. Taking the divergence of (53) and (54) and implementing the assumption of *incompressibility* for both the solid skeleton and the pore water (i.e., zero fluid and solid dilatation), we recover Laplace's equation (50) for the pore-pressure. This means that pore-pressure distribution would still be given by (51) for either a rigid porous bed or a deformable but incompressible porous bed. Beside the pore-pressure, there will be solid *effective* stresses σ_{xx},

σ_{yy}, σ_{xy} carried by the porous solid skeleton. In a poroelastic medium, these effective stresses will be related to the solid displacements X and Y through Hooke's law, i.e., equations (15), (16) and (17) after replacing τ_{xx}, τ_{yy}, τ_{xy} in the left-hand side by σ_{xx}, σ_{yy}, σ_{xy} respectively, and replacing ρ_s in the right-hand side by $(1-n)\rho_s$, which is now the solid density in this composite medium (proportion of space occupied by the solid phase is $1-n$). Adding the contribution from the solid effective stresses to the pore-pressure contribution, we obtain the *total* stress tensor in this two-phase medium

$$\tau_{xx} = \sigma_{xx} - p, \qquad \tau_{yy} = \sigma_{yy} - p \tag{55}$$

$$\tau_{xy} = \sigma_{xy} \,. \tag{56}$$

It can also be shown that the solid displacement X and Y in this incompressible poroelastic bed are given by (13), where the potential ϕ is still governed by Laplace's equation (14), and the function ψ is still governed by Kelvin's equation (18), after replacing ρ_s by $(1-n)\rho_s$. As for the boundary conditions, we note that the condition (23) on the continuity of normal stress across the interface will now have to be replaced by *two* separate continuity conditions: one on water pressure continuity, and the other on solid normal stress continuity, respectively requiring that

$$(p)^+ = (p)^- \tag{57}$$

$$\left(2G\frac{\partial Y}{\partial y} + (1-n)\rho_s\frac{\partial \phi}{\partial t} + (1-n)\rho_s g\eta_b\right)^- = 0 \,. \tag{58}$$

The other boundary conditions (19), (20), (22), (24) and (25) will remain identical to the elastic bed case. It is, therefore, of interest to compare the solution for a poroelastic bed to the already discussed elastic case. Mei and Foda (1981a) demonstrated that, for a poroelastic medium with a low permeability coefficient which is typical for many geophysical porous media, the leading-order harmonic behavior is *pure elastic,* plus a boundary-layer correction which provides the seepage-induced energy dissipation.

4. Boundary-Layer Processes

As stated above, the benthic boundary layer (BBL) constitutes an important component of the general wave-bed interaction problem. Typically, the response has a *sharp* water-to-bed transition across the boundary layer. Part of this boundary layer exists above the water-sediment interface, and the other part below the interface. As demonstrated in the solutions of the above two Sections, the wave length $L = 2\pi/k$ is clearly the primary length scale of the response above and below the boundary layer. Within the boundary layer, however, the thickness of the boundary layer δ, which is much shorter than L, becomes the representative length scale of the response there.

Embedded within the BBL, there is usually an even thinner layer of flowing solid grains extracted from the sediment bed. The granular flow is maintained by a tangential force from the fluid flow above. Bagnold (1954, 1956) postulated and verified experimentally that in addition to the net drift of the granular flow, random collisions among neighboring grains will give rise to granular oscillations in all three directions. Therefore, similar to the kinetic theory of gases, these collisions will give rise to a pressure field inside the granular flow, which Bagnold termed the dispersive-grain pressure tensor. Bagnold's "dispersive-grain" pressure tensor provides the necessary reaction force to balance gravity; or to support the submerged weight of this bed-load of flowing particles. Here, the grain concentration is found to attenuate progressively to zero with increasing height above the bed.

The fluid tangential stresses on the granular bed will be developed in a viscous or often turbulent boundary layer above the bed. Studies on the properties of oscillatory fluid boundary layers dates all the way back to the classical work of Stokes (1851). Stokes developed a solution to the Navier-Stokes equations for the flow around a flat plate immersed in a fluid at rest and oscillating in its own plane. Nearly a century and a half later, the extensive research on this particular area has made major inroads in improving our current understanding of the real processes taking place there. Extensions to turbulent boundary layers, nonlinear streaming effects, and modifications due to sediment ripples are some of the critical issues which received particular attention in the past (see, e.g., Sleath, 1984 for a review).

Inside the non-fluidized porous bed, another boundary layer structure has been identified by Mei and Foda (1981a) as part of a leading order solution to Biot's (1941, 1956) poroelasticity equations. This was motivated, in part, by the experimental observation by Yamamoto *et al.* (1978) that the length scale of pore-pressure attenuation in a soil bed under water waves was much shorter than the wave length. This is in contradiction to Equation (51) for pore-pressure distribution in either a rigid or a deformable but incompressible porous bed. Yamamoto *et al.* (1987) and Madsen (1987) solved the full poroelasticity equations for a general compressible bed and showed that the extra consolidation effect associated with the inclusion of compressibility would reproduce this fast pore-pressure attenuation with depth. Mei and Foda (1981a) showed that this attenuation length scale is in fact a general length scale in Biot's equations and, in physical terms, it represents the thickness of a thin boundary layer where seepage flow is important. Such a boundary layer would exist near the surface of the seabed where water above the bed is free to squeeze in and out of the porous bed. The enhanced seepage would clearly increase friction on the walls of the internal pores, and this in turn will cause faster attenuation in pore-pressure to balance this extra friction (Darcy's law).

In what follows, a more detailed discussion will be given on the mechanics of each one of these *three* sub-layers of the BBL. The separate discussions are motivated by the distinct features of these sub-layers. However, whenever possible, the natural coupling among the sub-layers will be emphasized as well.

4.1. *Fluid boundary layer*

Most of the early studies on waves interaction with the seabed were particularly concerned with the mechanics of the bottom boundary layer but under the simplifying assumptions of flat, rigid, nonporous, and nonerodible bottom. Even under such an idealization, two important complications would arise when attempting to extend the above-mentioned Stokes (1851) solution to the case of water waves over a stationary bed. One is related to the fact that, unlike Stokes's infinitely long flat plate, water waves have finite length and, therefore, there will be small but nonzero contribution from the nonlinear convective inertia terms in the equations of motion (i.e., the nonlinear terms in the left-hand sides of Equations (1) and (2)). The product of wave-harmonic terms will clearly give rise to *new* wave harmonics. The frequencies of the new harmonics will be either the multiples (super-harmonics) or the differences (subharmonics) of the frequencies of the generating harmonics. For example, the self-interaction (or self-product) of the primary wave harmonic will give rise to a second harmonic (twice the primary frequency) and a zeroth harmonic (zero frequency) constituents in the general response. Of particular interest is this zeroth-order harmonic constituent, since it represents a weak wave-induced streaming flow capable of transporting mass (e.g., water, sediment, pollutants, etc.) above the bed. The solution for mass transport velocity due to simple-harmonic propagating waves was first obtained by Longuet-Higgins (1953), and verified experimentally by Collins (1963), Mei *et al.* (1972) and others. Liu (1977) and Sleath (1978) investigated the effect of bed permeability on the mass transport velocity and found that the deviation from Longuet-Higgins solution could be significant. It is found that the mass transport just outside the boundary layer is

$$\bar{u}_m = \frac{U_o^2 k}{4\omega}\left(5 + \frac{4(\rho\omega/2\mu)^{3/2} K}{k}\right) \tag{59}$$

where $\bar{u}_m$ is the wave-average mass-transport velocity, and U_o is the amplitude of the wave oscillatory velocity just outside the boundary layer. The first term in (59) is the Longuet-Higgins (1953) solution and the second represents the effect of bed permeability.

Besides convective nonlinearity and associated mass transport applications, there is the other complication due to the possible transition of the flow from laminar (Stokes solution) to turbulent. Different models have been put forward to extend the analysis to account for turbulent effects. The treatment by Kajiura (1968) employed an equivalent eddy viscosity to replace the molecular viscosity μ. Bakker's (1974) model is instead based on the introduction of a turbulent mixing length, and Jonsson (1980) developed a velocity defect analysis. All the models have borrowed from the well-established steady-state turbulent flow models. Another group of models were developed to extend the analysis to boundary-layer flows over rippled but still rigid beds (e.g., Lyne, 1971; Sleath, 1975; Longuit-Higgins, 1981). An excellent review of these and other models of fluid boundary-layer flows over a rigid bed is given in Sleath (1984).

4.2. *Granular-flow boundary layer*

Bagnold, in his pioneering work in the mid-1950s, has successfully laid down many of the fundamental concepts of the theory of granular flow in fluids. For example, he identified two distinct categories of flowing grains. A bed load is defined as that portion of the moving grains whose weight is supported by solid stresses, through grain-to-grain collisions and slidings, and transmitted to the unfluidized bed as effective stresses. On the other hand, the suspended load is defined as the grains whose weight is instead supported by fluid pressure. This component of fluid pressure would necessarily be transmitted into the underlying bed as an excess pore-pressure. Bagnold (1955) was able to measure this excess pressure component in the laboratory. Bagnold attributed the source of this excess water pressure to the working of turbulent eddies on the flowing grains. Recent measurements, however, seem to suggest that there are cases where sediment suspension may be originated inside the porous bed itself, due to some fluidization processes (e.g., Hanes, 1992; Foda and Tzang, 1993). More discussion of this issue will be given later.

Following Bagnold, there have been many attempts to develop empirical formulae for the prediction of sediment transport rates. However, there are still many fundamental obstacles that need to be overcome before being able to confidently predict sediment transport in most cases of interest. One such obstacle is our poor understanding of the mechanical behavior of high-concentration fluid-sediment mixtures. This is even more so for oscillatory flows than for the much more studied steady flows. Recently, however, Dick and Sleath (1991) reported on an experimental study where they measured velocity and concentration distribution both within and above a granular bed in oscillatory flow. They found that the profile of velocity amplitude and phase across the bulk of the granular-flow layer is almost *linear* with height, i.e., the velocity distribution there may be expressed as

$$\frac{u}{U_o} = K_1 y \cos(\omega t - K_2 y + \phi_o) \tag{60}$$

where U_o is the velocity amplitude outside the viscous boundary layer, K_1 and K_2 are calibration constants, ϕ_0 is a phase constant, and y is measured from the base of the moving layer. They noted that the above solution is similar to the inner limit of Stokes' (1851) solution for oscillatory flow over a flat bed (with axes fixed in the bed) but of course with different numerical values for the solution constants. Further above the granular layer and into the fluid boundary layer, it was possible to fit logarithmic curves to segments of the velocity profile. However, they confirmed that the movement of the bed has a significant effect on the fluid boundary layer. Bed roughness height is increased and the amplitude of the velocity falls off more slowly than for a fixed bed under equivalent flow conditions. Another important observation from their measurement is that the shear stress inside the granular-flow layer is *not* in phase with the velocity gradient (see, e.g., Fig. 18 in their paper). This clearly rules out a Newtonian fluid model to describe the behavior of the grains-water mixture. Instead, a rather simple alternative would be to adopt a *viscoelastic* constitutive model to account for the observed phase lag. Recall that a general viscoelastic material subjected to harmonic loading would be characterized by a complex shear modulus G_e defined by (45) or equivalently a complex viscosity coefficient $\mu_e = iG_e/\omega$. The real part of μ_e would represent the coefficient for the in-phase (viscous) part of the response, while the imaginary part as the coefficient for the out-of-phase (elastic) part of the response.

Beside this *flowing* mode of sediment motion where shear is almost uniformly distributed across the moving sediment layer, there is also the important *sliding* mode which is observed primarily in larger-scale movement events. In the slide mode, shearing motion is confined to within a thin zone at the base of the layer with the rest of the mass moving at an almost constant bulk velocity with little or no mixing. Figure 2 sketches the difference between a flowing mode and a sliding mode. Note that a basal boundary layer in a sliding mode is in fact a *second-level* boundary-layer structure within the overall benthic boundary layer (BBL). This is an example of the intricate complexities that are often encountered when studying BBL processes.

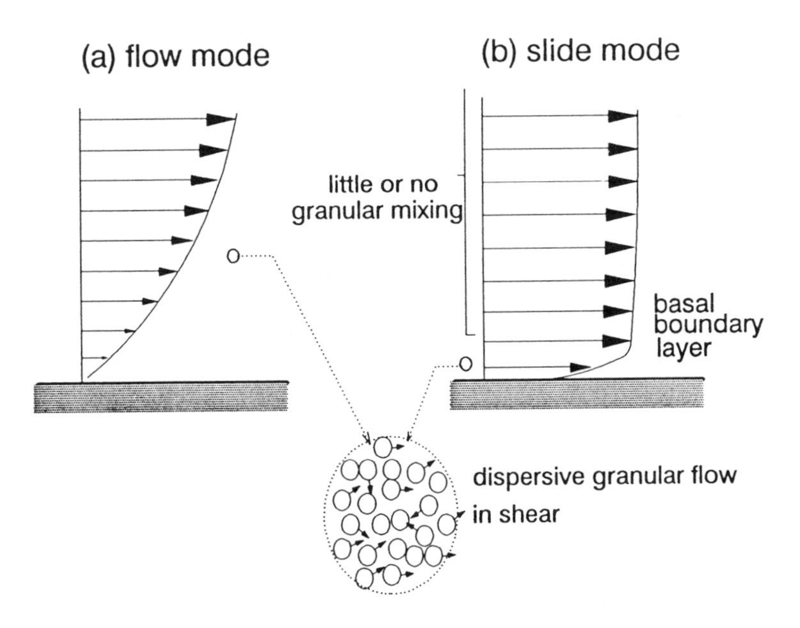

Fig. 2. Modes of granular flow.

Sliding is likely to be a preferred mode of motion for a variety of large-scale sediment transport phenomena such as, for example, sediment slumping off the edges of submarine slopes, submarine canyons or even continental shelves (Press and Siever, 1974); or earthquake or storm-induced submarine slides, avalanching, turbidity currents and mud flows (Saxov and Nieumenhuis, 1980). Foda (1993) and Dagang and Foda (1993) studied the dynamics of the basal boundary layer for relatively deep sliding motion and showed that the shear granular flow in a thin basal layer is unstable to internal wave perturbation of the kind recently proposed by Foda (1987). This instability wave will be excited when the bulk sliding velocity exceeds a certain threshold limit U_c. They then showed that the presence of such an instability wave will have important dynamical effects that would result in certain exchanges of *mass, energy and momentum* across the interface between the high-shear basal zone and the overlying bulk-sliding mass. Such exchanges may prove to be the means by which transition would take place from a sliding mode to a flowing mode and vice versa.

Finally, there is one particular property of the granular boundary layer in oscillatory flow which has received special attention lately. This is the phenomenon of wave-induced intermittent suspension of sediments. The phenomenon has been observed in the field by many investigators including Brenninkmeyer (1976), Thornton and Morris (1978) and Hanes *et al.* (1988). A typical observation is that of cloudy bursts or puffs of sediments sporadically ejecting from the seabed under shoaling waves. The strength of bursting seems to be greatly enhanced near the breaker line. Madsen (1974) reports divers' observations that the sediment bed seemingly "explodes" just prior to the passage of the crest of near-breaking waves. The precise mechanism for such rapid *vertical* entrainments of sediment from the seabed is not yet known. This is why there has not been systematic observation of the phenomenon in the laboratory. For example, Dick and Sleath (1992) did not report on similar observation of vertical motion in their experiments on artificial granular beds. On the other hand, Shibayama *et al.* (1986) reported on the generation, in a laboratory wave flume, of strong sediment jets under near-breaking water waves, somewhat confirming Madsen's observations (1974). These jets would lift a substantial amount of sediment into suspension over the whole water depth. Furthermore, Clukey *et al.* (1983) and Foda and Tzang (1993) reported on observations in the wave flume of sediment plumes emanating from a micro volcano-like structure on the surface of a fine silty bed under small-amplitude monochromatic waves.

There have been a number of mechanisms proposed to explain how the sediment is carried up from the bed into suspension. Much of the early work was based on the assumption that suspension is caused by fluid turbulence, and various mixing-length theories were put forward (e.g., Homma and Horikawa, 1962; Bakker, 1974). Later work, supported by careful measurements, indicates that most of the suspension is however associated with the periodic wave velocity near the bed rather than with turbulence. Nielson (1979) developed analytical models in support of this concept of a regular and repeatable sediment suspension in a wave field. Sleath (1982) demonstrated in the laboratory that sediment jets will be regularly ejecting off the crests of a rippled sand bed every time the wave flow reverses direction. This happens as the sediment-laden vortices in the lee of each ripple is pushed back towards the ripple's crest due to flow reversal. Even though this is a very-likely mechanism for sediment suspension over a rippled bed, it is certainly not the only one at work based on the variety of field and laboratory measurements. For example, Inman's (1957) observation of sediment suspension near the surf zone describes a nondiscrete, dense suspension layer above highly-transitional rippled seafloor. Hanes (1992) reviewed some of the field data from the Stanhope Lane Experiment, Prince Edward Island, Canada, and concluded that the time scales of many of the recorded intermittent suspension events were much longer (by a factor of five or more) than would be expected from a vortex-cloud mechanism. He suggested that these observations could be the result of some other processes involving direct vertical ejection from the bed. Shibayama *et al.* (1986) indicated that the observed strong suspension under a near-breaking wave is not related to the small-scale vortices near the bed, but may be associated with the much larger scale vortex structure of the plunging breaker.

Recently, Conley and Inman (1992) included a video camera in their field measurement scheme in order to visually inspect the details of the suspension process under waves. Their measurements were conducted in water depths of 1 to 3 meters, seaward of the surf zone off the ocean pier of Scripps Institution of Oceanography, La Jolla, California. They identified three distinct regimes in the development of the granular boundary layer, with sediment pluming being the ultimate outcome. The observed development appears to represent a transitional sequence from laminar to strong-turbulent flow. They identified three distinct regimes of streaking, roiling and pluming, with sediment pluming being characterized as a rather sudden and explosive lift off of sediment into the interior of the water column. They suggested that pluming may be associated with a breakup of turbulent vortices caused by the adverse pressure gradients of the wave field.

4.3. *Boundary layer in a poroelastic bed*

Underneath the base of the granular-flow layer, another boundary layer structure may develop in the unfluidized porous bed. This porous-media boundary layer owes its existence primarily to the usually small *compressibility* that may exist in the solid skeleton and/or the pore fluid. Initially, we have dropped the compressibility effects from the analysis. The rationale for this assumption, which is employed in many wave-seabed studies, is based on the fact that water waves are much slower (by several orders of magnitude) than sound (or compressional) waves in either water or the seabed solids. Therefore, in most applications, water waves are uncoupled from compressibility waves (the exception of microseism generation has been noted earlier). However, besides the generation of compressibility waves, there is the important *consolidation effect* that results from the coupling of the small compressibility in a porous bed with the viscosity of the pore fluid. Terzaghi (1945) was the first to identify this slow diffusion effect in his classical work on soil mechanics. The associated diffusion process may take on a boundary layer structure and it can be easily demonstrated as follows. We consider the simple case of a compressible pore-fluid with a bulk modulus β, so that the fluid continuity equation becomes

$$\frac{\partial U}{\partial x} + \frac{\partial V}{\partial y} = -\frac{1}{\beta}\frac{\partial p}{\partial t} \,. \tag{61}$$

Using this in the generalized Darcy's equations, we get

$$\nabla^2 p = \frac{n\mu}{K\beta}\frac{\partial p}{\partial t} + \frac{\rho}{\beta}\frac{\partial^2 p}{\partial t^2} \,. \tag{62}$$

Comparing (60) with (50), we see that the inclusion of compressibility (i.e., a finite β) has yielded the two new terms on the right-hand side of (62), representing the two different compressibility-induced effects. The last term in (62) is associated with fluid inertia and represents the possible compressibility-wave effect, while the other term, involving the product of the permeability coefficient and the bulk modulus, represents the consolidation effect. A simple order-of-magnitude comparison between these two terms on the right-hand side of (62) shows, for typical values of the parameters involved, that

$$\frac{n\mu}{K\beta}\frac{\partial p}{\partial t} \Big/ \frac{\rho}{\beta}\frac{\partial^2 p}{\partial t^2} \sim \frac{n\mu}{\rho\omega K} \gg 1 \tag{63}$$

confirming that the compressibility-wave effects may be neglected in comparison with the consolidation effects as stated above, so that (62) becomes the diffusion (consolidation) equation

$$\nabla^2 p = \frac{n\mu}{K\beta}\frac{\partial p}{\partial t} \,. \tag{64}$$

The diffusion length-scale $\delta\ell$ for a wave-frequency ω is seen from (64) to be

$$\delta\ell = \sqrt{\frac{K\beta}{n\omega\mu}} \,. \tag{65}$$

A more complete treatment is given by Mei and Foda (1981a). They started from the full Biot's poroelasticity equations and examined the general properties of the linear poroelastic wave solutions. One class of solutions, which Biot (1956) called *waves of the first kind*, can be reduced to well-known elastodynamic wave solutions by taking the limit of no pore fluid in the pore space. This limit, however, would lead to the vanishing of the other class of solution termed *waves of the second kind*. Biot (1956) found that the waves of the second kind have an attenuation distance that is much shorter than typical length scales of the waves of the first kind. Mei and Foda (1981a) argued that based on this general result, at least two vastly different length scales are inherent in Biot's equations, suggesting a boundary-layer analysis. The analysis has shown that the poroelastic problem may be reduced to first solving a conventional elasticity problem (e.g., bed response solution in Sec. 1) and then making a boundary-layer correction near the free surface. The validity condition for such a boundary-layer structure is that the derived scale for boundary-layer thickness $\delta\ell$ is much shorter than the problem's other primary length scale L, i.e., the wave length. For a poroelastic material with pore fluid bulk modulus β and solid skeleton Poisson's ratio ν, it is shown that the scale of the boundary-layer thickness is given by

$$\delta\ell = \sqrt{\frac{K}{\mu\omega}}\left[\frac{n}{\beta} + \frac{1}{G}\frac{1-2\nu}{2(1-\nu)}\right]^{-1/2} \,. \tag{66}$$

Thus, the boundary-layer thickness is very small if any of the following situations prevail, small permeability, high frequency, small bulk modulus of the pore fluid and small modulus in the solid. It is shown that $\delta\ell$ would range in value between a few centimeters to about ten meters for the case of a 10-s water wave and a wide range of geophysical porous media as the seabed (Mei and Foda, 1981a). This clearly makes $\delta\prime$ much shorter than the corresponding water wave length $L \sim O$ (100 m), satisfying the boundary-layer theory requirement. Approximate boundary-layer solutions were shown to be in agreement with the *exact* solutions by Yamamoto *et al.* (1978) and Madsen (1978). The method has also been used successfully to treat a variety of practical problems (e.g., Mei and Foda, 1981b; Mynett and Mei, 1982; Foda, 1982, 1985; Tsai *et al.*, 1990; Sakai *et al.*, 1992).

5. Damping Rates of Water Waves

Inside the viscous or viscoelastic boundary layer near the bottom, it can be shown that the mean rate of dissipation of wave energy $P_{\rm bl}$, per unit seafloor surface area, is given by (e.g., Dalrymple and Liu, 1978)

$$P_{\rm bl} = \mathrm{Re} \int_{\rm bl} \left(\mu_e \frac{\partial u}{\partial y} \right) \cdot \left(\frac{\partial u}{\partial y} \right)^* dy \tag{67}$$

where Re means the real part of, the asterisk denotes complex conjugate, and the integration is performed across the entire thickness of the boundary layer. In addition, the water wave will lose more energy if there is a net energy transfer into the seabed below the fluidized boundary layer. This net energy transfer is equal to the net working of the wave's traction stresses on the seabed surface

$$P_{\rm sb} = \mathrm{Re}(\tau_{yy} v^* + \tau_{xy} u^*) \; . \tag{68}$$

Note that (68) is calculated at the interface between the fluidized (viscous or viscoelastic) boundary layer and the underlying seabed. If the behavior in the seabed is assumed to be elastic, then Hooke's law dictates that stresses are 90° out-of-phase with corresponding velocities, implying zero dissipation from (68). For obvious reasons, (68) will also yield zero seabed dissipation in the cases of an inviscid liquid bed or, at the other extreme, a rigid bed. The steady-state wave energy balance requires that

$$\frac{d}{dx}(E c_g) = -(P_{\rm bl} + P_{\rm sb}) \tag{69}$$

where E is the wave energy density in the water layer and c_g is the group velocity, Re $(\partial\omega/\partial k)$. E is the combination of the wave's kinetic and potential energy and is defined as

$$E = \frac{1}{T}\mathrm{Re}\int_0^T \left[\int_{\eta_b}^{h+\eta} \frac{1}{2}\rho(u^2+v^2)dy + \frac{1}{2}\rho(\eta^2-\eta_b^2)\right] . \tag{70}$$

It is easy to show that the leading order contribution to E is quadratic in the wave's surface amplitude, so that from (47) we may approximate (69) to get another expression for the amplitude's damping rate k_i (e.g., Dalrymple and Liu 1978)

$$k_i = (P_{\mathrm{bl}} + P_{\mathrm{sb}})/(2c_g E) . \tag{71}$$

Recall that k_i is the imaginary part of the eigen-wavenumber of the problem which can also be obtained from the problem's dispersion relation. However, from a practical standpoint, k_i is usually much smaller in magnitude than k_r (the real part of the eigen-wavenumber), and it is normally more reliable, although approximate, to use the dispersion relation to get k_r and then use (71) to get k_i.

Calculations by MacPherson (1980) and Hsiao and Shemdin (1980) for a viscoelastic bed, and Gade (1958) and Dalrymple and Liu (1978) for a pure viscous bed have typically yielded energy dissipation rates which are much larger in magnitude than those obtained from rigid-bed's bottom friction models (e.g., Putnam and Johnson, 1949; Ippen, 1966). MacPherson (1980) made numerical comparisons among various viscous models and showed (e.g., his Fig. 7) that internal dissipation can be much larger than predictions from bottom-friction models. Table 1 lists the various analytical expressions for the wave amplitude's damping rate k_i as developed by different dissipative wave models.

Although the obtained damping rates from the compliant bed models are significantly higher than those predicted by the earlier rigid-bed bottom friction models, they are still too small, for realistic values of marine sediment properties (e.g., mud viscosity), to explain some of the observed spectacular dampings, especially over mudbank deposits. A more serious shortcoming of these above-mentioned linear damping models is that the observed enhancement of damping during severe wave loading (e.g., Jacob and Qasim, 1974; Wells, 1983) is a basic *nonlinear* feature of the problem that cannot be explained by a linear model. Foda (1989) developed a nonlinear damping model for water waves over a very soft bed. The elastic stiffness at the mudline is assumed to be very small, as is typically the case for upper-stratum gel-like marine mud, so that its effect is assumed comparable in magnitude with that

Table 1. Some analytical damping rates k_i from different dissipative wave models.

Mechanism	Damping rate k_i
• Deep water viscous wave, or inviscid wave above a viscous bed (Lamb 1932):	$\dfrac{4k_r^3\mu}{\rho\omega}$
• Bottom friction dissipation (Putnam and Johnson, 1949):	$\dfrac{4fA\omega^3}{3\pi g(\sinh k_r h)^3}$ (f: bottom friction coefficient)
• Laminar boundary-layer dissipation (Ippen, 1966):	$k_r^2\left[\sqrt{\rho\omega/2\mu}(2k_r h + \sinh 2k_r h)\right]^{-1}$
• Stiff viscoelastic bed (MacPherson, 1980):	$\dfrac{\rho g\mu}{4\omega(\mu^2 + G^2/\omega^2)}$
• Rigid permeable bed (Reid and Kajiura, 1957):	$\dfrac{K\rho\omega}{\mu}\dfrac{2k_r}{2k_r h + \sinh 2k_r h}$

of wave nonlinearity. The considered nonlinearity here is that due to the finite amplitude of the water wave. It is no longer assumed that the wave has an infinitesimal amplitude, which is the basis for the developed linear solutions of Secs. 2 and 3. In particular, the linear free-surface boundary conditions (19) and (20) would be replaced by nonlinear conditions involving quadratic and cubic terms in the unknown velocity potential ϕ. One well-known property of finite-amplitude, or nonlinear water waves is that the waves are unstable to sideband perturbations. Benjamin and Feir (1967) developed the sideband instability theory for water waves over a rigid bed and showed that this instability is possible only for water depth h larger than a critical value $h_c = 1.36/k$ where k is the wavenumber. The analysis has shown that the Benjamin and Feir's sideband instability works just as well for the case of a compliant bed. Notice that the critical depth h_c of Benjamin and Feir theory does not apply for the soft bed case and in fact the instability is possible for any water depth. This can easily be explained by referring to the leading-order linear dispersion relations (37) for a rigid bed versus (41) for a soft bed. It is clear that in the case of a soft bed the leading-order behavior would be that of a deep-water wave, and hence the wave would be exposed to the sideband instability for practically any depth of the surface water. Although the ensuing algebra is somewhat lengthy, the associated physics of the proposed sideband damping mechanism is rather straightforward and may be summarized as follows. The sideband instability will be responsible for distributing the incoming wave energy from a single carrier frequency to a narrow frequency band around this carrier frequency. In the presence of the bed's elasticity and viscosity, it is shown that these newly-generated sideband oscillations are *highly dissipative*. This is because these sidebands have weak but extremely short elastic shear waves in them. Because of their short length scales, these shear waves are capable of generating significant shear strains in the bed. Strong dissipation will therefore follow as these significant shear strains interact with the viscous boundary layer at the water-bed interface. Thus, in this damping model, energy is first lost to the sideband which in turn loses its acquired energy—via elastic shear waves—to boundary-layer friction. The model was shown to be able to reproduce the observed almost-total damping of water waves within a distance of only 4 to 8 wavelengths as reported by MacPherson and Kurup (1981) for monsoon waves over the mudbanks of Kerala, India. Unlike the linear models of MacPherson (1980) etc., where extremely high mud viscosity has to be assumed to reproduce the large damping, the sideband damping calculation required a kinematic viscosity of $10^{-5}\text{m}^2/\text{s}$ or only ten times that of water for the bottom mud to achieve similar results.

6. Fluidization Processes

Next, we turn our attention to the particular class of problems concerning the modes of mechanical failure in the seabed when subjected to water waves loading. We will divide the discussion into two almost separate groups, one regarding soil failures in a cohesive-sediment bed, and the other for a cohesionless-sediment bed. For a bed made of cohesive sediment, e.g., marine mud, we will discuss a soil fluidization mechanism caused by wave-induced *shear* stresses. A fluidization failure would take place inside the bed when the internal shear strains exceeds a certain *yield* limit, following a given nonlinear constitutive law for the bed material.

A different kind of fluidization failure will be discussed for a seabed of cohesionless sediment (e.g., sand or silt). This type of fluidization will be driven by wave-induced build up in *pore-pressure*.

6.1. *Fluidization of marine mud by waves*

Several recent interaction models have proposed different constitutive laws to describe the rheology of soft muddy bed in waves. These include a layered viscoelastic bed by Maa and Mehta (1988), a viscous bed under a finite amplitude wave by Jiang *et al.* (1990), and a Bingham-plastic bed by Mei and Liu (1987). Meanwhile, recent laboratory measurements by Chou *et al.* (1993) have shown that the viscoelastic properties of cohesive sediment in oscillatory shear are strongly dependent on the applied strain level. This makes the general material response to transient loading highly nonlinear. The measurements reported by Chou (1989) and Chou *et al.* (1993) confirmed that clay muds behave as elastic solids at low oscillatory strains and as viscous fluids at high oscillatory strains. This yield-like behavior is related to the break up of the soil's fragile aggregate matrix. Progressive break up at more aggregate joints will make the overall behavior more viscous and less elastic. Therefore, there will be an intermediate strain range where the sediment may be described as a viscoelastic material with variable viscosity and stiffness coefficients. The implication from this general rheological observation is that the material properties of the bed sediment cannot be set *a priori*, but should be made as functions of the imposed wave height. It is, for example, reasonable to expect that for the same bed material, larger waves will fluidize more of the bottom sediments than smaller waves would. The reason being that larger waves will induce larger strains in the mud, thus expanding the thickness of the mud layer where strains are large enough to make the response fluid-like. Beside strain magnitude, it is also found that mud rheology is a strong function of solid concentration or of the mud's relative density $\rho_r = \rho_s/\rho$. The measurements have shown that the material will have larger yield strain as it gets denser. Figure 3 shows a sample result for the rheology of one representative clay mineral, K1, called kaolin, for ranges of shear strain amplitude $|\gamma|$ and mud density (Chou *et al.*, 1993). The figure shows the least-square curve fits for the measured viscosity and shear modulus for the clay K1 at a salinity 35 ppt and an angular frequency of 1.5 rad/s. As stated above, there are three empirical regimes identified. At strain amplitudes less than some critical yield amplitude γ_e, the viscosity coefficient is zero and the behavior is elastic. For intermediate strains both viscosity and shear modulus are nonzero. At the higher strains there is no detectable shear modulus and the behavior is purely viscous.

Foda *et al.* (1993) incorporated some of these rheological findings into the development of an interactive model which predicts the depth of fluidization as a function of the imposed water-wave height. Their model was again made of an inviscid water layer of a depth h setting on a semi-infinite elastic bed as shown in Fig. 1. The formulation, however, differed from the basic non-

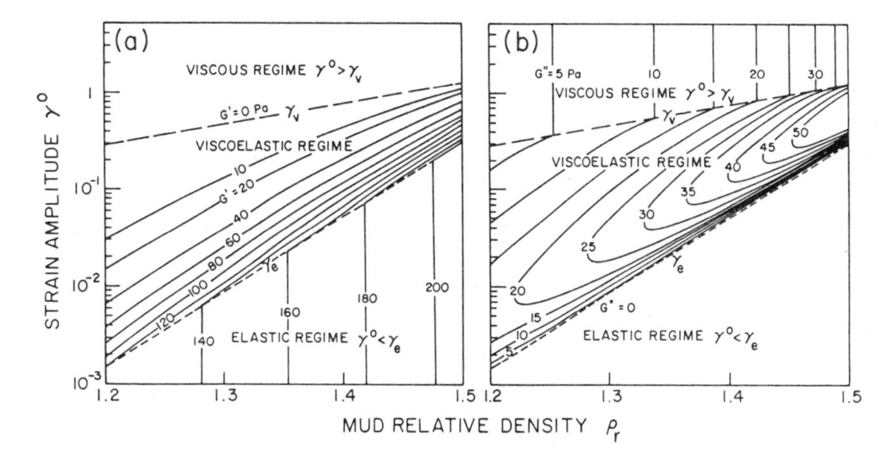

Fig. 3. Contours of the components of the complex viscoelastic shear modulus G_e Equation (45) for K1 kaolin at a salinity of 35 ppt and an angular frequency of 1.5 rad/s.
(a) The storage modulus $G\prime = G$ (real part of G_e).
(b) The loss modulus $G\prime\prime = \mu\omega$ (imaginary part). (Chou, 1989)

dissipative wave solution of Sec. 1 in two important aspects. They fitted a viscoelastic boundary layer of an unknown depth d at the water-bed interface, and they allowed the elastic bed to become slowly stratified with depth. The fluidized surface mud has a representative complex viscosity coefficient $\mu_e = iG_e/\omega$, where G_e is defined by (44). This viscosity coefficient is made to be a function of the imposed wave height, with the functional relationship being based on the experimentally-obtained rheological relation for the material. The stratification in the underlying elastic bed is assumed to be due to consolidation effects, with both density and elastic stiffness of the mud slowly increasing with depth. The formulation of the interaction problem proceeded in a manner similar to the treatment given in Sec. 1. With the inclusion of the viscoelastic boundary layer, it was also possible to satisfy the conditions of continuity of both tangential velocity and shear stress across the water-sediment interface. A WKB solution is developed for Kelvin's equation (18) which now has a variable coefficient $c_s(y)$ for the stratified bed. The depth d of the fluidized layer is determined by requiring that the elastic strain at the top of the elastic bed is equal to the limiting or yield strain γ_e for the given mud material. This produces the desired relationship between the depth of fluidization and the imposed wave height.

Figure 4 shows the assumed stratification profiles for three of the cases examined by Foda *et al.* (1993). These are cases which are relevant to a relatively deep and exposed open-coast mud bed. As shown in the figure, we fixed the density profile and considered three stiffness profiles I, II and III of similar shapes, with I being the softest and III the stiffest mud. This was done in order to examine the effect of relative mud stiffness on the resulting fluidization. Figure 5 shows the obtained fluidization relations for the three cases. The general trend is that of an increase in fluidization depth with increasing wave height. However, notice that under the same wave height, the weakest or softest bed I produces the *shallowest* depth of fluidization. This may roughly be explained as follows. As the elastic stiffness of the bed vanishes, its response approaches that of an inviscid liquid with diminishing shear strains and hence, less fluidization potential.

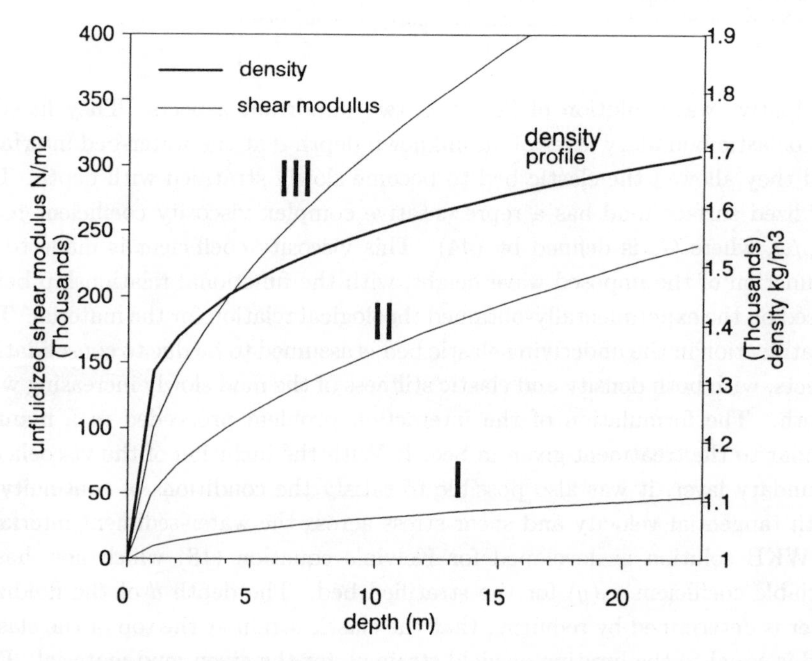

Fig. 4. Stratification profiles for three assumed cases of soft muddy beds. All three cases share the same density profile. Mud stiffness is lowest in case I as shown in the shear modulus profiles. These profiles correspond to the *unfluidized* elastic regime [cf. Fig. 3].

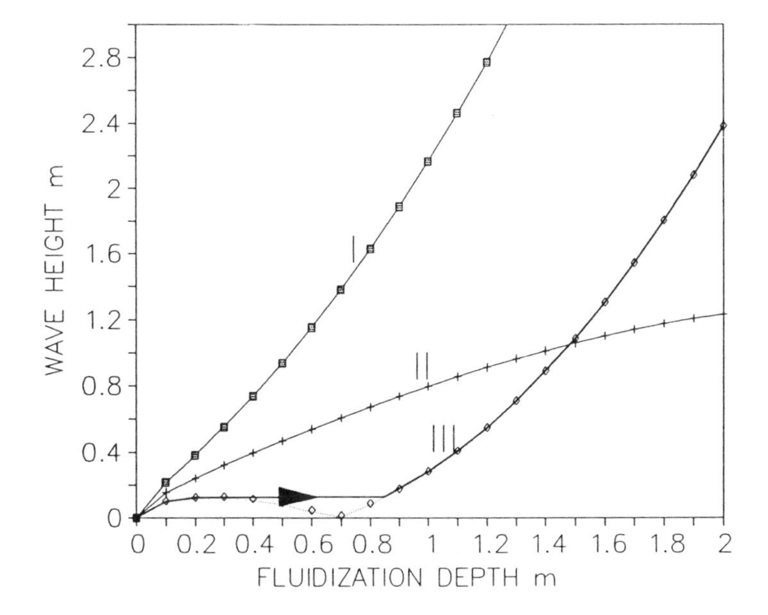

Fig. 5. Fluidization depth curves for the stratification cases of Fig. 4. (Foda *et al.*, 1993)

Another peculiar feature of the obtained solution in this range of parameters is the occurrence of a local maximum in the wave-height fluidization-depth relationship as shown in Fig. 5. This implies that sometimes the relationship is not single-valued. Case III in Fig. 5 demonstrate this property. The fluidization depth for this case has a local maxima at a wave height of about 0.15 meters and a corresponding fluidization depth of $d = 0.2$ m. The model calculation indicates that deeper fluidization depth for this case actually requires smaller wave heights, until a depth of about 0.85 meters is reached which requires the same wave height of 0.15 meters. This means that for a range of wave heights below 0.15 meters, each wave height will have not one, but two possible depths of fluidization depending on which branch the solution path is following. An important consequence of this property of the solution may be seen by considering what happens in the solution as the wave height increases gradually from zero, passing the critical local maximum at $H = 0.15$ m. The solution then indicates that, as the wave height increases infinitesimally above 0.15 meters, there will be a finite increase in the induced fluidization depth, from 0.2 meters to 0.85 meters; what might be termed a sudden *bulk fluidization* in the seabed. We demonstrated this in Fig. 4 by drawing a horizontal arrow which connects these two points through a discontinuous jump. As the wave increases further above this critical wave height, a gradual and monotonic increase in fluidization depth will resume. This was not limited to this particular case only but was reproduced for many other examined cases, suggesting that this observed phenomenon of bulk fluidization may be indeed a general property of the considered nonlinear system.

6.2. *Fluidization in cohesionless soil*

Recent experimental studies by Clukey *et al.* (1983), Foda *et al.* (1991), Tzang (1992) and Foda and Tzang (1993) have shown that water waves can sometimes fluidize cohesionless fine (silty) soil beds. The fluidization process is observed as a rather sudden rise in the *mean* pore-pressure inside the soil bed. Sometimes the magnitude of the rise approaches in some locations the value of the submerged weight per unit area of the soil column above. This corresponds to nearly reaching a stage of total fluidization where a block of soil particles becomes effectively in complete suspension, with its weight now supported almost entirely by the intervening fluid and thus accounting for the net rise in fluid pressure. Partial fluidization response corresponds to partial suspension of the soil particles and hence a smaller rise in mean fluid pressure inside the bed.

Here, the main findings reported by Foda and Tzang (1993) will be summarized. Figures. 6a, b and c show samples of pore-pressure records from their experiments on silty soils of a mean grain size $D_{50} = 0.05$ mm, placed in a 2-feet deep soil box under a 2-feet deep water layer. The shown measurements were taken at the same location inside the soil box, 8 inches below the soil-water interface but during different experimental runs. The figures clearly show three different types of pore-pressure response to loading from shallow water waves of heights from 5 to 12 centimeters and wave periods around 2 seconds. The details of the experimental set-up and experimental data are given in Tzang (1992). The first type of response is shown in Fig. 6a and clearly displays a resonant stage in a distinct three-phase response: initial, final and *resonant* transition phase. During the initial phase of the response, i.e., from the initiation of wave loading on the soil box until a time $t \sim 70$ seconds as shown in Fig. 6a, the excess pore-pressure at the point of measurement is oscillating around a mean value which is only *slightly* larger than the hydrostatic pressure and is slowly increasing with time. On the other hand, the final phase of this response (starting, say, from $t \sim 90$ seconds in Fig. 6a) is characterized by pore-pressure oscillation around a mean value which is *significantly* larger than the hydrostatic water pressure, i.e., close to a state of complete fluidization. The intermediate transition between these two phases (i.e., between time $t \sim 70$ and $t \sim 90$ seconds) is relatively short in duration and is characterized by a rather rapid amplification in the amplitude of oscillation — from about 0.08 kN/m^2 to nearly 0.8 kN/m^2 (tenfold) in a span of just five seconds or so. After that, there is a rather complex transition towards the steady harmonic state of the final state. Accompanying this amplification in oscillation is an equally significant rise in the mean value of the pore-pressure — about 1.0 kN/m^2 in the figure — to connect with the final phase. This type of response is referred to as a *resonant-fluidization* response.

Figure 6b shows an example of the observed second kind of response. Here, there is no clear evidence of a similar resonant transition as in the first kind. This kind of response was observed to take place *only* in preresonated soils or in soils which experienced responses similar to that shown in Fig. 6a. In fact, the record shown in Fig. 6b is from the experimental run immediately following that of Fig. 6a. The two runs were separated by a consolidation period (no wave loading) of twenty-four hours, and again the two records were taken from the same pressure transducer at a depth 8 inches below the soil surface. The wave height and period during this second run were very similar to those of the earlier run. In spite of that, the recorded responses are quite different as shown in Fig. 6a as against Fig. 6b. In this kind of response, it takes only a few cycles of wave loading to achieve a near complete collapse in the soil's solid skeleton and hence, the resulting significant rise in the mean pore-pressure. The fact that during this run a significant rise in the mean pore-pressure was achieved relatively shortly after the initiation of wave loading may suggest that the soil skeleton at the beginning of this run was very fragile and did not quite recover from the weakening effects of the earlier resonance event(s) of the previous run. This is perhaps why the soil skeleton after only a few cycles gave way and yielded almost entirely to the wave loading in a near complete fluidization failure as shown in Fig. 6b. This, in fact, was not restricted to just the run immediately following the resonant run but was observed to occur in a number of runs conducted in series following the resonant one, each separated by a few hours of consolidation time. This kind of response is referred to as a *nonresonant fluidization* response.

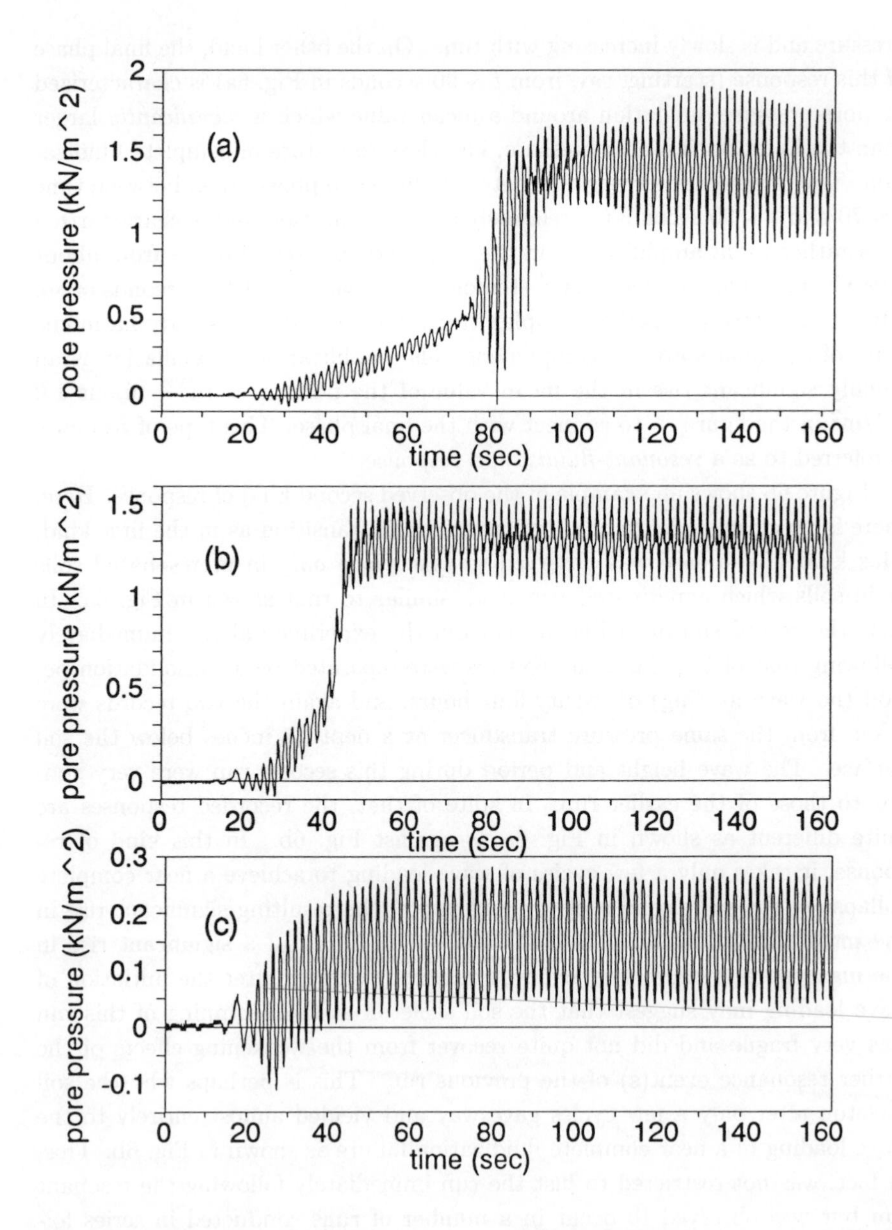

Fig. 6. Representative pore-pressure records taken at a depth $d = 8$ inches below the silt surface under monochromatic shallow water waves: (a) resonant fluidization response, (b) nonresonant fluidization response, and (c) unfluidized response. (Tzang, 1992)

An example of the third kind of response is shown in Fig. 6c. The behavior is similar to the second kind in that there is a relatively quick rise in the mean pore-pressure at the beginning of the run, asymptotically reaching a constant, larger-than-hydrostatic value. The main difference, however, is in the magnitude of this rise in the mean pore-pressure. The third kind of response is primarily characterized by a relatively small rise in the mean pore-pressure. For example, the mean pore-pressure rise in Fig. 6c is about 0.16 kN/m^2 which represents only 10% depth, compared to 1.33 kN/m^2 or an 86% rise (see Fig. 6b). We simply call this kind of response an *unfluidized* response.

6.2.1. *Sediment suspension observations*

The most conspicuous feature of sediment motion on the bed surface was the development of concentrated plumes of sediment rising from the soil bed upward into the water column. At any given point in time there would be a number of sporadically distributed plumes emanating from the soil surface. Continued visual observation of these sediment plumes will quickly become hindered by the increasing sediment suspension load in the water layer and the resulting very low visibility there. Such pluming of sediment was observed always during each cycle of wave loading on the silty soil. However, the intensity of sediment suspension was much more spectacular during a resonant or nonresonant fluidization response (e.g., Figs. 6a and b) than during an unfluidized response. Clukey *et al.* (1983) reported on similar observations of micro volcano-like structures at the bed surface which allowed soil particles to be ejected from within the bed into the overlying water column. There are similarities between these observed pluming and the intermittent sediment suspension events that were frequently observed *in the field* (see, e.g., Hanes, 1992) as discussed earlier in the granular-flow boundary-layer section. However, since in these field studies there was no concurrent measurement of pore-pressure in the bed, it is not possible to confirm if the underlying mechanics of the two observations are also similar.

Returning to the particular laboratory observations of fluidization induced pluming, it is interesting to note that no formation of bed ripples or dunes was observed during active generation of bed sediment plumes. On the other hand, control tests on a similar box filled with sand ($D_{50} = 0.29$ mm) resulted in the reverse observation: There were no sediment plumes ejecting from the bed but instead, there were small scale dunes or ripples covering the sand surface. In the silt tests, and associated with the occurrence of sediment plumes, there was a net transport of sediment moving in the same direction as that of wave propagation. Under the same waves, there was no significant net sediment transport during the sand tests.

Foda (1993) developed a theory to explain the observed internal resonances in Foda and Tzang experiments as a consequence of a strong *channeling* of the seepage flow within the silt bed. A strongly-channeled seepage flow is defined as a special case when the effective porosity is very low, so that the main seepage channels are sparsely distributed within the medium. A *new* slow-wave solution is developed for this special case and it is proposed that the resonances observed by Foda and Tzang (1993) are in the form of these strongly-channeled wave modes. Foda (1993) argued that this new wave mode is in fact an extension to Biot's (1956) wave of the second kind. The main difference is that Foda's solution produces a *propagating* wave solution capable of supporting the observed resonance, while the classical solution of Biot is for a *critically-damped* solution, effectively representing a diffusion process rather than a truly wave process. Because the wave speed in Foda's solution is much slower than the speed of the elastic shear wave, the response in the porous solid is governed by elastostatics. The inertia of the wave is essentially focused in the fluid flow. Furthermore, wave damping is caused by fluid viscous friction on the pores' elastic walls.

The developed solution also demonstrate that transient seepage flow can exhibit significant inertia, even for arbitrarily small permeability coefficient, in direct violation to Darcy's law. This is primarily because the seepage channels have flexible walls and that the compliance of these walls may significantly enhance fluid inertia. To illustrate, we may model the seepage flow as made of an ensemble of flows through *compliant* channels. For each one of these channels, transient pressure gradients working on the compliant walls will force moving contractions and expansions in the cross-sectional area, and this will force inertial flow from regions of contraction to regions of expansion. This compliance-induced inertial flow is seen from continuity to be *inversely* proportional to the cross-sectional area. The implication then seems to be that, with every thing else equal, the inertial flow will increase in importance as the channels' thickness decreases, or with *decreasing* permeability.

7. Behavior at a Structure-Seabed Interface

Marine pipes, caisson breakwaters, ocean submersibles, offshore terminals, oil-storage tanks, and the many different kinds of marine anchors and mooring systems are just a few examples of gravity marine structures resting under their own weight on the seabed. The seabed contact forces on such structures would, in general, be a combination of fluid and solid loadings. The fluid load comes from the integration of pore-pressure distribution over the seabed-structure interface. A similar integration of soil effective or "contact" stresses over the interface would give the solid loads. The distribution of these stresses and the associated intricate balance between the pore fluid and the soil skeleton will clearly influence the overall stability of the structure.

For example, the disappearance of the interface's effective stresses means that the structure would be effectively in a state of suspension, or that the soil at the interface would be in a state of liquefaction. This is usually a highly unstable situation even if it happens momentarily during a wave cycle, since at that time there is virtually no foundation resistance to shear loading. The slightest lateral perturbation carries with it the potential of causing the structure to *slide* on its foundation. If liquefaction is extended into the underlying soil, the structure may instead *sink* under its own weight into the fluidized soil. Aside from these types of global failure modes, localized liquefactions may lead with time to a fatigue-type foundation failure. For example, localized interface liquefaction will cause a limited loss of contact between the base of a structure and the solid soil skeleton but only over part of the seabed interface. This amounts to the development of a fluid gap over that part of the interface and hence, the possibility of mobilizing and pumping of surface sediment in the gap region. Toe erosion around the edges of large gravity structures is a prime example of such a process (Hoeg, 1986; Gerwick, 1986). The repeated occurrences of these localized detachment events would lead to a progressive undermining of the supporting soil, stress focusing, and an eventual foundation failure (Niedoroda *et al.*, 1981; Burland *et al.*, 1978). In other cases, like that of a marine pipe laid on the seafloor, the slow pumping of soil from under the pipe would lead to the slow erosion-induced settlement, or self-burial, of the pipe into the seabed (Grace, 1978; Chao and Hennessy, 1972; Mao, 1986; Husbergen, 1984; Work, 1987). The reverse phenomenon of a pipe *breakout* from the seafloor has also been observed, especially after the pipe has been subjected to severe storm or hurricane wave loading (Blumberg, 1964; Christian *et al.*, 1974; Nataraja and Gill, 1983). In this case, the hydrodynamic lift force on the pipe would be strong enough to cause a complete detachment and the eventual flotation of the pipe away from the seabed.

There have been a few analytical studies on wave-induced stresses in the seabed in the presence of marine structures. Liu and O'Donnell (1979), Monkmeyer *et al.* (1983) and McDougal *et al.* (1988) assumed a rigid porous bed and studied seepage forces on buried pipes. Biot's poroelasticity equations were the basis for the models by Mynett and Mei (1982) for stresses under a caisson breakwater in waves, and by Mei and Foda (1981b), Foda (1985) and Chang and Liu (1985) for wave-induced stresses in the seabed around exposed, half-buried, or fully-buried pipes. Most of the above studies were concerned with the *scattering* problem where the structure is assumed fixed in space and the solution is developed for the response due to the combination of an incident plus a scattered wave. Mei and Foda (1982) gave a simple solution, using complex functions, for a complementing radiation problem where a flat footing resting on the seabed is oscillating vertically, or heaving, in calm sea. If the motion is changed from heaving to a unidirectional pull-up, the problem then becomes that of a *body breakout* from the seabed. Early interest in the breakout problem was primarily for salvage engineering applications where the objective is to determine the magnitude of the pull-up force and the duration of the pull, which are necessary to lift sunken vessels, etc., from the seabed (Liu, 1969; DeHart and Ursell, 1967). Other more recent applications include wave-induced breakout of marine pipes from the seafloor, repositioning of construction caissons, response of arctic offshore platforms to ice uplift forces, and the potentially significant problem of reclaiming gravity structures from depleted or nonproducing offshore oil wells around the world. Foda (1982) developed poroelastic solutions for an object detaching itself from the seabed at a prescribed upward velocity or a prescribed pull force. Central to the developed solution is the behavior in the expanding fluid gap between the body and the seabed. A general situation is sketched in Fig. 7a. Water is shown to flow into the expanding gap from both the pores of the seabed as well as from the ambient water above the seabed through the gap periphery P. The pumping of pore water upward into the gap will necessarily be accompanied by pressure gradients into the seabed, resulting in the development of negative pore-pressure (below ambient) or *mud-suction* stresses at the bed-body interface. This mud suction will provide the main resistance against body breakout from the seabed. This will persist for some time (in some cases, a few hours or days in salvaging large sunken vessels) until a *breakout time* is reached. At this time, mud suction is suddenly dissipated and hence, a rather sharp release or detachment of the body from the seafloor would take place. The dissipation of mud suction happens when the water flux into the expanding gap becomes dominated by the supply from ambient water through the gap periphery P rather than from pore-water pumping. This detachment process was shown by Foda (1982) to be described in terms of an Abel's integro-differential equation relating applied lift force and gap thickness. Solving this equation produces a power-law relationship between the lift force F and the associated breakout time t_b

$$t_b = \gamma F^{-3/2} \tag{72}$$

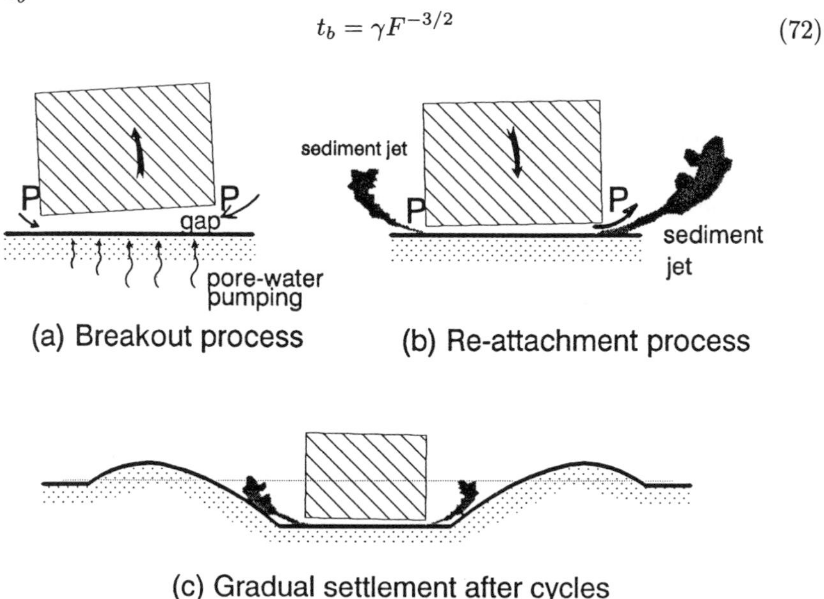

Fig. 7. The process of body breakout (a) and re-attachment (b) to the seafloor. Sediment depletion from underneath the body after many cycles of breakout and attachment leads to the gradual body settlement into the seabed as shown in (c).

where γ is a material constant containing properties of the poroelastic bed and of the body base (Foda, 1985). Mei *et al.* (1985) solved the breakout problem for a rigid porous bed and for a uniform as well as a pivoting body lift-off. Their breakout force-time power law differed from (72) in the expression for γ, and in the power exponent being -1, instead of $-3/2$. Experimental examination of the phenomenon and the theoretical predictions were carried out by Foda *et al.* (1989). They reproduced the phenomenon of pipe breakout from a soil bed in waves. The experimental breakout force-time relation was found to fit between the poroelastic and the rigid porous predictions but more closely to the rigid bed power law. However, a much larger structure in the prototype scale and/or softer bed material may make the response a little closer to the poroelastic model.

Foda (1990) extended the analysis to examine what happens next as the detached body begins to fall down under its own weight and re-establishes contact with the seabed (Fig. 7b). It is observed that just before contact, the interior gap pressure will reach extremely large magnitudes. Connecting with ambient pressure at the edge of the gap implies that strong pressure gradients will develop along the fluid gap. As a result, a very strong water jet will be generated and ejected from underneath the body just before it re-establishes seabed contact. This has been confirmed experimentally by a series of model runs at the Naval Civil Engineering Laboratory (NCEL), Port Hueneme, CA (Foda, 1990). The general response is sketched in Fig. 7b. The strength of the ejecting jet will mobilize and carry a fairly large load of sediment from underneath the body and deposit it on the side, as shown in Fig. 7c. Repeating the cycles of breakout and attachment will clearly result in the continuous pumping of sediment from under the body in what amounts to a slow and gradual process of self burial.

The support from the National Science Foundation Fluid, Particulate and Hydraulic Systems Program, through grants CTS-9215889 and MSM8718959, and EPA Grant R817170 are gratefully acknowledged.

References

Bagnold, R. A. (1954). Experiments on a gravity-free dispersion of large solid spheres in a Newtonian fluid under shear. *Proc. Roy. Soc. Lond.* **A225**:49-63.

Bagnold, R. A. (1955). *Proc. Instn. Civil Engrs.* pt. III.

Bagnold, R. A. (1956). The flow of cohesionless grains in fluids. *Philos. Trans. Roy. Soc.* **249**:235-297.

Bakker, W. T. (1974). Sand constration in oscillatory flow. *Proc. 14th Coastal Eng.* Copenhagen. 1129–1148.

Bear, J. (1972). *Dynamics of Fluids in Porous Media.* Elsevier.

Benjamin, T. B. and J. E. Feir (1967). The disintegration of wave trains on deep water. *J. Fluid Mech.* **27**:417–430.

Biot, M. A. (1941). General theory of three-dimensional consolidation. *J. Appl. Phys.* **12**:155–165.

Biot, M. A. (1956). Theory of propagation of elastic waves in a fluid-saturated porous solid, part I: low frequency range, and part II: higher frequency range. *J. Acoust. Sci. Am.* **28**:168–191.

Blumberg, R. (1964). Hurricane winds, waves and currents test, marine pipeline design. *Pipeline Industry*. June–Nov.

Brenninkmeyer, B. M. (1976). In situ measurements of rapidly fluctuating, high sediment concentrations. *Marine Geol.* **20**:117–128.

Burland, J. B., A. C. M. Penman, and K. A. Gallagher (1976). Behaviour of a gravity foundation under working and failure conditions. *European Offshore Petroleum Conference and Exhibition*, Vol. 1, London. 111–120.

Chang, A. H.-D. and P. L.-F. Liu (1986). Seepage force on pipelines buried in a poroelastic seabed under wave loadings. *Appl. Ocean Res.* **8**:22–32.

Chao, L. and P. V. Hennessy (1972). Local scour under ocean outfall pipelines. *J. Water Pollution Control Federation* **44**:1443–1447.

Chou, H.-T. (1989). Rheological response of cohesive sediments to water waves. Ph.D. thesis. University of California at Berkeley. 149.

Chou, H.-T., M. A. Foda, and J. R. Hunt (1993). Rheological response of cohesive sediment to oscillatory forcing, in *Nearshore and Estuarine Cohesive Sediment Transport,* Coastal Estuarine Sci. Ser., Vol. 42, ed. A. J. Mehta, AGU, Washington, D.C. 126–148.

Christian, J. T., P. K. Taylor, J. K. Yen, and D. R. Erali (1974). Large diameter underwater pipeline for nuclear power plant designed against soil liquefaction. *Proc. OTC.* 597–602.

Clukey, E. C., F. H. Kulhawy, and P. L.-F. Liu (1983). Laboratory and field investigation of wave-sediment interaction. Joseph H. DeFrees Hydraulics Laboratory, School of Civil and Environmental Engineering, Cornell University, Ithaca, New York, Report 83–1.

Collins, J. I. (1963). Inception of turbulence at the bed under periodic gravity waves. *J. Geophys. Res.* **69**:6007–6014.

Conley, D. C. and D. L. Inman (1992). Field observation of the fluid-granular boundary layer under near-breaking waves. *J. Geophys. Res.* **97**:9631–9643.

Dalrymple, R. A. and P. L-F. Liu (1978). Waves over soft muds: a two layer model. *J. Phys. Oceanog.* **8**:1121–1131.

De Hart, R. C. and C. R. Ursell (1967). Ocean bottom breakout forces, including field test data and the development of an analytical method. Southwest Res. Inst. report on Contract N-ONR-336300.

Dick, J. A. and J. F. A. Sleath (1991). Velocities and concentrations in oscillatory flow over beds of sediment. *J. Fluid Mech.* **233**:165–196.

Foda, M. A. (1982). On the extrication of large objects from the ocean bottom. *J. Fluid Mech.* **117**:211–231.

Foda, M. A. (1985). Pipeline breakout from seafloor under wave action. *Appl. Ocean Res.* **2**:79–84.

Foda, M. A. (1987). Internal dissipative waves in poroelastic waves. *Proc. Roy. Soc. Lond.* **A413**:383–405.

Foda, M. A. (1989). Side band damping of water waves over a soft bed. *J. Fluid Mech.* **201**:198–211.

Foda, M. A. (1990). Soil pumping underneath a rocking marine structure. Technical report submitted to Sea Floor Division, Naval Civil Engineering Laboratory, Port Hueneme, California. 70.

Foda, M. A. (1994). Inertial (non-Darcian) channeled seepage flow. *J. Geophys. Res. (Oceans)* (to appear).

Foda, M. A. (1994). Landslides riding on basal pressure waves. *J. Continuum Mech. and Thermo.* **6**:61–79.

Foda, M. A. and S.-Y. Tzang (1994). Resonant fluidization of silty soil by water waves. *J. Geophys. Res (Oceans)* (to appear).

Foda, M. A., J. R. Hunt, and H. T. Chen (1993). A nonlinear model for the fluidization of marine mud by waves. *J. Geophys. Res.* **98**:7039–7047.

Foda, M. A., J. Y.-H. Chang, and A. W.-K. Law (1990). Wave-induced breakout of half-buried marine pipes. *J. Wtrwy., Port., Coast., and Oc. Eng. ASCE.* **116**:267–286.

Gade, H. G. (1958). Effects of a non-rigid, impermeable bottom on plane surface waves in shallow water. *J. Mar. Res.* **16**:61–82.

Gerwick, B. C., Jr. (1986). *Construction of Offshore Structures.* Wiley, New York.

Grace, R. A. (1978). *Marine Outfall Systems: Planning, Design and Construction.* Prentice-Hall, New Jersey.

Graff, K. F. (1975). *Wave Motion in Elastic Solids.* Ohio State University Press, Cleveland, Ohio.

Hanes, D. M. (1992). The structure of events of intermittent suspension of sand due to shoaling waves, in *The Sea,* Vol. 9, part B, eds. B. Le Mehante and D. M. Hanes. Wiley Interscience. 941–952.

Hanes, D. M., C. E. Vincent, D. A. Huntley, and T. E. Clarke (1988). Acoustic measurements of suspended sand concentration in the C2S2 experiment at Stanhope Lane, Prince Edward Island. *Marine Geol.* **81**:185–196.

Hasselmann, K. *et al.* (1973). Measurements of wind-wave growth and swell decay during the joint North Sea Wave Project. *Deutsch. Hydrograph. Inst.* **A80**:1–95.

Hoeg, K. (1986). Geotechnical issues in offshore engineering, in *Marine Geotechnology and Nearshore/Offshore Structures,* ASTM STP 923, eds. R. C. Chaney and H. Y. Fang. American Society for Testing and Materials, Philadelphia. 7–50.

Homma, M. and K. Horikawa (1962). Suspended sediment due to wave action. *Proc. 8th Conf. Coastal Eng. ASCE.* 168–193.

Hsiao, S. V. and O. H. Shemdin (1980). Interaction of ocean waves with a soft bottom. *J. Phys. Oceanog.* **10**:605–610.

Husbergen, C. J. (1984). Stimulated self-burial of submarine pipelines, OTC-2967, Houston, TX, May 1984.

Inman, D. L. (1957). Wave generated ripples in nearshore sands. Tech. Memo. 11, U.S. Army Corps of Eng., Beach Erosion Board, 65.

Ippen, A. T. (1966). *Estuary and Coastline Hydrodynamics,* 6th edition. Cambridge University Press.

Jacob, P. G. and S. Z. Qasim (1974). Mud of a mudbank in Kerala, southwest coast of India. *Indian J. Mar. Sc.* **3**:115–119.

Jiang, L., W. Kioka, and A. Ishida (1990). Viscous damping of Cnoidal waves over fluid-mud seabed. *J. Wtrwy., Port., Coast., and Oc. Eng. ASCE.* **116**:470–491.

Jonsson, I. G. (1980). A new approach to oscillatory rough turbulent boundary layers. *Ocean Eng.* **7**:109–152.

Kajiura, K. (1968). A model of the bottom boundary layer in water waves. *Bull. Earthquake Res. Inst.* **46**:5–123.

Liu, C. L. (1969). Ocean sediment holding strength against body breakout. U.S. Naval Civ. Eng. Lab., Port Hueneme, California, Technical Report R635.

Liu, P. L.-F. (1974). Damping of water waves over porous bed. *Proc. ASCE. J. Hydraulic Div.* **99**:2263–2271.

Liu, P. L.-F. (1977). Mass transport in water waves propagated over a permeable bed, *Coastal Eng.* **1**:79–96.

Liu, P. L.-F. and P. O'Donnell (1979). Wave-induced forces on buried pipelines in permeable seabed. *Proc. Civ. Eng. in Oceans* **4**:111–121.

Longuet-Higgins, M. S. (1953). Mass transport in water waves. *Proc. Roy. Soc. Lond.* Ser. A **245**:535–581.

Longuet-Higgins, M. S. (1981). Oscillatory flow over steep sand ripples. *J. Fluid Mech.* **107**:1–35.

Longuet-Higgins, M. S. (1950). A theory on the origin of microseism. *Phil. Trans. Roy. Soc.* **A243**:1–35.

Lyne, W. H. (1971). Unsteady viscous flow over a wavy wall. *J. Fluid Mech.* **50**:33–48.

Maa, P. Y. and A. J. Mehta (1988). Soft mud properties: Voight model. *J. Wtrway., Port, Coast., and Oc. Eng. ASCE.* **114**:765–770.

MacPherson, H. (1980). The attenuation of water waves over a non-rigid bed. *J. Fluid Mech.* **97**:721–742.

MacPherson, H. and P. G. Kurup (1981). Wave damping at the Kerala mudbanks, southwest India. *Indian Jour. Mar. Sci.* **10**:154–160.

Madsen, O. S. (1974). Stability of a sand bed under breaking waves. *Proc. Conf. Coastal Eng.* **14**:794–581.

Madsen, O. S. (1978). Wave-induced pore-pressure and effective stresses in a porous bed. *Geotechnique* **28**:377–393.

Mao, Y. (1986). The interaction between a pipeline and an erodible bed. Institute of Hydrodynamics and Hydraulic Engineering, Technical University of Denmark, Research Series Paper No. 39.

Massel, S. R. (1976). Gravity waves propagated over permeable bottom. *Proc. ASCE. J. Wtrwy. Harbors Coastal Eng. Div.* **102**:111–121.

McDougal, W. G., S. H. Davidson, P. L. Monkmeyer, and C. K. Sollitt (1988). Wave-induced forces on buried pipelines. *J. Wtrwy., Port, Coast. and Oc. Eng. ASCE.* **114**:220–236.

Mei, C. C. and K. F. Liu (1987). A Bingham-plastic model for a muddy seabed under long waves. *J. Geophys. Res.* **92**:14581–14594.

Mei, C. C. and M. A. Foda (1981a). Wave-induced response in a fluid-filled poro-elastic solid with a free surface: A boundary-layer theory. *Geophys. J. Astron. Soc.* **66**:597–631.

Mei, C. C. and M. A. Foda (1981b). Wave-induced stresses around a pipe laid on a poroelastic seabed. *Geotechnique* **31**:311–318.

Mei, C. C., P. L-F. Liu, and T. G. Carter (1972). Mass transport in water waves. M.I.T. Dept. Civil Eng., Ralph M. Parson Laboratory Report. 146.

Mei, C. C., R. W. Yeung, and K. P. Liu (1985). Lifting of a large body from a porous seabed. *J. Fluid Mech.* **152**:203–215.

Monkmeyer, P. A., P. G. Manttovani, and H. Vincent (1983). Wave-induced seepage effects on buried pipelines. *Proc. Coastal Structures, Arch. ASCE.* 519–531.

Mynett, A. E. and C. C. Mei (1982). Wave-induced stresses in a poroelastic seabed beneath a rectangular caisson. *Geotechnique* **33**:235–247.

Nataraja, M. S. and H. S. Gill (1983). Ocean-wave induced liquefaction, *J. Geotech. Eng.* **109**:573–590.

Niedoroda, A. W., C. Dalton, and R. G. Bea (1981). The descriptive physics of scour in the ocean environment. OTC 13, Vol. 4, Houston, Texas. 297–304.

Packwood, A. R. and D. H. Peregrine (1980). The propagation of solitary waves and bores over a porous bed. *Coast. Eng.* **3**:221–242.

Press, F. and R. Siever (1974). *The Earth, Part II: Surface Processes.* W. H. Freeman and Co., San Francisco. 235.

Putnam, J. A. (1949). Loss of wave energy due to percolation in a permeable sea bottom. *Trans. A.G.U.* **30**(5):349–356.

Putnam, J. A. and J. W. Johnson (1949). The dissipation of wave energy by bottom friction. *Trans. Am. Geophys. Union* **30**:67–74.

Reid, R. O. and K. Kajiura (1957). On the damping of gravity waves over a permeable sea bed. *Trans. A.G.U.* **38**:662–666.

Sakai, T., K. Hatanaka, and H. Mase (1992). Wave-induced effective stress in seabed and its momentary liquefaction. *J. Wtrwy., Port, Coast. and Oc. Eng. ASCE.* **118**:202–206.

Saxov, S. and J. K. Nieumenhuis, eds. (1980). *Marine Slides and Other Mass Movements.* Plenum, New York.

Scheidegger, A. E. (1974). *The Physics of Flow Through Porous Media.* 3rd edition. University of Toronto Press.

Shibayama, T., A. Higuchi, and K. Horikawa (1986). Sediment suspension due to breaking waves. *Proc. 20th Coastal Eng. Conf. ASCE.* 1509–1522.

Sleath, J. F. A. (1970). Wave-induced pressure in beds of sand. *Proc. ASCE J. Hydraulic Div.* **96**:367–378.

Sleath, J. F. A. (1975). Transition in oscillatory flow over rippled beds. *Proc. I.C.E.* Pt.2 **59**:309–322.

Sleath, J. F. A. (1978). Discussion of "Mass transport in water waves propagated over a permeable bed" (by P. L.-F. Liu 1977). *Coast. Eng.* **2**:169–171.

Sleath, J. F. A. (1982). The suspension of sand by waves. *J. Hyd. Res.* **20**:439–452.

Sleath, J. F. A. (1984). *Sea Bed Mechanics.* Wiley, New York.

Stokes, G. G. (1847). On the theory of oscillatory waves. *Trans. Cambridge Phil. Soc.* Vol. 8, and *Supplement, Sci. Papers*, Vol. 1.

Stokes, G. G. (1851). On the effect of the internal friction of fluids on the motion of pendulums. *Trans. Camb. Philos. Soc.* **9**:20–21.

Terzaghi, K. (1945). *Theoretical Soil Mechanics.* Wiley, New York.

Thornton, E. B. and W. D. Morris (1978). *Suspended Sediments Within the Surf Zone. Coastal Sediments '77.* ASCE. 655–668.

Ting, F. C.-K. (1989). Interaction of water waves with a density-stratified fluid in a rectangular trench. W. M. Keck Laboratory of Hydraulics and Water Resources, California Institute of Technology, Report KH-R-50.

Ting, F. C.-K. and F. Raichlen (1986). Wave interaction with a rectangular trench. *J. Wtrwy., Port, Coast. and Oc. Eng. ASCE.* **112**:454–460.

Tsai, Y. T., W. G. McDougal, and C. K. Sollitt (1990). Response of finite depth seabed to waves and caisson motion. *J. Wtrwy., Port, Coast. and Oc. Eng. ASCE.* **116**:1–20.

Tubman, M. W. and J. N. Suhayda (1976). Wave action and bottom movements in fine sediments. *Proc. 15th Coastal Eng. Conf.*, Honolulu, Hawaii. **2**:1168–1183.

Tzang, S.-Y. (1992). Water wave-induced soil fluidization in a cohesionless fine grained seabed. Ph.D. thesis. University of California at Berkeley.

Wells, J. T. (1983). Fluid mud dynamics in low, moderate and high tide range environments. *Canadian J. Fish. Aquatic Sci.*, Supplement 1. 130–142.

Wells, J. T., Y. A. Park and J. H. Choi (1985). Storm-induced fine sediment transport, west coast of South Korea. *Geo-Marine Letters.* 4:177–180.

Work, P. A. (1987). Self-burial of marine pipelines. UCB/HEL-87/05, Hydraulic Engineering Laboratory, Department of Civil Engineering, University of California at Berkeley, Berkeley, CA.

Yamamoto, T. H., L. Koning, H. Sellmeijer, and E. Van Hijum (1978). On the response of a poro-elastic bed to water waves. *J. Fluid Mech.* **87**:193–206.

Zhang, D. and M. A. Foda (1993). A model for low-drag landslides. *Proc. Hyd. Eng. 1993, ASCE*, San Francisco, California. **1**:322–327.

MODEL EQUATIONS FOR WAVE PROPAGATIONS FROM DEEP TO SHALLOW WATER

PHILIP L.-F. LIU

This paper reviews two different approaches for deriving shallow water equations. The Hamiltonian approach is first used to obtain the Boussinesq equations in terms of the horizontal velocity on the free surface. A direct perturbation method is introduced to derive a general nonlinear shallow water equation. Various forms of Boussinesq equations are discussed. Some of these Boussinesq equations are shown to be unstable subject to short wave disturbances. The dispersion characteristics, both linear and nonlinear, of the shallow water equations are compared with those of the Stokes' wave theory. As far as the linear dispersion is concerned, an optimal form of the Boussinesq equations is identified, which is applicable even in deep water. However, if the nonlinearity is important in deep water, none of the equations discussed in this paper can provide an adequate description of nonlinear dispersion in deep water.

1. Introduction

To design a coastal structure in the nearshore region, engineers must have a means to estimate wave climate. Numerical models are often used to calculate wave propagation from an offshore location, where wave data are available, to the nearshore area of concern. Waves, approaching the surf zone from offshore, experience changes caused by combined effects of bathymetric variations, interference of man-made structures, and nonlinear interactions among wave trains. Inside the surf zone, where wave breaking is a dominating feature, waves undergo a much more rapid transformation. Early efforts to model this wave evolution process were based primarily on the geometrical ray theory, which ignores both diffraction and nonlinearity.

Significant research accomplishments were made in the 1970s and 1980s to overcome the shortcoming imbedded in the ray theory. To include the wave diffraction, mild-slope equations are derived from the linear wave theory by assuming that evanescent modes can be ignored and the bathymetric changes are small within a typical wavelength. Furthermore, the vertical structure of the velocity field is assumed to be the same as that for a progressive wave over a constant depth. Therefore, mild-slope equations are two-dimensional (in the horizontal space) partial differential equations of the elliptic type, which require boundary conditions along the entire boundaries of the computational domain. Numerical solutions have demonstrated that mild-slope equations give adequate description of combined refraction and diffraction for small amplitude waves. The general mild-slope equation for nonlinear waves is still not available.

In applying mild-slope equations to a large coastal region, one encounters the difficulty in defining the location of the breaker line *a prioi.* One of the most important developments in wave modeling during the last decade was the application of the parabolic approximation to mild-slope equations. The parabolic approximation can be viewed as a modification of the ray theory. While waves propagate along "rays", wave energy is allowed to "diffuse" across the "rays". Therefore, the effects of diffraction are included approximately in the parabolic approximation. Although the parabolic approximation models have been used primarily for forward propagation, weak backward propagation modes can be included by an iterative procedure (Liu and Tsay, 1983; Chen and Liu, 1993). Nonlinearity can also be included in the forward propagation mode. More detailed discussions on mild-slope equations and the associated parabolic approximations can be found in Liu (1990) and Mei and Liu (1993).

As waves approach the surf zone, wave amplitudes become large and the Stokes' wave theory is no longer valid. A more relevant approach is based on Boussinesq equations for weak nonlinear and weak dispersive waves. Peregrine (1967, 1972) provided several versions of the Boussinesq equations, written in terms of either the depth-averaged velocity or the velocity along the bottom or the velocity on the free surface. Because the dispersive terms in the Boussinesq equations are of higher order, they can be further manipulated by replacing the time derivative or the spatial derivatives by the lower order relations (e.g., see Mei, 1989). Although all these different forms of Boussinesq equations have the same order of magnitude of accuracy, their dispersion relations and the associated phase velocity are different.

The major restriction of the Boussinesq equations is their depth limitation. The best forms of the Boussinesq equations, using the depth-averaged velocity, break down when the depth is larger than one-fifth of the equivalent deep water wavelength, corresponding to a five percent phase velocity error (McCowan, 1987). For many engineering applications, a less restrictive depth limitation is desirable. Moreover, when the Boussinesq equations are solved numerically, high frequency numerical disturbances related to the grid size could cause instability. The search for a new set of two-dimensional governing equations which can describe the wave propagation from a deeper depth to a shallow depth is currently an active area of research (Witting, 1984; McCowan and Blackman, 1989; Murray, 1989; Madsen, Murray and Sorensen, 1991; Nwogu 1993).

In this paper, we first review the derivation of Boussinesq equations using the Hamiltonian approach and the direct perturbation procedure. In the direct perturbation approach, shallow water equations for highly nonlinear waves are also derived. The Boussinesq equations become a special case. The dispersive characteristics of nonlinear shallow water equations and Boussinesq equations are then compared with those of Stokes' waves. The possibility of extending the range of applicability of nonlinear shallow water equations into deeper water is discussed. Different methods for improving the dispersive characteristics are also discussed.

2. Governing Equation and Boundary Conditions

In this section, the governing equation and boundary conditions for water waves propagating in a varying depth are summarized. Denoting $\mathbf{x}' = (x', y')$ as the horizontal coordinates and z' the vertical coordinate, we define the flow domain as a layer of water bounded by a free surface $z' = \eta'(\mathbf{x}', t)$ and a solid bottom $z' = -h'(\mathbf{x}')$. Using the characteristic wavelength, $(k')^{-1}$ as the horizontal length scale, the characteristic depth, h'_o as the vertical length scale, and $(k'\sqrt{gh'_o})^{-1}$ as the time scale, we introduce the following dimensionless variables:

$$\begin{aligned} \mathbf{x} = k'\mathbf{x}', \quad & z = z'/h'_o \,, \\ h = h'/h'_o, \quad & t = k'\sqrt{gh'_o}t' \,. \end{aligned} \tag{2.1}$$

Assuming that the flow field is irrotational, we represent the velocity field by the gradient of a velocity potential, Φ'. Denoting a'_o as the characteristic wave amplitude, we normalize the free surface displacement η' and the associated potential function and obtain the following dimensionless variables:

$$\eta = \eta'/a'_o \,, \quad \Phi = k'\frac{\sqrt{gh'_o}}{ga'_o}\Phi'. \tag{2.2}$$

The dimensionless continuity equation becomes

$$\mu^2\left(\frac{\partial^2\Phi}{\partial x^2}+\frac{\partial^2\Phi}{\partial y^2}\right)+\frac{\partial^2\Phi}{\partial z^2}=0\ ,\quad -h<z<\varepsilon\eta \tag{2.3}$$

where

$$\varepsilon = a_o'/h_o'\ ,\quad \mu^2=(k'h_o')^2 \tag{2.4}$$

are parameters representing nonlinearity and frequency dispersion, respectively. The no-flux boundary condition along the bottom requires

$$\mu^2\nabla h\cdot\nabla\Phi+\frac{\partial\Phi}{\partial z}=0\ ,\quad z=-h \tag{2.5}$$

in which $\nabla=(\partial/\partial x,\partial/\partial y)$ denotes the gradient vector on a horizontal plane. On the free surface, both the kinematic and the dynamic boundary conditions must be satisfied. The dynamic condition specifies the continuity of pressure across the free surface. Setting the atmospheric pressure at zero, the Bernoulli equation applied on the free surface becomes

$$\frac{\partial\Phi}{\partial t}+\frac{1}{2}\varepsilon\left[\left(\frac{\partial\Phi}{\partial x}\right)^2+\left(\frac{\partial\Phi}{\partial y}\right)^2+\mu^{-2}\left(\frac{\partial\Phi}{\partial z}\right)^2\right]$$

$$+\eta=0, z=\varepsilon\eta\ . \tag{2.6}$$

The kinematic boundary condition states that the free surface is a material surface. Thus, following the free surface movement, the rate of change of the free surface, $F=z-\eta=0$, must vanish. The dimensionless kinematic boundary condition is expressed as

$$\frac{\partial\eta}{\partial t}+\varepsilon\nabla\eta\cdot\nabla\Phi=\mu^{-2}\frac{\partial\Phi}{\partial z}\ ,\quad z=\varepsilon\eta\ . \tag{2.7}$$

3. Approximate Governing Equations on the Horizontal Plane

The main objective of deriving approximate equations is to reduce the three-dimensional governing equations and boundary conditions to two-dimensional forms so that lesser computational efforts are required in modeling wave propagation in a large domain. Several different methods can be used to achieve this goal. In this section, we will discuss two of them: a Hamiltonian approach and a direct perturbation approach.

3.1. *Hamiltonian approach*

The total energy of a flow field is the sum of kinetic and potential energy. Denoting Ω as the projection of flow domain on the horizontal plane, we can express the total energy, which is also called Hamiltonian, in the following dimensionless form:

$$\mathcal{H} = \frac{1}{2}\int\int_{\Omega}\left\{\int_{-h}^{\varepsilon\eta}\left[\left(\frac{\partial\Phi}{\partial x}\right)^2 + \left(\frac{\partial\Phi}{\partial y}\right)^2 + \mu^{-2}\left(\frac{\partial\Phi}{\partial z}\right)^2\right]dz + \eta^2\right\}dxdy \quad (3.1)$$

in which the total energy has been normalized by a factor $\rho g a_o'^2/k'^2$. Because the total energy (Hamiltonian) must be finite, we assume that η, Φ and their derivatives vanish along the lateral boundaries of the horizontal domain.

The canonical theorem states that the free surface boundary conditions, (2.6) and (2.7), are equivalent to the following canonical equations (Broer, 1974; Zakharov, 1968; Miles, 1977):

$$\frac{\partial\eta}{\partial t} = \frac{\delta\mathcal{H}}{\delta\phi} \quad , \quad \frac{\partial\phi}{\partial t} = -\frac{\delta\mathcal{H}}{\delta\eta} \quad (3.2)$$

in which δ denotes a variational derivative and $\phi(\mathbf{x}, t)$ represents the potential evaluated on the free surface

$$\phi(\mathbf{x}, t) = \Phi(\mathbf{x}, \varepsilon\eta(\mathbf{x}, t), t) \ . \quad (3.3)$$

To use the canonical equations, we need to know the relation between Φ and ϕ. In other words, the vertical distribution of the potential function must be derived first. This task can only be achieved approximately. In the following section, an approximated Hamiltonian will be obtained by adopting the Boussinesq approximation, i.e., $0(\varepsilon) \approx 0(\mu^2) << 1$.

3.1.1. *An approximate Hamiltonian*

The Hamiltonian can be rewritten as the sum of kinetic energy, E_k, and the potential energy E_p, i.e.,

$$\mathcal{H} = E_k + E_p \quad (3.4)$$

where

$$E_p = \frac{1}{2}\int\int_\Omega \eta^2 dxdy \tag{3.5}$$

$$E_k = E_{ko} + E_{k\eta} \tag{3.6}$$

with

$$E_{ko} = \frac{1}{2}\int\int_\Omega\int_{-h}^{0}\left[\left(\frac{\partial\Phi}{\partial x}\right)^2 + \left(\frac{\partial\Phi}{\partial y}\right)^2 + \mu^{-2}\left(\frac{\partial\Phi}{\partial z}\right)^2\right] dzdxdy \tag{3.7}$$

$$E_{k\eta} = \frac{1}{2}\int\int_\Omega\int_{0}^{\varepsilon\eta}\left[\left(\frac{\partial\Phi}{\partial x}\right)^2 + \left(\frac{\partial\Phi}{\partial y}\right)^2 + \mu^{-2}\left(\frac{\partial\Phi}{\partial z}\right)^2\right] dzdxdy\,. \tag{3.8}$$

Applying the Green's theorem to the volume integral on the right-hand side of (3.7) and using the continuity equation (2.3) as well as the no-flux boundary condition on the bottom, (2.5), we obtain

$$E_{ko} = \frac{1}{2}\int\int_\Omega\left[\mu^{-2}\left(\Phi\frac{\partial\Phi}{\partial z}\right)\right]_{z=0} dxdy \tag{3.9}$$

where the integrand is evaluated on the still water level, $z = 0$. We remark here that up to this point, no approximation has been made.

The kinetic energy above the still water level, $z = 0$, given by (3.8) can be approximated by using the Taylor's series expansion

$$E_{k\eta} = \frac{1}{2}\int\int_\Omega \varepsilon\eta\left[\left(\frac{\partial\Phi}{\partial x}\right)^2 + \left(\frac{\partial\Phi}{\partial y}\right)^2 + \mu^{-2}\left(\frac{\partial\Phi}{\partial z}\right)^2\right]_{z=0} dxdy + 0(\varepsilon^2). \tag{3.10}$$

The approximate Hamiltonian can be written as

$$\mathcal{H} = \frac{1}{2}\int\int_\Omega\left\{\mu^{-2}\left(\Phi\frac{\partial\Phi}{\partial z}\right)_{z=0} + \varepsilon\eta\left[\left(\frac{\partial\Phi}{\partial x}\right)^2 + \left(\frac{\partial\Phi}{\partial y}\right)^2 + \mu^{-2}\left(\frac{\partial\Phi}{\partial z}\right)^2\right]_{z=0} + \eta^2\right\} dxdy + 0(\varepsilon^2)\,. \tag{3.11}$$

To continue the derivation of Boussinesq-type equations, we must find the vertical structure of the potential function, $\Phi(\mathbf{x}, z, t)$, such that the approximate Hamiltonian can be evaluated at the still water surface, $z = 0$.

3.1.2. *Vertical structure of the potential function*

Many approaches can be taken to find the vertical structure of the potential function. Here, we follow the procedure originally developed by Lin and Clark (1959). Expanding the potential function in a power series in terms of $(z + h)$

$$\Phi(\mathbf{x}, z, t) = \sum_{n=0}^{\infty} (z + h)^n \phi^{(n)}(\mathbf{x}, t) \tag{3.12}$$

we can obtain recursive relations among $\phi^{(n)}(\mathbf{x}, t)$ by substituting (3.12) for the continuity equation (2.3), and the bottom boundary condition (2.5)

$$\phi^{(1)} = -\frac{\mu^2 \nabla h \cdot \nabla \phi^{(0)}}{1 + \mu^2 |\nabla h|^2} \tag{3.13}$$

$$\phi^{(n+2)} = -\frac{\mu^2 [\nabla^2 \phi^{(n)} + 2(n+1)\nabla h \cdot \nabla \phi^{(n+1)} + (n+1)\nabla^2 h \phi^{(n+1)}]}{(n+1)(n+2)(1 + \mu^2 |\nabla h|^2)} \tag{3.14}$$

where $n = 0, 1, 2, \ldots$. From the recursive relations, $\phi^{(n)}$ can be expressed in terms of $\phi^{(0)}$, which is the potential along the bottom. For instance, up to $0(\mu^4)$ the potential function can be written as

$$\begin{aligned}\Phi(\mathbf{x}, z, t) = \phi^{(0)} - \mu^2 \bigg[& h\nabla h \cdot \nabla \phi^{(0)} + \frac{h^2}{2} \nabla^2 \phi^{(0)} + z \nabla \cdot (h \nabla \phi^{(0)}) \\ & + \frac{z^2}{2} \nabla^2 \phi^{(0)} \bigg] + 0(\mu^4) \, .\end{aligned} \tag{3.15}$$

For later use, we introduce $\Phi_\alpha(\mathbf{x}, t)$ as the potential at an arbitrary elevation $z = z_\alpha(\mathbf{x})$. From (3.15), Φ_α can be expressed as

$$\begin{aligned}\Phi_\alpha = \phi^{(0)} - \mu^2 \bigg[& h \nabla h \cdot \nabla \phi^{(0)} + \frac{h^2}{2} \nabla^2 \phi^{(0)} \\ & + z_\alpha \nabla \cdot (h \nabla \phi^{(0)}) + \frac{z_\alpha^2}{2} \nabla^2 \phi^{(0)} \bigg] + 0(\mu^4) \, .\end{aligned} \tag{3.16}$$

Subtracting (3.16) from (3.15) and using $\Phi_\alpha = \phi^{(0)} + 0(\mu^2)$, we can write the potential function in terms of Φ_α

$$\Phi(\mathbf{x}, z, t) = \Phi_\alpha(\mathbf{x}, t) - \mu^2 \left[(z - z_\alpha)\nabla \cdot (h\nabla\Phi_\alpha) + \frac{(z^2 - z_\alpha^2)}{2}\nabla^2\Phi_\alpha \right] + 0(\mu^4) . \quad (3.17)$$

From (3.16), we can also express $\phi^{(0)}$ in terms of Φ_α

$$\phi^{(0)} = \Phi_\alpha + \mu^2 \left[h\nabla h \cdot \nabla\Phi_\alpha + \frac{h^2}{2}\nabla^2\Phi_\alpha + z_\alpha\nabla \cdot (h\nabla\Phi_\alpha) + \frac{z_\alpha^2}{2}\nabla^2\Phi_\alpha \right] + 0(\mu^4) . \quad (3.18)$$

Along the still water surface, $z_\alpha = 0$, and $\Phi_\alpha(\mathbf{x}, 0, t) = \Phi_o$. From (3.18), we obtain

$$\phi^{(0)} = M^{-1}\Phi_o \quad (3.19)$$

where

$$M^{-1} = 1 + \mu^2 \left[h\nabla h \cdot \nabla + \frac{h^2}{2}\nabla^2 \right] + 0(\mu^4) . \quad (3.20)$$

To evaluate the integrand in (3.11), we need to find the expression for the gradient of Φ on $z = 0$. From (3.12) $\sim$ (3.14) and (3.19), we have

$$\left.\frac{\partial\Phi}{\partial z}\right|_{z=0} = \sum_{n=0}^{\infty}(n+1)h^n\phi^{(n+1)} = SM^{-1}\Phi_o + 0(\mu^6) \quad (3.21)$$

where

$$S = -\mu^2 \Big\{ (1 - \mu^2|\nabla h|^2)\nabla(h \cdot \nabla) - \mu^2 \Big[2h\nabla h \cdot \nabla(\nabla h \cdot \nabla) + h\nabla^2 h(\nabla h \cdot \nabla) + h^2 \left(\frac{1}{2}\nabla^2(\nabla h \cdot \nabla) + (\nabla h \cdot \nabla)\nabla^2 + \frac{1}{2}\nabla^2 h\nabla^2 \right) + \frac{1}{6}h^3\nabla^2\nabla^2 \Big] \Big\} . \quad (3.22)$$

The first term in the Hamiltonian or (3.9) can be evaluated as

$$\begin{aligned} E_{ko} &= \frac{\mu^{-2}}{2}\int\int_{\Omega} \Phi_o S M^{-1}\Phi_o dxdy \\ &= \frac{1}{2}\int\int_{\Omega}\left\{h\nabla\Phi_o\cdot\nabla\Phi_o + \frac{\mu^2 h^2}{2}\nabla\cdot[\nabla\Phi_o(\nabla h\cdot\nabla\Phi_o)]\right. \\ &\quad \left. + \frac{\mu^2 h^3}{3}\nabla\Phi_o\cdot(\nabla^2\nabla\Phi_o)\right\} dxdy + 0(\mu^4)\,. \end{aligned} \tag{3.23}$$

Finally, the approximated Hamiltonian (3.11), can be expressed in terms of Φ_o and η in the following form:

$$\begin{aligned} \mathcal{H} &= \frac{1}{2}\int\int_{\Omega}\left\{(h+\varepsilon\eta)|\nabla\Phi_o|^2 + \frac{\mu^2 h^2}{2}\nabla\cdot[\nabla\Phi_o(\nabla h\cdot\nabla\Phi_o)]\right. \\ &\quad \left. + \frac{\mu^2 h^3}{3}\nabla\Phi_o\cdot\nabla^2(\nabla\Phi_o) + \eta^2\right\} dxdy + 0(\mu^4, \varepsilon\mu^2, \varepsilon^2)\,. \end{aligned} \tag{3.24}$$

We note that from (3.17) with $z_\alpha = 0$,

$$\frac{\partial\Phi}{\partial x} = \frac{\partial\Phi_o}{\partial x} + 0(\mu^2),\ \frac{\partial\Phi}{\partial y} = \frac{\partial\Phi_o}{\partial y} + 0(\mu^2),\ \frac{\partial\Phi}{\partial z} = 0(\mu^2)$$

which have been used in deriving the Hamiltonian (3.24).

To apply the canonical theorem, we first recognize that the potential on the actual free surface ($z = \varepsilon\eta$) , ϕ, is different from that on the still water level, Φ. However, the difference is small and can be shown by using the Taylor's series expansion

$$\phi = \Phi(x, \varepsilon\eta, t) = \Phi(x, 0, t) + \left.\frac{\partial\Phi}{\partial z}\right|_{z=0}\varepsilon\eta + \cdots\,.$$

Hence,

$$\Phi_o = \phi + 0(\varepsilon\mu^2)\,. \tag{3.25}$$

Therefore, within the limit of accuracy for the Hamiltonian, ϕ and Φ_o are exchangeable.

Applying the canonical theorem (3.2), to the Hamiltonian, (3.24), yields

$$\begin{aligned} \frac{\partial\eta}{\partial t} &= -\nabla\cdot[(h+\varepsilon\eta)\nabla\Phi_o] \\ &\quad - \frac{1}{2}\mu^2\nabla\cdot\left[h^2\nabla(\nabla\cdot(h\nabla\Phi_o)) - \frac{h^3}{3}\nabla(\nabla\cdot\nabla\Phi_o)\right] \end{aligned} \tag{3.26}$$

$$\frac{\partial\Phi_o}{\partial t} = -\frac{\varepsilon}{2}|\nabla\Phi_o|^2 - \eta\,. \tag{3.27}$$

Introducing the horizontal velocity vector on the still water surface as

$$\mathbf{u}_o = \nabla \Phi_o \tag{3.28}$$

we can rewrite (3.26) in the following form

$$\frac{\partial \eta}{\partial t} + \nabla \cdot [(h + \varepsilon \eta)\mathbf{u}_o] + \frac{\mu^2}{2} \nabla \cdot \left[h^2 \nabla (\nabla \cdot h\mathbf{u}_o) - \frac{h^3}{3} \nabla (\nabla \cdot \mathbf{u}_o) \right] = 0(\varepsilon \mu^3, \mu^4, \varepsilon^2) \tag{3.29}$$

which represents the continuity equation. Taking the gradient of (3.27), we obtain the momentum equation in terms of the velocity $\mathbf{u}_o$

$$\frac{\partial \mathbf{u}_o}{\partial t} + \varepsilon \mathbf{u}_o \cdot \nabla \mathbf{u}_o + \nabla \eta = 0(\varepsilon \mu^3, \mu^4, \varepsilon^2) \, . \tag{3.30}$$

Equations (3.29) and (3.30) are the conventional Boussinesq equations expressed in terms of the horizontal velocity on the still water surface. These equations can be rewritten in terms of horizontal velocity on the bottom or the depth-averaged velocity. We will discuss these alternative forms in Subsec. 3.2.2.

3.2. *A direct perturbation approach*

3.2.1. *Governing equations for finite amplitude waves*

In the Hamiltonian approach, we have to employ the Boussinesq approximation, i.e., $0(\varepsilon) \approx 0(\mu^2) << 1$. Moreover, the knowledge of the vertical structure of the potential function is essential and is obtained via a perturbation method. In this section, we present a direct perturbation approach, which allows the parameter, ε, representing the nonlinearity, to be arbitrary.

To facilitate the perturbation procedure efficiently, we integrate the continuity equation (2.3) from the bottom, $z = -h$, to the free surface, $z = \varepsilon\eta$. Using the kinematic boundary conditions (2.5) and (2.7), one may obtain the depth-integrated continuity equation

$$\nabla \cdot \left[\int_{-h}^{\varepsilon\eta} \nabla \Phi dz \right] + \frac{\partial \eta}{\partial t} = 0 \tag{3.31}$$

which is an exact equation. Following the perturbation procedure given in Subsec. 3.1.2 and substituting (3.17) into the continuity equation (3.31), we obtain

$$\begin{aligned}
&\frac{\partial \eta}{\partial t} + \nabla \cdot [(\varepsilon\eta + h)\nabla\Phi_\alpha] \\
&\quad + \mu^2 \nabla \cdot \{(\varepsilon\eta + h)\,[\nabla(z_\alpha \nabla \cdot (h\nabla\Phi_\alpha)) \\
&\quad + \frac{1}{2}(h - \varepsilon\eta)\nabla(\nabla \cdot (h\nabla\Phi_\alpha)) + \frac{1}{2}\nabla(z_\alpha^2 \nabla^2 \Phi_\alpha) \\
&\quad - \frac{1}{6}(\varepsilon^2\eta^2 - \varepsilon\eta h + h^2)\nabla\nabla^2\Phi_\alpha\Big]\Big\} = 0(\mu^4)\,. \qquad (3.32)
\end{aligned}$$

We reiterate here that the parameter, ε, is assumed to be an arbitrary constant. Substitution of (3.17) for the dynamic free surface boundary condition, (2.6), yields

$$\begin{aligned}
&\frac{\partial \Phi_\alpha}{\partial t} + \eta + \frac{\varepsilon}{2}|\nabla\Phi_\alpha|^2 + \mu^2 \Bigg[(z_\alpha - \varepsilon\eta)\nabla \cdot (h\nabla\frac{\partial \Phi_\alpha}{\partial t}) \\
&\quad + \frac{1}{2}(z_\alpha^2 - \varepsilon^2\eta^2)\nabla^2(\frac{\partial \Phi_\alpha}{\partial t})\Bigg] - \varepsilon\mu^2 \nabla\Phi_\alpha \cdot \Bigg[-\nabla z_\alpha(\nabla \cdot (h\nabla\Phi_\alpha)) \\
&\quad + (\varepsilon\eta - z_\alpha)\nabla(\nabla \cdot (h\nabla\Phi_\alpha)) - z_\alpha \nabla z_\alpha \nabla^2\Phi_\alpha \\
&\quad + \frac{1}{2}(\varepsilon^2\eta^2 - z_\alpha^2)\nabla(\nabla^2\Phi_\alpha)\Bigg] + \frac{\varepsilon\mu^2}{2}[(\nabla \cdot (h\nabla\Phi_\alpha))^2 \\
&\quad + 2\varepsilon\eta\nabla \cdot (h\nabla\Phi_\alpha)\nabla^2\Phi_\alpha + \varepsilon^2\eta^2(\nabla^2\Phi_\alpha)^2] = 0(\mu^4)\,. \qquad (3.33)
\end{aligned}$$

From (3.17), the horizontal velocity at $z = z_\alpha$ can be defined as

$$\begin{aligned}
\mathbf{u}_\alpha = (\nabla\Phi)\Big|_{z=z_\alpha} &= \nabla\Phi_\alpha \\
&+ \mu^2 \left[\nabla z_\alpha \nabla \cdot (h\nabla\Phi_\alpha) + z_\alpha \nabla z_\alpha \nabla^2\Phi_\alpha\right] + 0(\mu^4)\,. \qquad (3.34)
\end{aligned}$$

Equivalently,

$$\begin{aligned}
\nabla\Phi_\alpha = \mathbf{u}_\alpha &- \mu^2[\nabla z_\alpha \nabla \cdot (h\nabla\mathbf{u}_\alpha) \\
&+ z_\alpha \nabla z_\alpha \nabla \cdot \mathbf{u}_\alpha] + 0(\mu^4)\,. \qquad (3.35)
\end{aligned}$$

Replacing $\nabla\Phi_\alpha$ by the right-hand side member in (3.35), we can rewrite the continuity equation (3.32) as

$$\frac{\partial\eta}{\partial t}+\nabla\cdot[(\varepsilon\eta+h)\mathbf{u}_\alpha]+\mu^2\nabla\cdot\left\{\left(\frac{z_\alpha^2}{2}-\frac{h^2}{6}\right)h\nabla(\nabla\cdot\mathbf{u}_\alpha)\right.$$
$$\left.+\left(z_\alpha+\frac{h}{2}\right)h\nabla[\nabla\cdot(h\mathbf{u}_\alpha)]\right\}+NL1=0(\mu^4) \tag{3.36}$$

where

$$NL1=\mu^2\nabla\cdot\left\{\varepsilon\eta\left[\left(z_\alpha-\frac{1}{2}\varepsilon\eta\right)\nabla(\nabla\cdot h\mathbf{u}_\alpha)\right.\right.$$
$$\left.\left.+\frac{1}{2}\left(z_\alpha^2-\frac{1}{3}\varepsilon^2\eta^2\right)\nabla(\nabla\cdot\mathbf{u}_\alpha)\right]\right\}. \tag{3.37}$$

The terms in $NL1$, (3.37), are of the order of magnitude of $0(\mu^2)$. Notice that these terms are the combination of cubic and quadratic nonlinearity.

Taking the gradient of (3.33) and using (3.35) in the resulting equation, we obtain

$$\frac{\partial\mathbf{u}_\alpha}{\partial t}+\nabla\eta+\varepsilon\mathbf{u}_\alpha\cdot\nabla\mathbf{u}_\alpha+\mu^2\left\{z_\alpha\nabla\left[\nabla\cdot(h\frac{\partial\mathbf{u}_\alpha}{\partial t})\right]\right.$$
$$\left.+\frac{1}{2}z_\alpha^2\nabla(\nabla\cdot\frac{\partial\mathbf{u}_\alpha}{\partial t})\right\}+NL2=0(\mu^4) \tag{3.38}$$

where

$$NL2=\mu^2\varepsilon\left\{(\mathbf{u}_\alpha\cdot\nabla z_\alpha)\nabla(\nabla\cdot h\mathbf{u}_\alpha)+z_\alpha\nabla[\mathbf{u}_\alpha\cdot\nabla(\nabla\cdot h\mathbf{u}_\alpha)]\right.$$
$$+z_\alpha(\mathbf{u}_\alpha\cdot\nabla z_\alpha)\nabla(\nabla\cdot\mathbf{u}_\alpha)+\frac{z_\alpha^2}{2}\nabla[\mathbf{u}_\alpha\cdot\nabla(\nabla\cdot\mathbf{u}_\alpha)]$$
$$\left.+\nabla\cdot(h\mathbf{u}_\alpha)\nabla(\nabla\cdot h\mathbf{u}_\alpha)-\nabla\left[\eta\nabla\cdot\left(h\frac{\partial\mathbf{u}_\alpha}{\partial t}\right)\right]\right\}$$
$$-\mu^2\varepsilon^2\left\{\frac{1}{2}\nabla\left[\eta^2\left(\nabla\cdot\frac{\partial\mathbf{u}_\alpha}{\partial t}\right)\right]+\nabla[\mathbf{u}_\alpha\cdot(\eta\nabla(\nabla\cdot h\mathbf{u}_\alpha)]\right.$$
$$\left.-\nabla[\eta(\nabla\cdot h\mathbf{u}_\alpha)\nabla\cdot\mathbf{u}_\alpha]\right\}$$
$$-\mu^2\varepsilon^3\nabla\left\{\frac{\eta^2}{2}[\nabla(\nabla\cdot\mathbf{u}_\alpha)-(\nabla\cdot\mathbf{u}_\alpha)^2]\right\}. \tag{3.39}$$

We reiterate that similar to those terms in $NL1$, the terms in $NL2$ are nonlinear in both quadratic and cubic forms. By allowing the wave parameter, ε, to be arbitrary, we permit finite amplitude waves in very shallow water.

3.2.2. *Summary of conventional Boussinesq equations*

In the Boussinesq approximation, the Ursell number is assumed to be of the order of magnitude of one. In other words, $0(\varepsilon) \approx 0(\mu^2) << 1$. Consequently, the members of $NL1$ and $NL2$ in (3.37) and (3.39) are in the same order of magnitude as or smaller than $0(\mu^4)$. Therefore, the Boussinesq equations written in terms of the horizontal velocity components, $\mathbf{u}_\alpha$, and the free surface displacement, η, are given in (3.36) and (3.38) without $NL1$ and $NL2$. These equations are accurate up to $0(\mu^4, \mu^2\varepsilon)$. When the velocity is evaluated on the still water surface, $z_\alpha = 0$ and $\mathbf{u}_\alpha = \mathbf{u}_o$, the Boussinesq equations can be reduced to (3.29) and (3.30) which were derived from the Hamiltonian approach.

The limitations of the Boussinesq equations are two-fold: In very shallow waters where waves are close to breaking, the nonlinearity is important and the nonlinearity parameter, ε, could become as large as $0.3 \sim 0.4$. At the same time the dispersion parameter, μ^2, becomes smaller as the depth decreases. Therefore, free surface profiles for a near breaking wave obtained from the Boussinesq equation are usually more symmetric with respect to the wave crest than that observed in the laboratory (Liu, 1990). This shortcoming can be overcome by relaxing the restriction on nonlinearity parameter, ε. In other words, some terms in $NL1$ and $NL2$ given in (3.37) and (3.39), such as $0(\mu^2\varepsilon)$ terms, can be included in the Boussinesq equations. The second limitation of the Boussinesq equations is their inadequate behavior in the intermediate depth region. We will illustrate this point by examining the dispersion relations corresponding to different forms of the Boussinesq equations in the following section.

The conventional Boussinesq equations appear in different forms. They can be expressed in terms of the velocity on the free surface, as shown in (3.29) and (3.30). They can also be written in terms of velocity vectors on the bottom or the depth-averaged velocity (Peregrine, 1972). For later uses, we present these well-known Boussinesq equations here. In terms of the velocity along the bottom, $z_\alpha = -h$, (3.36) and (3.38) becomes

$$\frac{\partial \eta}{\partial t} + \nabla \cdot [(\varepsilon\eta + h)\mathbf{u}_b] + \mu^2 \left\{ \nabla \cdot \frac{1}{3} h^3 \nabla(\nabla \cdot \mathbf{u}_b) - \frac{1}{2} h^2 \nabla [\nabla \cdot (h\mathbf{u}_b)] \right\} = 0(\mu^4) \tag{3.40}$$

$$\frac{\partial \mathbf{u}_b}{\partial t} + \nabla\eta + \varepsilon \mathbf{u}_b \cdot \nabla \mathbf{u}_b + \mu^2 \left\{ -h\nabla\left(\nabla \cdot h \frac{\partial \mathbf{u}_b}{\partial t}\right) + \frac{1}{2}h^2 \nabla\left(\nabla \cdot \frac{\partial \mathbf{u}_b}{\partial t}\right)\right\} = 0(\mu^4) \tag{3.41}$$

where $\mathbf{u}_b$ denotes the velocity on the sea bottom. To write the Boussinesq equations in terms of the depth-averaged velocity

$$\bar{\mathbf{u}} = \frac{1}{h+\varepsilon\eta}\int_{-h}^{\varepsilon\eta} \nabla\Phi dz \, . \tag{3.42}$$

We first rewrite the continuity equation (3.31) as

$$\frac{\partial \eta}{\partial t} + \nabla \cdot [(h+\varepsilon\eta)\bar{\mathbf{u}}] = 0 \tag{3.43}$$

which is exact. Secondly, from (3.17), we obtain

$$\nabla\Phi_\alpha = \frac{\bar{\mathbf{u}}}{h} - \frac{\varepsilon\eta}{h^2}\bar{\mathbf{u}} + \mu^2 \left\{ \frac{h}{6}(h^2 - 3z_\alpha^2)\nabla \cdot \left(\frac{\bar{\mathbf{u}}}{h}\right) - \frac{1}{2}h(h + 2z_\alpha \nabla \cdot \bar{\mathbf{u}})\right\} + 0(\mu^4, \varepsilon^2) \, . \tag{3.44}$$

Substituting (3.34) and (3.44) for (3.38), we obtain after lengthy manipulation,

$$\frac{\partial \bar{\mathbf{u}}}{\partial t} + \nabla\eta + \varepsilon \bar{\mathbf{u}} \cdot \nabla \bar{\mathbf{u}} + \mu^2 \left[\frac{h^2}{6} \nabla\left(\nabla \cdot \frac{\partial \bar{\mathbf{u}}}{\partial t}\right) - \frac{h}{2}\nabla\nabla \cdot \left(h \frac{\partial \bar{\mathbf{u}}}{\partial t}\right)\right] = 0(\mu^4, \varepsilon^2) \, . \tag{3.45}$$

We remark here that different forms of the Boussinesq equations can be further deduced by replacing the higher order time derivative terms in the momentum equations, (3.41) and (3.45) by the spatial derivative of η.

3.3. *Dispersion relation and phase velocity of nonlinear shallow-water equations*

The major difference among the conventional Boussinesq equations is in the higher order derivative terms. These higher order derivative terms affect the dispersive properties and the stability of the equations. In this section, we examine the dispersion relations for the nonlinear shallow-water equations as well as the conventional Boussinesq equations. To simplify the discussion, we only investigate one-dimensional constant depth cases. From (3.36) and (3.38), we can write the continuity and the momentum equation as

$$\begin{aligned}&\frac{\partial \eta}{\partial t}+\frac{\partial}{\partial x}\left[(\varepsilon\eta+h)u_\alpha\right]+\mu^2\left(\alpha+\frac{1}{3}\right)h^3\frac{\partial^3 u_\alpha}{\partial x^3}\\&\quad+\beta\left\{\mu^2\varepsilon\alpha h^2\frac{\partial}{\partial x}\left(\eta\frac{\partial^2 u_\alpha}{\partial x^2}\right)-\frac{\mu^2\varepsilon^2}{2}h\frac{\partial}{\partial x}\left(\eta^2\frac{\partial^2 u_\alpha}{\partial x^2}\right)\right.\\&\quad\left.+\frac{\mu^2\varepsilon^3}{6}\frac{\partial}{\partial x}\left(\eta^3\frac{\partial^2 u_\alpha}{\partial x^2}\right)\right\}=0\end{aligned}\tag{3.46}$$

$$\begin{aligned}&\frac{\partial u_\alpha}{\partial t}+\frac{\partial \eta}{\partial x}+\mu^2\alpha h^2\frac{\partial^3 u_\alpha}{\partial x^2\partial t}+\varepsilon u_\alpha\frac{\partial u_\alpha}{\partial x}\\&\quad+\beta\mu^2\varepsilon\left\{h^2\alpha\frac{\partial}{\partial x}\left(u_\alpha\frac{\partial^2 u_\alpha}{\partial x^2}\right)+h^2\frac{\partial u_\alpha}{\partial x}\frac{\partial^2 u_\alpha}{\partial x^2}-h\frac{\partial}{\partial x}\left(\eta\frac{\partial^2 u_\alpha}{\partial x\partial t}\right)\right\}\\&\quad-\beta\mu^2\varepsilon^2\frac{\partial}{\partial x}\left[\frac{\eta^2}{2}\frac{\partial^2 u_\alpha}{\partial x\partial t}+u_\alpha\eta h\frac{\partial^2 u_\alpha}{\partial x^2}-\eta h\left(\frac{\partial u_\alpha}{\partial x}\right)^2\right]\\&\quad-\beta\mu^2\varepsilon^3\frac{\partial}{\partial x}\left\{\frac{\eta^2}{2}\left[\frac{\partial^2 u_\alpha}{\partial x^2}-\left(\frac{\partial u_\alpha}{\partial x}\right)^2\right]\right\}=0\end{aligned}\tag{3.47}$$

where $\alpha=(z_\alpha/h)^2/2+(z_\alpha/h)$ and $\beta=1$. If the terms in the order of $\mu^2\varepsilon$, $\mu^2\varepsilon^2$ and $\mu^2\varepsilon^3$ are ignored (i.e., $\beta=0$), the above equations are reduced to the Boussinesq equations.

We look for a solution of (3.46) and (3.47) in the expansion

$$\eta=\eta_1(\theta)+\varepsilon\eta_2(\theta)+\varepsilon^2\eta_3(\theta)+\dots\tag{3.48}$$

$$u_\alpha=u_1(\theta)+\varepsilon u_2(\theta)+\varepsilon^2 u_3(\theta)+\dots\tag{3.49}$$

$$\theta=kx-wt\tag{3.50}$$

$$\omega=\omega_0+\varepsilon^2\omega_2+\dots\tag{3.51}$$

where $\varepsilon = a/h$ denotes the small parameter for nonlinearity. Substituting (3.48) $\sim$ (3.51) into (3.46) and (3.47), we obtain the hierarchy

$$-\omega_o\eta_1' + khu_1' + \mu^2(\alpha + \frac{1}{3})k^3h^3u_1''' = 0 \tag{3.52}$$

$$-\omega_o u_1' + k\eta_1' - \mu^2\alpha k^2h^2\omega_o u_1''' = 0 \tag{3.53}$$

$$-\omega_o\eta_2' + khu_2' + \mu^2(\alpha + \frac{1}{3})k^3h^3u_2''' = -k(\eta_1 u_1)' - \beta\mu^2\alpha k^3h^2(\eta_1 u_1'')' \tag{3.54}$$

$$\begin{aligned} -\omega_o u_2' + k\eta_2' - \mu^2\alpha k^2h^2\omega_o u_2''' &= -ku_1u_1' \\ -\beta\mu^2 \left\{\alpha k^3h^2(u_1u_1'')' + k^3h^2u_1'u_1'' + \omega_o k^2h(\eta_1u_1'')'\right\} \end{aligned} \tag{3.55}$$

$$\begin{aligned} -\omega_o\eta_3' + khu_3' + \mu^2(\alpha + \frac{1}{3})k^3h^3u_3''' &= -k(\eta_1u_2 + \eta_2u_1)' + \omega_2\eta_1' \\ -\beta\mu^2\alpha k^3h^2[(\eta_1u_2'')' + (\eta_2u_1'')'] &+ \beta\frac{\mu^2}{2}k^3h(\eta_1^2u_1'')' \end{aligned} \tag{3.56}$$

$$\begin{aligned} -\omega_o u_3' + k\eta_3' - \mu^2\alpha k^2h^2\omega_o u_3''' &= -k(u_1u_2)' + \omega_2u_1' \\ -\beta\mu^2\Big\{k^3h^2\alpha[(u_1u_2'')' + u_2u_1''] &+ k^3h^2\left[u_1'u_2'' + u_2'u_1''\right] + \omega_o k^2h[(\eta_1u_2'')' \\ +(\eta_2u_1'')'] + \frac{1}{2}\omega_o k^2(\eta_1^2u_1'')' &- k^3h(u_1\eta_1u_1'')' + [\eta_1(u_1')^2]'\Big\}\,. \end{aligned} \tag{3.57}$$

The solution for the leading order equations (3.52) and (3.53) represents a periodic wave

$$\eta_1 = \cos\theta \quad , \quad u_1 = \frac{k}{\omega_o}\left(\frac{1}{1 - \mu^2\alpha k^2h^2}\right)\cos\theta \tag{3.58}$$

with the linear dispersion relation

$$C^2 = \frac{\omega_o^2}{k^2} = h\left[\frac{1 - \mu^2(\alpha + \frac{1}{3})k^2h^2}{1 - \mu^2\alpha k^2h^2}\right]\,. \tag{3.59}$$

If the above equation becomes negative, the frequency and the phase velocity become imaginary. This implies that the solution grows in time and becomes unstable. Hence, the instability condition requires

$$\left[1 - \mu^2\left(\alpha + \frac{1}{3}\right)k^2h^2\right](1 - \mu^2\alpha k^2h^2) > 0\,.$$

However, because $-\frac{1}{2} < \alpha < 0$, the above condition becomes

$$1 - \mu^2 \left(\alpha + \frac{1}{3}\right) k^2 h^2 > 0 \,. \tag{3.60}$$

For $\alpha \leq -1/3$, the stability condition is always satisfied for all kh. On the other hand, if $0 > \alpha > -1/3$, the relative depth is limited by the following condition

$$\mu kh < \frac{1}{\sqrt{\alpha + \frac{1}{3}}} \tag{3.61}$$

for the stability requirement. For example, the conventional Boussinesq equations using the velocity on the free surface, $u_o(\alpha = 0)$, become unstable when $\mu(kh)$ is greater than $\sqrt{3}$.

As we mentioned before, the higher derivative terms (dispersive terms) in the Boussinesq equations can appear in different forms; we can replace the spatial derivative by the time derivative and vice versa through the leading order approximation. For instance, (3.47) can be rewritten as

$$\frac{\partial u_\alpha}{\partial t} + \frac{\partial \eta}{\partial x} - \mu^2 \alpha h^2 \frac{\partial^3 \eta}{\partial x^3} + \varepsilon u_\alpha \frac{\partial u_\alpha}{\partial x} = 0 \,. \tag{3.62}$$

The dispersive properties of the new set of Boussinesq equations (3.46) with $\beta = 0$ and (3.62) are different from those of (3.46) and (3.47). Substituting (3.48) $\sim$ (3.51) into (3.48) and (3.62), we find the following linear dispersion relation

$$C^2 = \frac{\omega^2}{k^2} = h \left[1 - \mu^2 \left(\alpha + \frac{1}{3}\right) k^2 h^2\right] (1 + \mu^2 \alpha k^2 h^2) \,. \tag{3.63}$$

Comparing (3.59) and (3.63), we observe that (3.63) is the truncated binomial expansion of (3.59) for small k^2h^2. To ensure that the system is stable, the kh value must satisfy the following condition

$$\left[1 - \mu^2 \left(\alpha + \frac{1}{3}\right) k^2 h^2\right] (1 + \mu^2 \alpha k^2 h^2) > 0 \,. \tag{3.64}$$

This is a more restrictive condition than (3.60). With $\alpha = -1/3$, the conventional Boussinesq equations using the depth-averaged velocity also become unstable when μkh is greater than $\sqrt{3}$. Hence, the form of Boussinesq equations expressed in (3.46) and (3.47) are preferred as far as the instability is concerned.

Substituting (3.58) for (3.54) and (3.55), we find that the right-hand side terms are proportional to $\sin 2\theta$. Therefore, the solution for (3.54) and (3.55) can be written as

$$\eta_2 = B\cos 2\theta \quad , \quad u_2 = D\cos 2\theta \tag{3.65}$$

where

$$B = \frac{\omega_o}{k}(1-4\mu^2\alpha k^2h^2)D - \frac{1}{4}\frac{k^2}{\omega_o^2}\frac{1}{(1-\mu^2\alpha k^2h^2)^2}$$
$$\left\{1-\beta\mu^2\left[(2\alpha+1)k^2h^2+2\omega_o^2h(1-\mu^2\alpha k^2h^2)\right]\right\} \tag{3.66}$$

$$D = \frac{1}{4\mu^2k^3h^3}\frac{k^2}{\omega_o}\,\frac{1}{1-\mu^2\alpha k^2h^2}\left\{3-2\mu^2\alpha k^2h^2\right.$$
$$\left.-\beta\mu^2(4\alpha+3)k^2h^2+2\beta\mu^4\left(\frac{1}{3}+\alpha+4\alpha^2\right)k^4h^4\right\}. \tag{3.67}$$

The right-hand side of the equations for η_3 and u_3 can be written as

$$RHS\text{ of }(3.56) = \left\{-\omega_2+\frac{1}{2}k(D+BU)+\beta\mu^2\left[\frac{k^3h}{4}U\right.\right.$$
$$\left.\left.-\frac{k}{2}\left(4\alpha k^2h^2D+\alpha k^2h^2BU-\frac{1}{4}k^2hU\right)\right]\right\}\sin\theta + L_1\sin 3\theta \tag{3.68}$$

$$RHS\text{ of }(3.57) = \left\{-\omega_2U+\frac{1}{2}UDk-\beta\mu^2\left[\frac{1}{2}k^2h\omega_oBU\right.\right.$$
$$+\left(\frac{k^4h^2}{2\omega_o}\frac{5\alpha-2}{1-\mu^2\alpha k^2h^2}+2k^2h\omega_o\right)D+\frac{3}{8}k^3\frac{1}{1-\mu^2\alpha k^2h^2}$$
$$\left.\left.-\frac{1}{2}\frac{k^5h}{\omega_o^2}\frac{1}{(1-\mu^2\alpha k^2h^2)^2}\right]\right\}\sin\theta + L_2\sin 3\theta \tag{3.69}$$

where

$$U = \frac{k}{\omega_o}\frac{1}{1-\mu^2\alpha k^2h^2} \tag{3.70}$$

and L_1 and L_2 are some complicated coefficients. The terms in $\sin 3\theta$ can be accommodated by solutions η_3 and u_3 which are proportional to $\cos 3\theta$, but the sin θ terms resonate with the operators on the left. There is a secular solution which is proportional to θ sin θ and is unbounded in θ. To eliminate the resonant term, we first combine (3.56) and (3.57) by eliminating η_3'. Thus, the right-hand side of the resulting equation becomes

$$k\,(RHS \text{ of } (3.56)) + \omega_o\,(RHS \text{ of } (3.57)) = \Bigg\{ -\omega_2(k+\omega_o U)$$
$$+\frac{1}{2}k^2(D+BU) + \frac{1}{2}k\omega_o UD - \beta\mu^2\Bigg[\Bigg(\frac{k^4h^2}{2}\frac{5\alpha-2}{1-\mu^2\alpha k^2h^2}$$
$$+2k^2h\omega_o^2 + 2\alpha k^4h^2\Bigg)D + \frac{k^2}{2}(\alpha k^2h^2+\omega_o^2h)BU - \frac{3}{8}k^4hU$$
$$+\frac{3k^3\omega_o}{8(1-\mu^2\alpha k^2h^2)} - \frac{k^5h}{2\omega_o(1-\mu^2\alpha k^2h^2)^2}\Bigg]\Bigg\}\sin\theta + (kL_1+\omega_oL_2)\sin 3\theta\,.$$

We now make the coefficient of sin θ be zero so that the resonant term vanishes. The final result for ω_2 is

$$\omega_2 = \frac{k}{2}D + \frac{k^2}{2\omega_o}\frac{1}{2-\mu^2\alpha k^2h^2}B - \beta\mu^2\frac{k^2h^2}{2-\mu^2\alpha k^2h^2}$$
$$\Bigg\{2k\left[\frac{5}{4}\alpha - 1 + \left(\alpha+\frac{\omega_o^2}{k^2h}\right)(1-\mu^2\alpha k^2h^2)\right]D$$
$$+\frac{\omega_o}{2h}\left(1+\alpha\frac{k^2h}{\omega_o^2}\right)B + \frac{3}{8}\frac{\omega_o}{h^2}\left(1-\frac{k^2h}{\omega_o^2}\right) - \frac{k^2}{2\omega_oh(1-\mu^2\alpha k^2h^2)}\Bigg\} \tag{3.71}$$

where B and D are defined in (3.66) and (3.67). ω_2 represents the dependence of the dispersion relation on amplitude. If the Boussinesq approximation is adopted (i.e., $\beta = 0$), ω_2 can be simplified to

$$\omega_2 = \Bigg\{\frac{1}{8}\frac{k^3}{\omega_o}\frac{1}{\mu^2k^3h^3}(3-2\mu^2\alpha k^2h^2)(3-5\mu^2\alpha k^2h^2)$$
$$-\frac{k^4}{8\omega_o^3}\frac{1}{1-\mu^2\alpha k^2h^2}\Bigg\}\frac{1}{(1-\mu^2\alpha k^2h^2)(2-\mu^2\alpha k^2h^2)}\,. \tag{3.72}$$

Returning to dimensional variables, we can express the dispersion relation (3.51) as

$$\frac{\omega}{\sqrt{ghk}} = \left[1-\left(\alpha+\frac{1}{3}\right)k^2h^2\right]^{1/2}(1-\alpha k^2h^2)^{-1/2}$$
$$+\left(\frac{a}{h}\right)^2\left\{\frac{1}{8k^2h^2}(3-\alpha k^2h^2)(3-5\alpha k^2h^2)-\frac{1}{8}\left[1-\left(\alpha+\frac{1}{3}\right)k^2h^2\right]^{-1}\right\}$$
$$\left[1-\left(\alpha+\frac{1}{3}\right)k^2h^2\right]^{-1/2}(1-\alpha k^2h^2)^{-1/2}(2-\alpha k^2h^2)^{-1}\,. \tag{3.73}$$

For the special case $\alpha = 0$, i.e., the Boussinesq equations written in terms of the free surface velocity, the dispersion relation can be expressed as

$$\omega = k\sqrt{h}\left(1-\frac{\mu^2}{3}k^2h^2\right)^{1/2}+\varepsilon^2\left(\frac{9}{16}\frac{k^3}{\omega_o}\frac{1}{\mu^2k^3h^3}-\frac{1}{16}\frac{k^4}{\omega_o^3}\right) \tag{3.74}$$

in the dimensionless form, and

$$\frac{\omega}{\sqrt{ghk}} = \left(1-\frac{1}{3}k^2h^2\right)^{\frac{1}{2}}+\left(\frac{a}{h}\right)^2\left\{\frac{9}{16k^2h^2}\right.$$
$$\left.-\frac{1}{16}\left(1-\frac{1}{3}k^2h^2\right)^{-1}\right\}\left(1-\frac{1}{3}k^2h^2\right)^{-\frac{1}{2}} \tag{3.75}$$

in the dimensional form.

4. Extension of Shallow-Water Equations to Deep Water

From the analysis shown in the previous section, we know that the Boussinesq equations are unstable subject to high frequency disturbances if $0 > \alpha > -1/3$ (see (3.60)). Even if the Boussinesq equations are stable for all kh (for example, $\alpha = -1/3$), the accuracy of these equations in intermediate and deep water is in doubt. A simple way to evaluate the accuracy of various forms of the nonlinear shallow water equations is to compare the dispersion relation derived from the these equations to that derived from Stokes' wave theory in the intermediate and deep water.

4.1. *Comparison between shallow-water equations and Stokes' wave theory*

For a uniform Stokes' wave train in a constant depth, the dispersion relation can be written as (Whitham, 1974):

$$\frac{\omega^2}{gk \tanh kh} = 1 + \left(\frac{9 \tanh^4 kh - 10 \tanh^2 kh + 9}{8 \tanh^4 kh}\right) k^2 a^2 + \dots \tag{4.1}$$

where ka denotes the wave slope and is considered as a small parameter. The dispersion relation can be further approximated as

$$\frac{\omega}{\sqrt{gk \tanh kh}} = 1 + \frac{1}{2}\left(\frac{9 \tanh^4 kh - 10 \tanh^2 kh + 9}{8 \tanh^4 kh}\right) k^2 a^2 \,. \tag{4.2}$$

The first term on the right-hand side of the above equation represents the linear wave dispersion relation, while the second term denotes the amplitude effect on the dispersion. In the shallow water limit, $kh << 1$, the dispersion relation becomes

$$\frac{\omega}{\sqrt{ghk}} = 1 - \frac{1}{6}k^2h^2 + \left(\frac{a}{h}\right)^2 \frac{9}{16\ k^2h^2} + \dots \,. \tag{4.3}$$

Once again, the shallow water limit of the Stokes' wave theory is valid only if $0(a/h) << 0(kh)^2$ (Whitham, 1974). Hence, the Stokes' theory works well in the shallow water only for extremely small amplitude waves. Because the Boussinesq equations are derived from the assumption that $0(a/h) = 0(kh)^2$ which covers the special case when the nonlinearity is weak, therefore, it is not surprising to observe that the dispersion relation for the conventional Boussinesq equations (3.75), after the higher-order terms in kh are dropped, is the same as (4.3).

However, we are more interested in knowing if the dispersion relation derived from the nonlinear shallow-water equations can be matched with that from the Stokes' wave theory in intermediate and deep water. Denoting ω and ω^s as the frequencies associated with the nonlinear shallow-water equations and the Stokes' theory respectively, we can define the ratios between these frequencies at two separate orders:

$$\frac{\omega_o}{\omega_o^s} = \sqrt{\frac{kh}{\tanh kh}}\left[\frac{1 - (\alpha + \frac{1}{3})k^2h^2}{1 - \alpha k^2h^2}\right]^{1/2} \tag{4.4}$$

$$\frac{\omega_2}{\omega_2^s} = \omega_2^* \sqrt{\frac{kh}{\tanh kh}} \left(\frac{1}{k^2h^2}\right) \frac{2 \tanh^4 kh}{9 \tanh^4 kh - 10 \tanh^2 kh + 9} \tag{4.5}$$

where

$$\omega_2^* = \frac{1}{2}D^* + \frac{1}{2}B^*\left[\frac{1-\alpha k^2h^2}{1-(\alpha+\frac{1}{3})k^2h^2}\right]\frac{1}{2-\alpha k^2h^2}$$
$$-\frac{\beta k^2h^2}{2-\alpha k^2h^2}\left\{2\left[\frac{5\alpha-2}{4}+\left(\alpha+\frac{1-(\alpha+\frac{1}{3})k^2h^2}{1-\alpha k^2h^2}\right)(1-\alpha k^2h^2)\right]D^*\right.$$
$$+\frac{1}{2}\left[1+\alpha\frac{1-\alpha k^2h^2}{1-(\alpha+\frac{1}{3})k^2h^2}\right]B^* + \frac{3}{8}\left[1-\frac{1-\alpha k^2h^2}{1-(\alpha+\frac{1}{3})k^2h^2}\right]$$
$$\left.-\frac{1}{2}\left[\frac{1-\alpha k^2h^2}{1-(\alpha+\frac{1}{3})k^2h^2}\right]^{\frac{1}{2}}\frac{1}{1-\alpha k^2h^2}\right\} \tag{4.6}$$

$$D^* = \frac{1}{4k^2h^2}\frac{[1-(\alpha+\frac{1}{3})k^2h^2]^{\frac{1}{2}}}{(1-\alpha k^2h^2)^{\frac{3}{2}}}\left\{3-2\alpha k^2h^2\right.$$
$$\left.-\beta\left[(4\alpha+3)k^2h^2+2\left(4\alpha^2+\alpha+\frac{1}{3}\right)k^4h^4\right]\right\} \tag{4.7}$$

$$B^* = \left[\frac{1-(\alpha+\frac{1}{3})k^2h^2}{1-\alpha k^2h^2}\right]^{\frac{1}{2}}D^* - \frac{1}{4[1-(\alpha+\frac{1}{3})k^2h^2]}$$
$$\frac{1}{1-\alpha k^2h^2}\left\{1-\beta k^2h^2[(2\alpha+3)-2(\alpha+\frac{1}{3})k^2h^2]\right\}\ . \tag{4.8}$$

When $\beta = 0$, the leading order frequency ratio remains the same as (4.4). But the second order frequency ratio can be simplified to

$$\omega_2^* = \left\{\frac{(3-\alpha k^2h^2)(3-5\alpha k^2h^2)}{k^2h^2} - \frac{1}{1-(\alpha+\frac{1}{3})k^2h^2}\right\}$$
$$\left[1-(\alpha+\frac{1}{3})k^2h^2\right]^{-1/2}(1-\alpha k^2h^2)^{-1/2}(2-\alpha k^2h^2)^{-1}\ . \tag{4.9}$$

These two frequency ratios are calculated for different kh, α and β values. In Fig. 1, we show the leading order frequency ratio (4.4) for the conventional Boussinesq equations with $\alpha = 0$. The agreement degenerates quickly for $kh \approx 1.5$. On the other hand, the agreement becomes much better if the velocity along the bottom ($\alpha = -0.5$) is used in the Boussinesq equations;

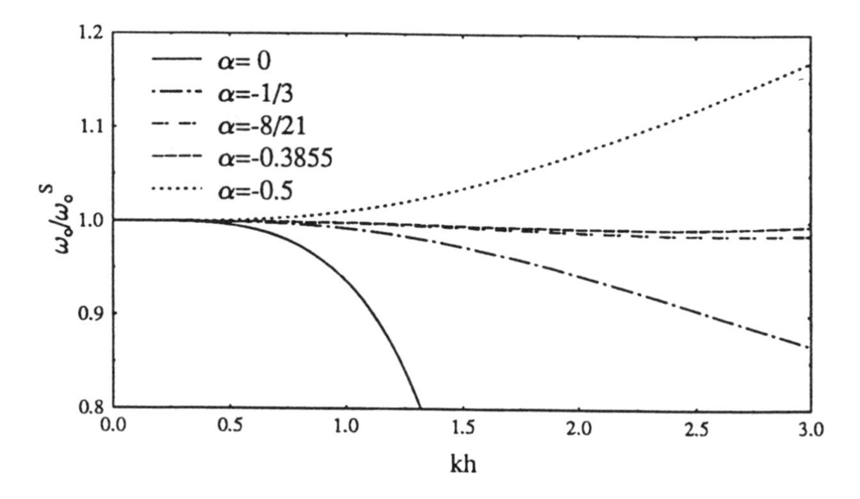

Fig. 1. First order frequency ratios between the Stokes' wave theory and Boussinesq-type equations.

the differences in deep water is roughly 20%. As far as the conventional Boussinesq equations are concerned, the best agreement occurs when the mean velocity is used, i.e., $\alpha = -1/3$. Since any value between 0 and -0.5 can be used for α without reducing the accuracy in the Boussinesq equations, one can find the optimal α value such that the differences in phase velocities and group velocities derived from ω_o and ω_o^s respectively, are minimized over the entire range of water depths. Chen and Liu (1993) reported that the optimal value is $\alpha = -0.3855$. Using a different approach, Madsen *et al.* (1991) derived a set of Boussinesq-type equations whose linear dispersion relation has the same form as that given in (3.59). Madsen *et al.* suggest that the optimal value for α should be $-8/21 = -0.381$. The approach used by Madsen *et al.* will be discussed in Subsec. 4.3.

The second order frequency ratio is also calculated for the conventional Boussinesq equations ($\beta = 0$) and the nonlinear shallow water equations ($\beta = 1$). Figure 2 shows that the conventional Boussinesq equations underestimate the second order frequency in intermediate and deep water. The second order frequency for the Boussinesq equation is insensitive to the α value when it is less than $-1/3$. Moreover, the second order frequency is practically zero for $kh > 1.5$. On the other hand, the nonlinear shallow water equations give

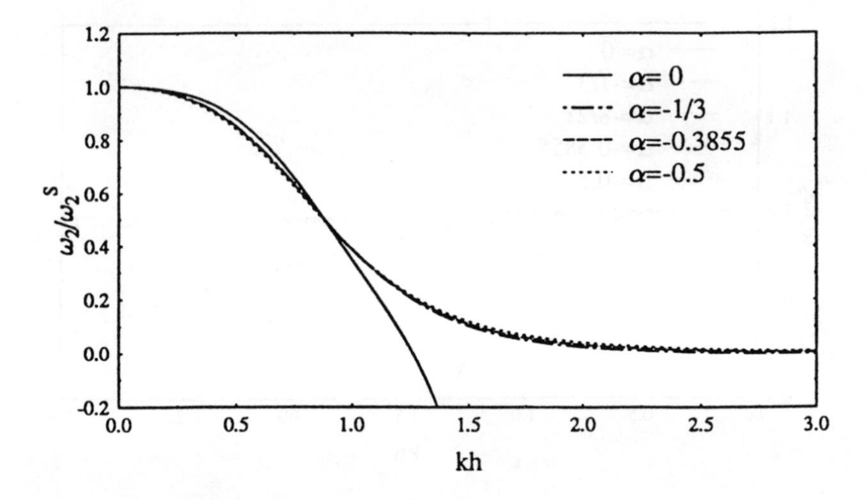

Fig. 2. Second order frequency ratios between the Stokes' wave theory and conventional Boussinesq equations ($\beta = 0$).

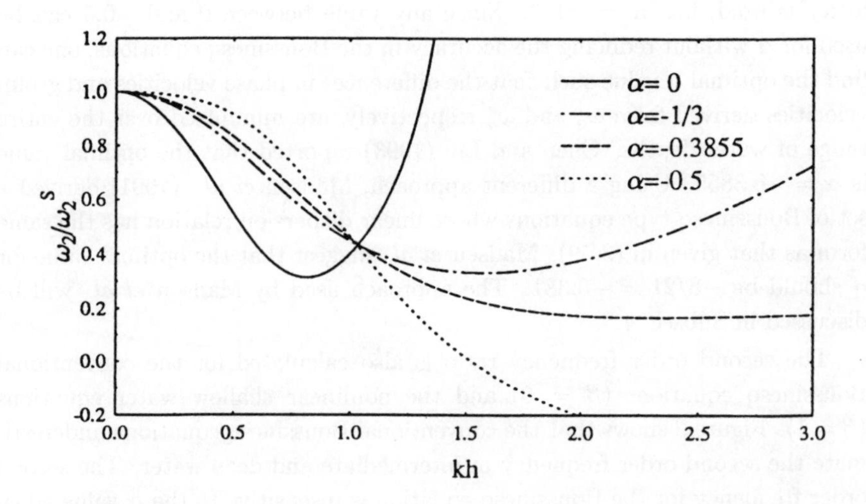

Fig. 3. Second order frequency ratios between the Stokes' wave theory and the nonlinear shallow water equations ($\beta = 1$).

slightly better estimations of the second order frequency (Fig. 3). Using the optimal α value determined from the first order frequency comparison, we show that the second order frequency for the nonlinear shallow water equation is about 25% of that from the Stokes' wave theory.

4.2. *Hamiltonian approach*

As shown in Sec. 3.1, the Boussinesq equations derived from the Hamiltonian given in (3.24) becomes unstable when μkh is greater than $\sqrt{3}$. The instability occurs when the Hamiltonian becomes negative. The remedy is to reconstruct the Hamiltonian into a quadratic form, which is always positive and definite. First, the Hamiltonian, (3.24), is rewritten as

$$\mathcal{H} = \frac{1}{2}\int\int_{\Omega} \left[\nabla\Phi_o \cdot (dR\nabla\Phi_o) + \eta^2\right] dxdy + 0(\mu^4, \varepsilon\mu^2, \varepsilon^2) \tag{4.10}$$

where R is a two by two symmetric tensor operator defined by

$$R = \begin{bmatrix} 1+\mu^2 a & \mu^2 b \\ \mu^2 b & 1+\mu^2 c \end{bmatrix} \tag{4.11}$$

$$a = \frac{h}{2}\frac{\partial^2}{\partial x^2}h - \frac{h^2}{6}\frac{\partial^2}{\partial x^2} \tag{4.12a}$$

$$b = \frac{h}{2}\frac{\partial^2}{\partial x\partial y}h - \frac{h^2}{6}\frac{\partial^2}{\partial x\partial y} \tag{4.12b}$$

$$c = \frac{h}{2}\frac{\partial^2}{\partial y^2}h - \frac{h^2}{6}\frac{\partial^2}{\partial y^2} \tag{4.12c}$$

and $d = \varepsilon\eta + h$ is the total depth. Following the approach suggested by Broer *et al.* (1987) and Mooiman (1991), we seek for a positive definite Hamiltonian in a quadratic form:

$$\mathcal{H} = \frac{1}{2}\int\int_{\Omega} \left[d(F\nabla\Phi_o)^2 + \eta^2\right] dxdy + 0(\varepsilon^2\ ,\ \varepsilon\mu^2\ ,\ \mu^4) \tag{4.13}$$

where F is another tensor operator to be found from the following relationship

$$\nabla\Phi_o \cdot (R\nabla\Phi_o) = (F\nabla\Phi_o)\cdot(F\nabla\Phi_o) + 0(\mu^4)\,.$$

Using (4.11) and (4.12), we obtain

$$F = \begin{bmatrix} F_{11} & F_{12} \\ F_{21} & F_{22} \end{bmatrix} \tag{4.14}$$

$$F_{11} = \frac{1}{\sqrt{2}}\left[1 + \frac{\mu^2}{2}(a+b)\right] \tag{4.15a}$$

$$F_{12} = \frac{1}{\sqrt{2}}\left[1 + \frac{\mu^2}{2}(b+c)\right] \tag{4.15b}$$

$$F_{21} = -\frac{1}{\sqrt{2}}\left[1 + \frac{\mu^2}{2}(a-b)\right] \tag{4.15c}$$

$$F_{22} = \frac{1}{\sqrt{2}}\left[1 - \frac{\mu^2}{2}(b-c)\right] \tag{4.15d}$$

where a, b, and c are operators defined in (4.12). The Hamiltonian defined in (4.13) is always positive and has the same accuracy as those given in (4.10) and (3.24). However, the Hamiltonian, (4.13), becomes unbounded for short waves, $\mu^2 \to \infty$. We must further approximate the operator F in the following manner

$$F_{11}^* = \frac{1}{\sqrt{2}[1 - \frac{\mu^2}{2}(a+b)]} = F_{11} + 0(\mu^4) \tag{4.16a}$$

$$F_{12}^* = \frac{1}{\sqrt{2}[1 - \frac{\mu^2}{2}(b+c)]} = F_{12} + 0(\mu^4) \tag{4.16b}$$

$$F_{21}^* = -\frac{1}{\sqrt{2}[1 - \frac{\mu^2}{2}(a-b)]} = F_{21} + 0(\mu^4) \tag{4.16c}$$

$$F_{22}^* = \frac{1}{\sqrt{2}[1 - \frac{\mu^2}{2}(b-c)]} = F_{22} + 0(\mu^4)\ . \tag{4.16d}$$

The canonical equations gives the final form of the stable Boussinesq-type equations:

$$\begin{aligned} \frac{\partial \eta}{\partial t} &= -\nabla \cdot \left[F^T(dF\nabla\Phi_o)\right] \\ &= -\frac{\partial}{\partial x}[F_{11}^* d(F_{11}^* u_o + F_{12}^* v_o) + F_{21}^* d(F_{21}^* u_o + F_{22}^* v_o)] \\ &\quad - \frac{\partial}{\partial y}[F_{12}^* d(F_{11}^* u_o + F_{12}^* v_o) + F_{22}^* d(F_{21}^* u_o + F_{22}^* v_o)] \end{aligned} \tag{4.17}$$

for continuity equation and

$$\frac{\partial \Phi_o}{\partial t} = -\frac{\varepsilon}{2}(F\nabla\Phi_o)^2 - \eta \tag{4.18}$$

for momentum equations. Taking the gradient of the momentum equation, we obtain

$$\begin{aligned}\frac{\partial u_o}{\partial t} = -\varepsilon \bigg[&(F_{11}^* u_o + F_{12}^* v_o)\frac{\partial}{\partial x}(F_{11}^* u_o + F_{12}^* v_o) \\ &+ (F_{21}^* u_o + F_{22}^* v_o)\frac{\partial}{\partial x}(F_{21}^* u_o + F_{22}^* v_o)\bigg] - \frac{\partial \eta}{\partial x}\end{aligned} \tag{4.19}$$

$$\begin{aligned}\frac{\partial v_o}{\partial t} = -\varepsilon \bigg[&(F_{11}^* u_o + F_{12}^* v_o)\frac{\partial}{\partial y}(F_{11}^* u_o + F_{12}^* v_o) \\ &+ (F_{21}^* u_o + F_{22}^* v_o)\frac{\partial}{\partial y}(F_{21}^* u_o + F_{22}^* v_o)\bigg] - \frac{\partial \eta}{\partial y} \, .\end{aligned} \tag{4.20}$$

Equations (4.17), (4.19) and (4.20) represent the modified Boussinesq equations which are stable for short waves.

Because the differential operators appear in the denominators on the right-hand side of (4.17), (4.19) and (4.20), numerical schemes for solving these equations must be designed with special care (Mooiman, 1991). To demonstrate the improved characteristics of the modified Boussinesq equations, we examine one-dimensional waves ($v_o = 0$, $\partial/\partial y = 0$) over a constant depth. From (4.17) and (4.19), we obtain

$$\frac{\partial \eta}{\partial t} = -\frac{\partial}{\partial x}\left[F_{11}^*(dF_{11}^* u_o) + F_{21}^*(dF_{21}^* u_o)\right] \tag{4.21}$$

$$\frac{\partial u_o}{\partial t} = -\varepsilon\left[F_{11}^* u_o \frac{\partial}{\partial x}(F_{11}^* u_o) + F_{21}^* u_o \frac{\partial}{\partial x}(F_{21}^* u_o)\right] - \frac{\partial \eta}{\partial x} \tag{4.22}$$

where

$$F_{11}^* = -F_{21}^* = \frac{1}{\sqrt{2}}\frac{1}{1 - \frac{\mu^2 h^2}{6}\frac{\partial^2}{\partial x^2}} \, . \tag{4.23}$$

The linearized version of the above equations becomes

$$\left(1 - \frac{\mu^2 h^2}{6}\frac{\partial^2}{\partial x^2}\right)^2 \frac{\partial \eta}{\partial t} = -h\frac{\partial u_o}{\partial x} \tag{4.24}$$

$$\frac{\partial u_o}{\partial t} = -\frac{\partial \eta}{\partial x} \,. \tag{4.25}$$

For a small amplitude periodic wave given in (3.58) the dispersion relation can be determined from (4.24) and (4.25) as

$$c^2 = \frac{\omega_o^2}{k^2} = h\left(\frac{1}{1+\frac{1}{6}\mu^2k^2h^2}\right)^2 . \tag{4.26}$$

It is quite obvious that the quantity on the right-hand side is always positive. Therefore, the system is stable for all kh values. The ratio of the frequency calculated from (4.26) to that from Stokes' wave can be expressed as

$$\frac{\omega_o}{\omega_o^s} = \sqrt{\frac{kh}{\tanh kh}}\left(\frac{1}{1+\frac{1}{6}k^2h^2}\right) . \tag{4.27}$$

Following the same procedure presented in Subsec. 3.3, we derive the second order frequency ratio defined in (4.5) with the following parameters:

$$\begin{aligned}\omega_2^* = {} & \frac{D^*}{4}\left[2\left(1+\frac{2k^2h^2}{3}\right)^{-2} - \left(1+\frac{k^2h^2}{6}\right)^{-2}\right] \\ & + \frac{1}{4}\left(1+\frac{k^2h^2}{6}\right)^{-1}\left[B^*\left(1+\frac{k^2h^2}{6}\right)^{-1} + D^*\left(1+\frac{2k^2h^2}{3}\right)^{-1}\right]\end{aligned} \tag{4.28}$$

$$D^* = \frac{1}{4k^2h^2}\left(1+\frac{15}{36}k^2h^2\right)^{-1}\left(1+\frac{k^2h^2}{6}\right)(3+k^2h^2)\left(1+\frac{2k^2h^2}{3}\right) \tag{4.29}$$

$$B^* = \left(1+\frac{k^2h^2}{6}\right)\left(1+\frac{2k^2h^2}{3}\right)^{-1}\left[\frac{1}{2}+\left(1+\frac{2k^2h^2}{3}\right)^{-1}D^*\right] . \tag{4.30}$$

In Fig. 4 the first order frequency ratio (4.27) is plotted for different kh values. As a reference the frequency ratio derived from the original Hamiltonian ($\alpha = 0$) is also plotted. The improvement made by the modified Hamiltonian is rather significant. However, the behavior of the shallow water equation with the optimal α value is still slightly better than that of the modified Boussinesq equations in the deep water limit (see Fig. 1). The characteristics of the second order frequency for the modified Boussinesq equations are more or less the same as those of the original Boussinesq equations with $\alpha < -1/3$ (see Figs. 2 and 5).

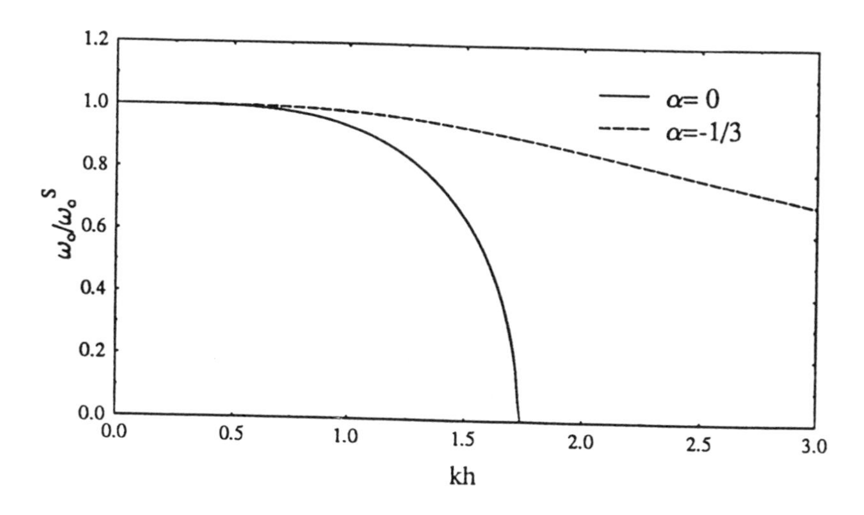

Fig. 4. First order frequency ratios between Stokes' wave theory and Boussinesq equation derived from the modified Hamiltonian.

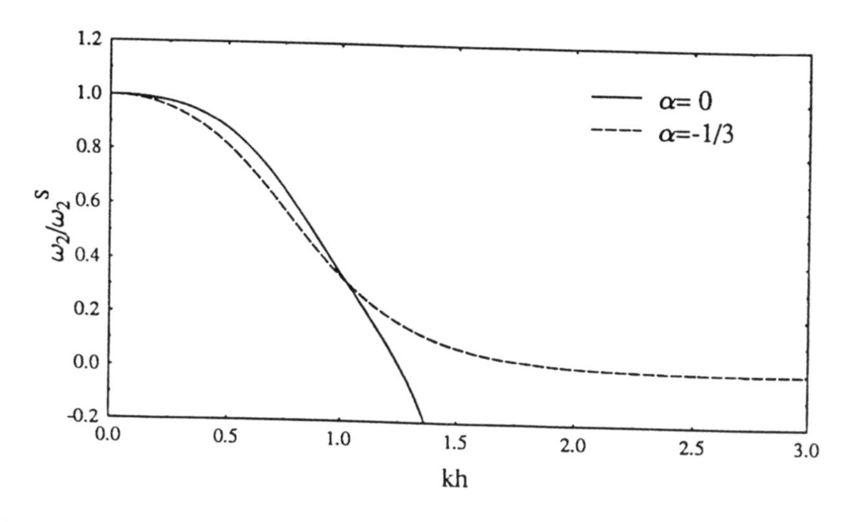

Fig. 5. Second order frequency ratios between Stokes' wave theory and the modified Boussinesq equations.

4.3. *Other approaches*

Madsen *et al.* (1991) took a different approach and derived a set of Boussinesq-type equations. They rewrote the conventional Boussinesq equations in terms of the depth-averaged velocity (3.43) and (3.45) in the following conservative form:

$$\frac{\partial \eta}{\partial t} + \frac{\partial P}{\partial x} + \frac{\partial Q}{\partial y} = 0 \tag{4.31}$$

$$\begin{aligned} &\frac{\partial P}{\partial t} + \varepsilon \frac{\partial}{\partial x}\left(\frac{P^2}{d}\right) + \varepsilon \frac{\partial}{\partial y}\left(\frac{PQ}{d}\right) + d\frac{\partial \eta}{\partial x} \\ &\quad - \frac{\mu^2}{3} h^2 \left(\frac{\partial^3 P}{\partial x^2 \partial t} + \frac{\partial^3 Q}{\partial x \partial y \partial t}\right) = 0 \end{aligned} \tag{4.32}$$

$$\begin{aligned} &\frac{\partial Q}{\partial t} + \varepsilon \frac{\partial}{\partial x}\left(\frac{PQ}{d}\right) + \varepsilon \frac{\partial}{\partial y}\left(\frac{Q^2}{d}\right) + d\frac{\partial \eta}{\partial y} \\ &\quad - \frac{\mu^2}{3} h^2 \left(\frac{\partial^3 P}{\partial x \partial y \partial t} + \frac{\partial^3 Q}{\partial y^2 \partial t}\right) = 0\,. \end{aligned} \tag{4.33}$$

where $d = \varepsilon\eta + h$ is the total depth and $P = \bar{u}(\varepsilon\eta + h)$ and $Q = \bar{v}(\varepsilon\eta + h)$ are the volume flux component in the x- and y-direction, respectively. From the leading order terms in the momentum equations we obtain

$$\frac{\partial^3 P}{\partial x^2 \partial t} + \frac{\partial^3 Q}{\partial x \partial y \partial t} + h\left(\frac{\partial^3 \eta}{\partial x^3} + \frac{\partial^3 \eta}{\partial x \partial y^2}\right) = 0(\varepsilon, \mu^2) \tag{4.34}$$

$$\frac{\partial^3 Q}{\partial y^2 \partial t} + \frac{\partial^3 P}{\partial x \partial y \partial t} + h\left(\frac{\partial^3 \eta}{\partial y^3} + \frac{\partial^3 \eta}{\partial x^2 \partial y}\right) = 0(\varepsilon, \mu^2)\,. \tag{4.35}$$

Madsen *et al.* (1991) argued that because the above quantities are in the same order of magnitude as the truncation errors in the Boussinesq equations, one can add a portion of these quantities into the Boussinesq equations without affecting the accuracy of the resulting equations. Hence, they multiplied (4.34) and (4.35) by $-\mu^2 B h^2$ and added them to the momentum equations (4.32) and (4.33) respectively. The resulting model equation becomes

$$\frac{\partial \eta}{\partial t} + \frac{\partial P}{\partial x} + \frac{\partial Q}{\partial y} = 0 \tag{4.36}$$

$$\frac{\partial P}{\partial t} + \varepsilon \frac{\partial}{\partial x}\left(\frac{P^2}{d}\right) + \varepsilon \frac{\partial}{\partial y}\left(\frac{PQ}{d}\right) + d\frac{\partial \eta}{\partial x} - \mu^2\left(B + \frac{1}{3}\right)h^2\left(\frac{\partial^3 P}{\partial x^2 \partial t} + \frac{\partial^3 Q}{\partial x \partial y \partial t}\right) - \mu^2 B h^3\left(\frac{\partial^3 \eta}{\partial x^3} + \frac{\partial^3 \eta}{\partial x \partial y^2}\right) = 0 \tag{4.37}$$

$$\frac{\partial Q}{\partial t} + \varepsilon \frac{\partial}{\partial x}\left(\frac{PQ}{d}\right) + \varepsilon \frac{\partial}{\partial y}\left(\frac{Q^2}{d}\right) + d\frac{\partial \eta}{\partial y} - \mu^2\left(B + \frac{1}{3}\right)h^2\left(\frac{\partial^3 Q}{\partial y^2 \partial t} + \frac{\partial^3 P}{\partial x \partial y \partial t}\right) - \mu^2 B h^3\left(\frac{\partial^3 \eta}{\partial y^3} + \frac{\partial^3 \eta}{\partial x^2 \partial y}\right) = 0 \tag{4.38}$$

where B is an empirical coefficient. The linear dispersion relation for the above equations can be written as

$$\frac{\omega_o}{\omega_o^s} = \sqrt{\frac{kh}{\tanh kh}}\left[\frac{1 + Bk^2h^2}{1 + (B + \frac{1}{3})k^2h^2}\right]^{1/2} . \tag{4.39}$$

Comparing (4.39) and (4.4), we find $B = -(\alpha + 1/3)$. Madsen *et al.* suggested that choosing the value $B = 1/21$ (or $\alpha = -8/21 = -0.381$) leads to phase velocity errors less than 3 % for the entire range $0 < h/\lambda_o < 0.75$ and to group velocity errors of less than 6% for the range $0 < h/\lambda_o < 0.55$, where λ_o is the wave length in deep water. The frequency ratio (4.39) is plotted in Fig. 1 and is very close to the curve created by using the optimal α value, -0.3855.

5. Concluding Remarks

Nonlinear shallow water equations are derived based on the assumption that the frequency parameter, μ, is small, while the nonlinearity parameter, ϵ, is an order one quantity. The Boussinesq equations become a subset of the nonlinear shallow water equations, i.e., $0(\varepsilon) \approx 0(\mu^2)$. It is shown that the Boussinesq equations can take on several different forms with the same order magnitude of accuracy. However, some of these equations are unstable in the range of short waves, which not only limits the applications of these Boussinesq-type equations to very shallow water, but also makes these equations vulnerable to any numerical disturbances. Several different methods for extending the shallow water equations to deep water are discussed.

Based on analyses for the constant depth, the linear dispersion characteristics of the Boussinesq-type equations can be matched with those of Stokes' waves in deep water by either using an appropriate velocity variable in the governing equations, or adding some of the higher derivative terms in the equations. A modified Hamiltonian has also been derived, which is always finite and positive. The associated modified Boussinesq equations also show significant improvements in the linear dispersion characteristics. However, all of these Boussinesq-type equations have very poor nonlinear dispersion characteristics in deep water. The nonlinear shallow water equations seem to give a better match with Stokes' wave theory in deep water. Nevertheless, the second order frequency (amplitude dispersion) is still significantly under-estimated.

The future research efforts should continue on identifying the most suitable model equation describing wave propagation from deep water to shallow water. The nonlinearity needs to be properly included. Moreover, the issue concerning the vertical structure of velocity field in the deep water should also be addressed.

Acknowledgement

The research reported here has been supported by a research grant from the U.S. Army Research Office (DAAL 03-92-G-0116) and by a research grant from the National Oceanic and Atmospheric Administration (NA90AA-D-SG078) to the Research Foundation of State University for the New York Sea Grant Institute.

References

Broer, L. J. F. (1974). On the Hamiltonian theory of surface waves. *Applied Scientific Research* **30**:430–446.

Broer, L. J. F., E. W. C. van Groessen and J. M. W. Timmers (1976). Stable model equations for long water waves. *Applied Scientific Research* **32**:609–636.

Chen, Y. and P. L.-F. Liu (1993). Scattering of linear and weakly nonlinear waves: Pseudospectral approaches. Center for Applied Research, University of Delaware, Research Report No. CACR-93-05.

Lin, C. C. and Clark A. (1959). On the theory of shallow water waves. *Tsing Hua J. of Chinese Studies, Special* **1**:54–62.

Liu, P. L.-F. (1990). Wave transformation, in *The Sea.* Vol. 9, eds. B. LeMehaute and D. M. Hanes. Wiley Interscience. 27–63.

Liu, P. L.-F. and Tsay, T. -K. (1983). On weak reflection of water waves, *J. Fluid Mech.* **131**:59–71.

Liu, P. L.-F., S. B. Yoon and J. T. Kirby (1985). Nonlinear refraction-diffraction of waves in shallow water. *J. Fluid Mech.* **153**:185–201.

Madsen, P. A., R. Murray and O. R. Sorensen (1991). A new form of the Boussinesq equations with improved linear dispersion characteristics. *Coastal Engineering* **15**:371–388.

McCowan, A. D. (1987). The range of application of Boussinesq-type numerical short wave models. *Proc. XXII IAHR Congress.* Lausanne, Switzerland, 379–384.

McCowan, A. D. and D. R. Blackman (1989). The extension of Boussinesq-type equations to modelling short waves in deep water, *Proc. 9th Australian Conf. Coastal and Ocean Engineering.* Adelaide, Australia. 412–416.

Mei, C. C. (1989). *The Applied Dynamics of Ocean Surface Waves.* World Scientific Publishing.

Mei, C. C. and P. L.-F. Liu (1993). Surface waves and coastal dynamics. *Annu. Rev. Fluid Mech.* **25**:215–240.

Miles, J. W. (1977). On Hamilton's principle for surface waves, *J. Fluid Mech.* **83**:153–158.

Mooiman, J. (1991). Boussinesq equations based on a positive definite Hamiltonian. Delft Hydraulics Report Z294.

Murray, R. J. (1989). Short waves modelling using new equations of Boussinesq type, *Proc. 9th Australian Conf. Coastal and Ocean Engineering.* Adelaide, Australia. 331–336.

Nwogu, O. (1993). Alternative form of Boussinesq equations for nearshore wave propagation. *J. Wtrwy., Port, Coast. and Oc. Eng. ASCE* **119**:618–638.

Peregrine, D. H. (1967). Long waves on a beach. *J. Fluid Mech.* **27**:815–827.

Peregrine, D. H. (1972). Equations for water waves and the approximations behind them, in *Waves on Beaches and Resulting Sediment Transport* (ed. Meyer, R.E.) Academic Press. 95–121.

Whitham, G. B. (1974) *Linear and Nonlinear Waves.* Wiley-Interscience.

Witting, J. M. (1984). A unified model for the evolution of nonlinear water waves. *J. Computational Physics* **56**:203–236.

Yoon, S. B. and P. L.-F. Liu (1993). A note on Hamiltonian for long water waves in varying depth, in *Wave Motion* (to appear).

Zakharov, V. E. (1968). Stability of periodic waves of finite amplitude on the surface of a deep fluid. *J. Appl. Mech. Tech. Phys.* **9**(2):190–194.

CROSS-SHORE SEDIMENT TRANSPORT PROCESSES

ROBERT G. DEAN

Cross-shore sediment transport is relevant to several coastal engineering problems, yet this process is poorly understood, in part due to the relatively short time that it has been under study. Applications range from the rapid seaward transport associated with beach and dune erosion due to episodic storm events to the extremely slow response to relative sea level rise. At any time and location in the nearshore, there are active forces which tend to displace sediment seaward and landward. One of these forces, gravity, can be quantified readily; however, the others are known only qualitatively. If the forces were held constant, the profile and cross-shore sediment transport would approach equilibrium and zero, respectively. Although equilibrium may never be achieved in nature due to the time varying forcing, a review of concepts and methodology is presented since they can be applied effectively to many coastal engineering problems. There have been accelerated efforts over the last decade to model profiles which are out of equilibrium. Two types of dynamic models are defined as "closed loop", which converge to a target profile, and "open loop", which do not necessarily approach an equilibrium. The formulation of these models includes the sediment continuity equation and a transport equation. The transport equation for the closed loop models is chosen to ensure equilibration for fixed forcing conditions, whereas various representations have been proposed for the open loop models including formulations of the bed load and/or suspended load transport dynamics. Based on published results and considerations of the dynamics, available models of the two types are examined and evaluated. Of the three published closed loop models examined, only one (SBEACH) can represent a bar. The model's performance, based on available comparisons with laboratory and field data, is considered reasonably good. However, the transport direction is independent of local beach slope, an effect that would disqualify the model from representing profile equilibration following a nourishment event, and possibly other applications. Also, because the model calibration is based on laboratory data using monochromatic waves, the bar feature is more accentuated than in nature. Previously published results including those from six open loop models tested against the same data sets are presented and reviewed. Wide ranges of behavior are exhibited by the various models with the transport differing by more than a factor of four and some transport predictions displaying instability effects. The tendency for a pronounced bar under the action of regular waves and a subdued bar for irregular waves was well predicted by most of the open loop models. It is not known whether or not these models are stable for long run times. Because of the uncertainties in convergence and transport magnitudes associated with the open loop models, it is concluded that closed loop models are more appropriate for engineering applications and that open loop models remain in the research arena.

1. Introduction

1.1. *Overview/Purpose*

Sediment transport at a point in the nearshore zone is a vector with both longshore and cross-shore components (see Fig. 1). It appears that under a number of coastal engineering scenarios of considerable interest, the transport is dominated by either the longshore or cross-shore component and this, in part, has led to a history of separate investigative efforts for each of these two components. The subject of total *longshore* sediment transport has been studied for approximately five decades. There is still considerable uncertainty regarding certain aspects of this transport component including the effects of grain size, barred profiles and the cross-shore *distribution* of the longshore transport; however, for many situations of interest, predictability of total longshore sediment transport is probably on the order of ±50%. A focus on *cross-shore* sediment transport is relatively recent, having commenced approximately one decade ago and uncertainty in prediction capabilities including the effects of all variables may be considerably greater. In some cases the limitations on prediction accuracy of both components may be due to a lack of knowledge of wave characteristics rather than an inadequate understanding of transport processes.

Cross-shore sediment transport encompasses both offshore transport such as those occuring during storms, and onshore transport which dominates during mild wave activity. Transport in these two directions appears to occur in significantly distinct modes and with markedly disparate time scales; as a result, the difficulties in predictive capabilities differ substantially. Offshore transport is the simpler of the two and tends to occur with greater rapidity and as a more regular process with transport more or less in phase over the entire active profile. This is fortunate since there is a considerably greater engineering relevance and interest in offshore transport due to the potential for damage to structures and loss of land. Onshore sediment transport often occurs in "wave-like" motions referred to as "ridge and runnel" systems in which individual packets of sand sequentially move toward, merge with and widen the dry beach. A complete understanding of cross-shore sediment transport is complicated by the contributions of both bed and suspended load components.

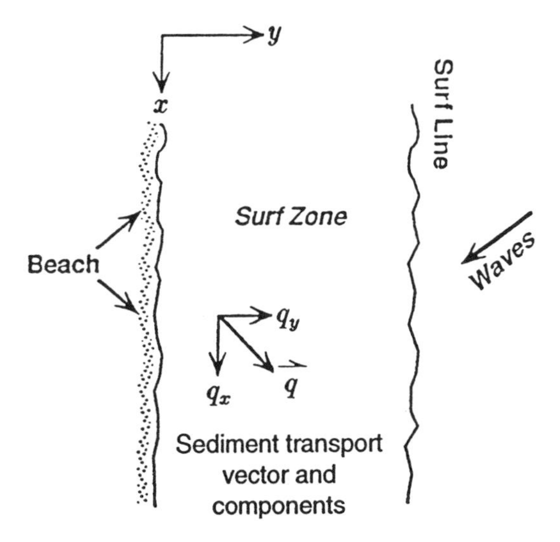

Fig. 1. Longshore (q_x) and cross-shore (q_y) sediment transport components.

The partitioning between the two components depends, in a presently unknown way, on grain size, local wave energy and other variables.

Cross-shore coastal sediment transport is relevant to a number of coastal engineering problems including: a) beach and dune response to storms, b) the equilibration of a beach nourishment project that is placed on slopes steeper than equilibrium, c) so-called "profile nourishment" in which the sand is placed in the nearshore with the expectation that it will move landward nourishing the beach (this involves the more difficult problem of onshore transport), d) shoreline response to sea level rise, e) seasonal changes of shoreline positions which can amount to 30 to 40 meters, f) overwash, the process of landward transport due to overtopping of the normal land mass due to high tides and waves, g) scour immediately seaward of shore parallel structures, and h) the three-dimensional flow of sand around coastal structures in which the steeper and milder slopes on the updrift and downdrift sides induce seaward and landward components respectively. These problems are schematized in Fig. 2.

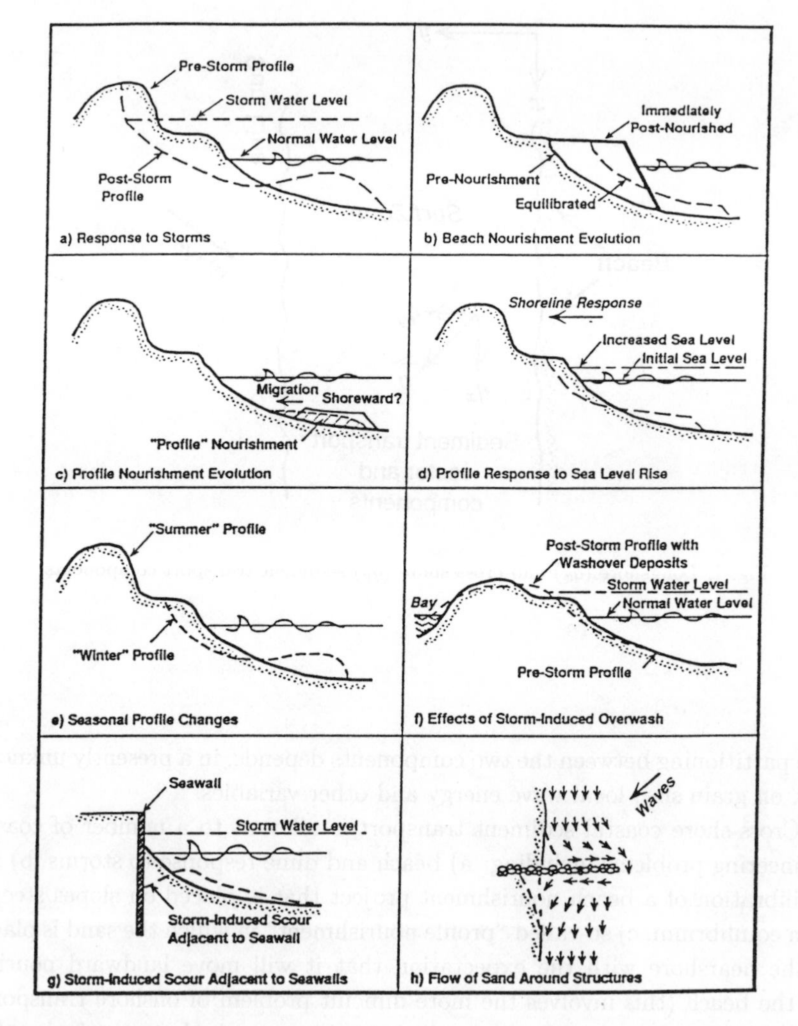

Fig. 2. Problems and processes in which cross-shore sediment transport is relevant.

1.2. *Scope*

This review consists of the following sections. Section 2, describing characteristics of natural and engineered nearshore systems, commences with a qualitative description of the forces acting within the nearshore zone, the characteristics of an equilibrium beach profile, and a discussion of conditions of equilibrium when the forces are balanced and the ensuing sediment transport when conditions change causing an imbalance. Survey capabilities to quantify profiles are reviewed. Section 3 addresses quantitative aspects of equilibrium beach profiles, the governing equations for profile evolution and several cross-shore transport formulations. Section 4 examines published results from several models and provides an evaluation of their capabilities. The summary and conclusions are provided in Sec. 5 and references are cited at the end.

2. Characteristics of the System

2.1. *Forces acting in the nearshore*

There are identifiable forces that occur within the nearshore active zone which includes and extends beyond the surf zone. The magnitudes of these forces can be markedly different inside and seaward of the surf zone. Under equilibrium conditions, these forces are by definition in overall balance and although there is motion of individual sand grains under even low wave activity, the profile remains more or less constant. Cross-shore sediment transport occurs when hydrodynamic conditions within the nearshore zone change, thereby modifying one or more of the forces resulting in an imbalance leading to transport gradients and profile change. Because cross-shore transport occurs when an imbalance occurs, it is essential herein to examine the case of equilibrium. Established terminology refers to onshore- and offshore-directed forces as "constructive" and "destructive" respectively. A brief review of these two types of forces is presented below; however, as will be noted, the term "forces" is used in the generic sense. Moreover, it will be evident that some forces could be constructive under certain conditions and destructive under others.

As noted, constructive forces are those that tend to cause onshore sediment transport. For classic nonlinear wave theories (Stokes, Cnoidal, Solitary, Stream Function, etc.), the wave crests are higher and of less duration than the troughs. This feature is most pronounced just outside the breaking point and also applies to the water particle velocities. For oscillatory water particle velocities expressed as a sum of phase-locked sinusoids such as for the Stokes or Stream Function wave theories, even though the time mean of the water particle velocity is zero, the average of the bottom shear stress, τ_b, expressed as

$$\overline{\tau_b} = \rho \frac{f}{8} \overline{|v_b| v_b} \tag{1}$$

can be shown to be directed onshore. In the above, ρ is the mass density of water, f is the Darcy–Weisbach friction coefficient which, for purposes here is considered constant over a wave period, and v_b is the instantaneous wave-induced bottom water particle velocity. A definition sketch is provided in Fig. 3. An example of the time-varying shear stress due to a near-breaking nonlinear (Stream Function) wave is shown in Fig. 4. Dean (1987a) has developed the average bottom shear stress based on the Stream Function wave theory and presented the results in the nondimensional form shown in Fig. 5.

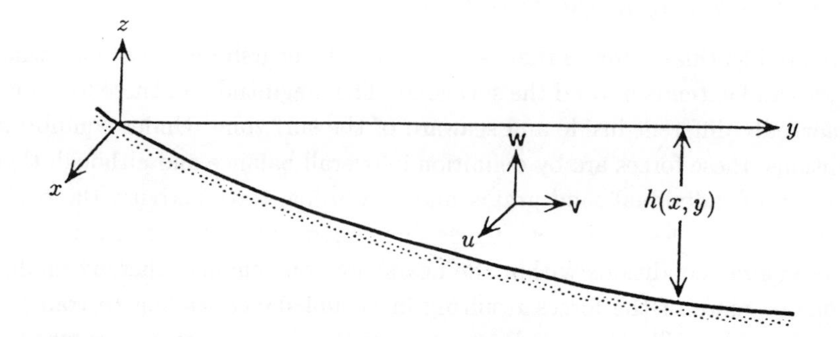

Fig. 3. Definition sketch.

A second constructive force originates within the bottom boundary layer causing a net mean velocity in the direction of propagating water waves. This streaming motion was first observed in the laboratory by Bagnold (1940) and has been quantified by Longuet–Higgins (1953) as due to the local transfer of momentum associated with friction–induced energy losses within the bottom boundary layer. For the case of laminar flows, the maximum value of this steady velocity, v_s, is surprisingly independent of the value of the viscosity and is given by (Phillips, 1977)

$$v_s = -\frac{3\sigma k H^2}{16 \sinh^2 kh} \tag{2}$$

which for the case of shallow water and a wave height proportional to the breaking depth will be shown to be 1.5 times the *average* of the return flow due to mass transport. In Eq. (2), σ is the wave angular frequency, k is the wave number and H, the wave height. Although the maximum velocity

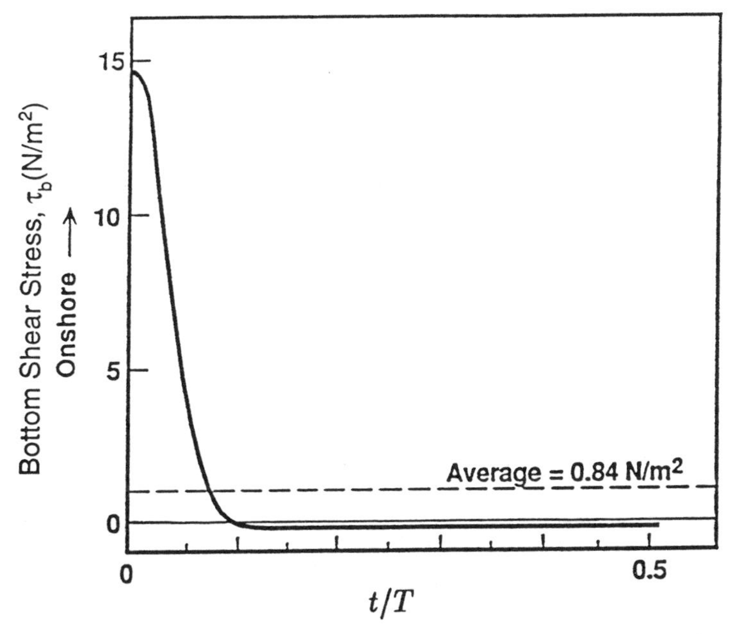

Fig. 4. Variation with time of the bottom shear stress under a breaking nonlinear wave. $H = 0.78$ m, $h = 1.0$ m, $T = 8.0$ s, and $D = 0.2$ mm, $f = 0.08$, critical shear stress $= 0.1\ N/\text{m}^2$.

is independent of the viscosity, the bottom shear stress, τ_{bs}, induced by the streaming velocity is not and is given by

$$\tau_{bs} = -\frac{\rho \varepsilon^{1/2} \sigma^{3/2} H^2 k}{8\sqrt{2}\sinh^2 kh} \tag{3}$$

in which ε is the eddy viscosity.

Within the surf zone, cross-shore transport may be predominantly due to sediment in suspension. If the suspension is intermittent, occurring in each wave period, the average water particle velocity during the time that the sediment particle is suspended determines the direction of cross-shore transport. Although this mode of sediment transport is not a true force, it does represent a mechanism which contributes to cross-shore sediment transport.

Turbulence, although also not a true force can be effective in mobilizing sediment and, depending on whether the net forces are shoreward or seaward at the time of mobilization, can be constructive or destructive respectively.

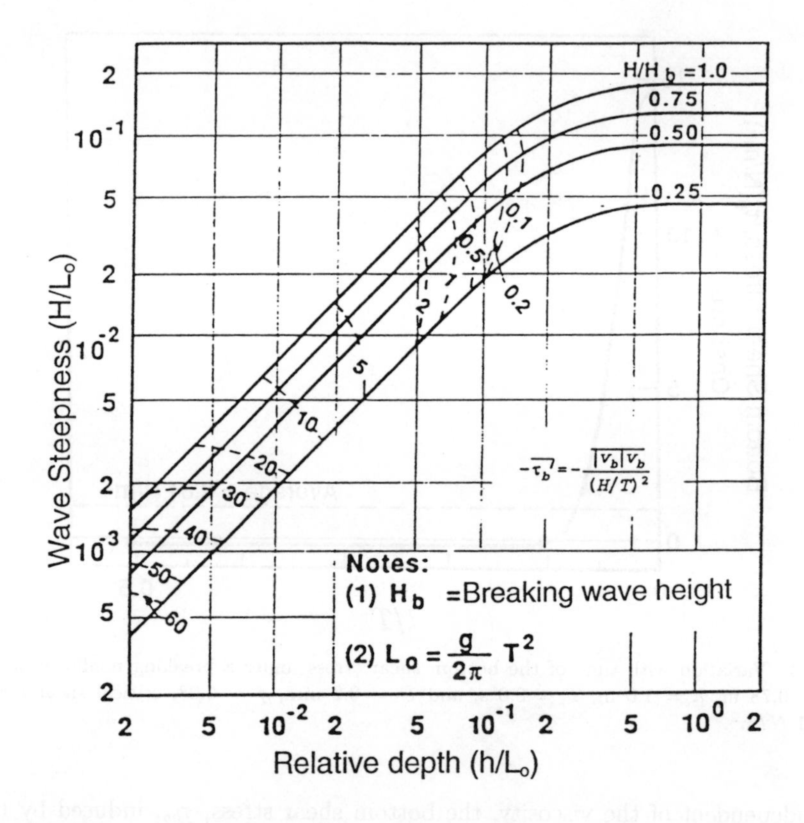

Fig. 5. Isolines of nondimensional average bottom shear stress, $\overline{\tau_b}$, versus relative depth and wave steepness. The shear stress is directed onshore. (Dean, 1987a)

Gravity, the most obvious destructive force, acts downslope which is generally seaward for a monotonic profile. However, for the case of a barred profile, gravity can act in the shoreward direction over portions of the profile. It will be shown later that gravity tends to "smooth" any irregularities that occur in the profile. It is enlightening to note that if gravity were the only force acting, the only possible equilibrium profile would be horizontal and sandy beaches as we know them would not exist.

The seaward return of wave mass transport induces a seaward stress on the bottom sediment particles. If the return flow due to mass transport were distributed uniformly over the water depth, based on linear shallow water wave

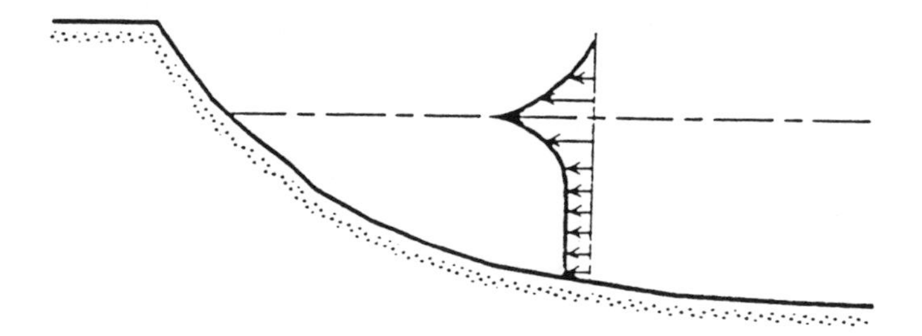

Fig. 6. Distribution over depth of the onshore flux of the onshore component of momentum.

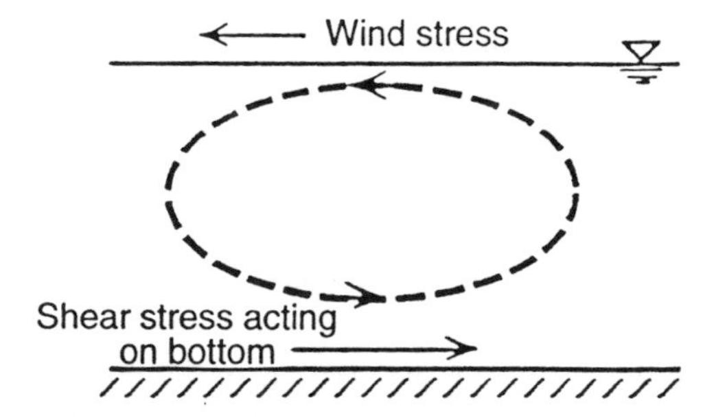

Fig. 7. Bottom stresses caused by surface winds.

theory, the return velocity is

$$\overline{V} = \frac{\sqrt{g}H^2}{8h^{3/2}} \tag{4}$$

or if, within the surf zone, the wave height is considered to be proportional to the local depth, h ($H = \kappa h$, and $\kappa \approx 0.78$)

$$\overline{V} = \frac{\kappa^2}{8}\sqrt{gh} \approx 0.08\sqrt{gh} \tag{5}$$

which as discussed earlier is two-thirds of the maximum streaming velocity and it is recognized that $\sqrt{gh}$ is the wave celerity in shallow water. In storm events in which there is overtopping of the barrier island, a portion or all of the potential return flow due to mass transport can be relieved through strong landward flows resulting in constructive forces.

It is well known that associated with wave propagation toward shore is a shoreward flux of linear momentum (Longuet–Higgins and Stewart, 1964). When waves break, the momentum is transferred to the water column resulting in a shoreward-directed thrust and thus a wave-induced setup within the surf zone, the gradient of which is proportional to the local bottom slope. This momentum is distributed over depth as shown in Fig. 6. In shallow water, linear water wave theory predicts that one-third of the momentum flux originates between the trough and crest levels and has its centroid at the mean water level and that the remaining two-thirds which originates between the bottom and the mean water level, is uniformly distributed over this dimension and thus has its centroid at the mid-depth of the water column. Because of the contribution within the free surface region, breaking waves induce an equivalent shear force on the water surface, thus causing a seaward-directed current termed "undertow" and an associated bottom shear stress within the breaking zone. The bottom shear stress, which will be quantified later, is dependent on the rate of energy dissipation and the relationship between the velocity gradient and shear stress. This effective surface shear force due to momentum transfer must be balanced by the bottom shear stress and the pressure forces due to the slope of the water surface.

Often during major storm events, strong onshore winds will be present in the vicinity of the shoreline. These winds cause a shoreward-directed surface flow and a seaward-directed bottom flow as shown in Fig. 7. Of course *seaward*-directed winds would cause shoreward-directed bottom velocities and thus constructive forces.

Considering linear wave theory and a linear shear stress relationship, with uniform and constant eddy viscosity, ε, the distribution of the mean velocity over depth without including the contribution of bottom streaming can be shown to be

$$\begin{aligned} v(z) = \frac{h}{\rho\varepsilon}\left[2\tau_\eta - \frac{\partial E}{\partial y}\right]\left[\frac{3}{8}\left(\frac{z}{h}\right)^2 + \frac{1}{2}\left(\frac{z}{h}\right) + \frac{1}{8}\right] \\ + \frac{3}{2}\frac{Q}{h}\left[1 - \left(\frac{z}{h}\right)^2\right] \end{aligned} \tag{6}$$

in which the term, $\frac{\partial E}{\partial y}$, reflects the effect of wave breaking and is zero outside the surf zone. The term Q represents return flow due to mass transport or other causes. For linear waves, the return flow due to mass transport is

$$Q = \frac{E}{\rho C} \tag{7}$$

in which C is the wave celerity. The associated wave setup gradient within the surf zone is

$$\frac{\partial \eta}{\partial y} = -\frac{3\varepsilon Q}{gh^3} + \frac{1}{\rho g h}\left(\frac{3}{2}\tau_\eta - \frac{7}{4}\frac{\partial E}{\partial y}\right) \tag{8}$$

where η is the mean water level.

The shear stress distribution associated with the velocity distribution in Eq. (6) is

$$\tau = \rho\varepsilon\frac{\partial v}{\partial z} = \left[2\tau_\eta - \frac{\partial E}{\partial y}\right]\left[\frac{3}{4}\frac{z}{h} + \frac{1}{2}\right] - 3\rho\varepsilon Q\frac{z}{h^2} \,. \tag{9}$$

Velocity and shear stress distributions inside and outside the surf zone are shown in Fig. 8 for no surface wind stress and cases of no overtopping and full overtopping. Reasonable profile conditions have been assumed for this example. The effects on the velocity profile of momentum transfer by breaking waves within the surf zone are evident.

Table 1. Constructive and Destructive Cross-Shore "Forces".

Constructive or Destructive	Description of force	Magnitude of force (N/m^2)	
		Inside Surf Zone	Outside Surf Zone
Constructive	Average bottom shear stress due to nonlinear waves (a)	0.84	0.84
	Streaming velocities (b)	1.19	1.19
	Overtopping	28.6	28.6
Destructive	Gravity (c)	0.046	0.046
	Undertow due to mass transport	28.6	28.6
	Undertow due to momentum flux transfer	7.9	0
Constructive or Destructive	Intermittent suspension	?	?
	Turbulence	?	?
	Wind effects (d)	0.47	0.47

Note: For the calculations resulting in the values in this table: $H = 0.78$ m, $h = 1.0$ m, $T = 8$ s. (a) $f = 0.08$, (b) $\varepsilon = 0.04$ m^2/s, (c) $D = 0.2$ mm, (d) Wind speed = 20 m/s.

Table 1 summarizes the forces identified as contributing to constructive or destructive forces or a combination of the two, and provides a quantitative estimate of their magnitudes for the following fairly representative conditions: $h = 1$ m, $H = 0.78$ m, $T = 8$ s, $D = 0.2$ mm, $\varepsilon = 0.04$ m^2/s, wind speed $= 20$ m/s. It is seen that of the bottom stresses that can be quantified, those associated with undertow due to mass transport and momentum flux transfer are dominant.

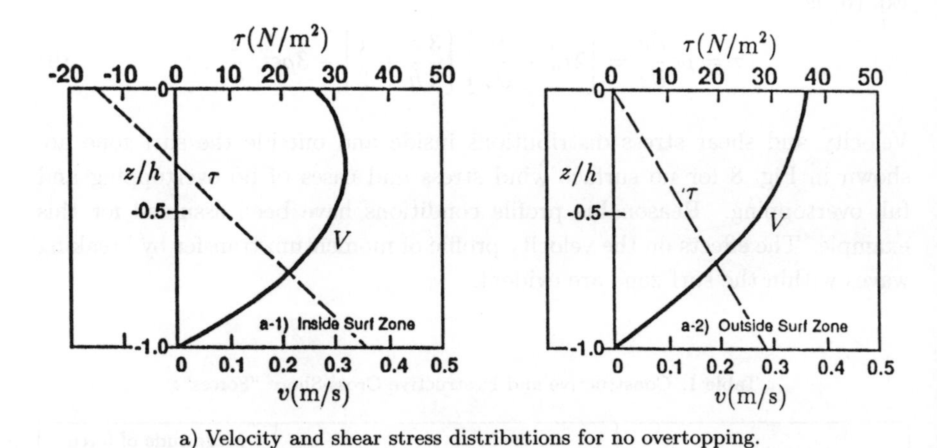

a) Velocity and shear stress distributions for no overtopping.

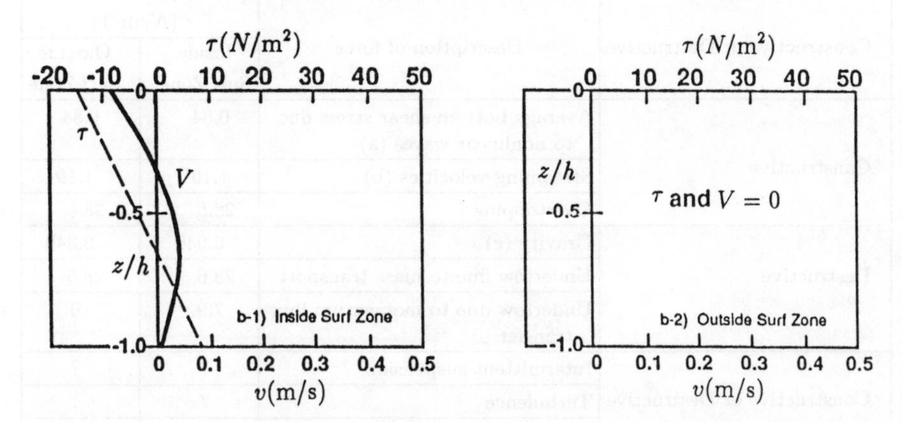

b) Velocity and shear stress distributions for overtopping such that no mass transport return flow exits.

Fig. 8. Velocity and shear stress distributions inside and outside surf zone for cases of no overtopping. (case a) and complete overtopping of mass transport (case b). $H = 0.78$ m, $h = 1$ m, $T = 8$ s, $D = 0.2$ mm, $\varepsilon = 0.04$ m^2/s.

2.2. *Equilibrium beach profile characteristics*

In consideration of cross-shore sediment transport, it is useful to first examine the case of equilibrium in which there is no net cross-shore sediment transport. The competing forces elucidated in the previous section can be fairly substantial, exerting tendencies for both onshore and offshore transport. A change will bring about a disequilibrium that causes cross-shore sediment transport. The concept of an equilibrium beach profile has been criticized, since in nature the forces affecting equilibrium are always changing with the varying tides, waves, currents and winds. Although this is true, the concept of an equilibrium profile is still one of the coastal engineer's most valuable tools in providing a framework to consider disequilibrium and thus cross-shore sediment transport. Also, many useful and powerful conceptual and design relationships are based on profiles of equilibrium. In this section, we first review the general properties of an equilibrium beach profile and the changes that cause an imbalance resulting in a transport in either the seaward or landward direction.

Generally, observed properties of equilibrium profiles are as follows:

1. they tend to be concave upwards,
2. the slopes are milder when composed of finer sediments,
3. the slopes tend to be flatter for steeper waves, and
4. the sediments tend to be sorted with coarser and finer sediments residing in shallower and deeper waters respectively.

The effects of changes that induce cross-shore sediment transport can be deduced from these known general characteristics. For example, an increase in water level will cause a disequilibrium: Due to the concave-upward nature of the profile, the depth at a particular reference distance from the new shoreline is now greater than it was before the increase. If the equilibrium profile had been planar, then a change in water level would not change the depth at a distance from the new shoreline and there would be no disequilibrium. Thus, this disequilibrium is a result of the concave-upward nature of the profile. It will be shown that without the introduction of additional sediment into the system, the only way in which the profile can reattain equilibrium is to recede, thus yielding sediment to fill the bottom to a depth consistent with the equilibrium profile. Also, since profiles are generally flatter for steeper waves, an increase or decrease in wave steepness will induce seaward or landward sediment flows respectively. Naturally, onshore and offshore winds will cause seaward and landward sediment transport respectively.

As an example of the shoreline response to storms, Fig. 9 presents results from Katoh and Yanagishima (1988) in which the offshore waves, shoreline position and beach face slope were measured over a period of approximately seven months. It is seen that the shoreline retreats abruptly during the higher wave events and advances more gradually during periods of milder wave activity. The beach slope and shoreline changes are of course correlated with the slope becoming milder during periods of shoreline retreat. Also of interest was the finding that the rate of shoreline advancement during the recovery phase was almost constant at 0.68 m/day.

Many beaches in nature have one or more longshore bars present. At some locations, these bars are seasonal and at some, they are more or less permanent. Figure 10a presents a profile from Chesapeake Bay in which at least six bars are evident and Fig. 10b shows profiles measured in a monitoring program to document the evolution of a beach nourishment project at Perdido Key, Florida. This project included both beach and profile nourishment. As seen from Fig. 10b, a bar was present before nourishment and gradually reformed as the nourished profile equilibrated during the two-year period shown in Fig. 10b. It will be shown later that the presence of bars depends on wave and sediment conditions and at a particular beach, the bars tend to form or move farther seaward during storms. It appears that the outer bars on some profiles are relict and may have been caused by a previous large storm which deposited the sand in water too deep for fair weather conditions to return the sand to shore. At some beaches with more than one bar, the inner bar will exhibit more rapid response to changing wave conditions than those farther offshore. Figure 11 presents results from Duck, North Carolina in which surveys for two profiles were conducted over a period of approximately eleven years. It is seen that at this location both the inner and outer bars undergo substantial changes whereas the shoreline remains relatively fixed, possibly due to coarser sediment in shallow water and at the shoreline.

Many studies have been carried out to observe and characterize the forms of beach profiles in nature and as produced in the laboratory. It is generally tacitly assumed that these profiles, on average, represent an equilibrium form. Equilibrium profiles will be discussed quantitatively later.

2.3. *Interaction of structures with cross-shore sediment transport*

The structure which interacts most frequently with cross-shore sediment transport is a shore parallel structure such as a seawall or revetment. During storm

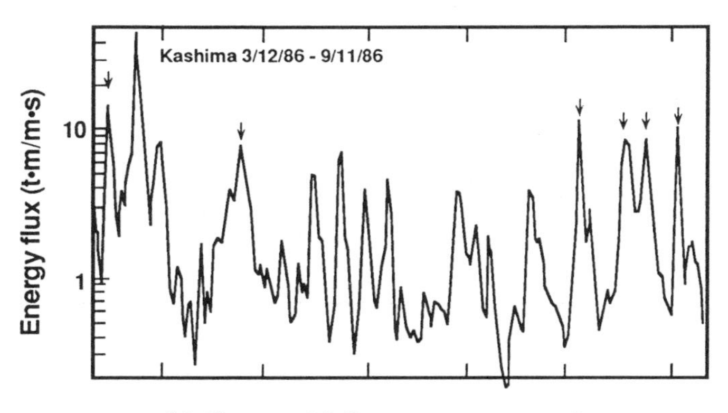

(a) Changes of daily mean wave energy flux.

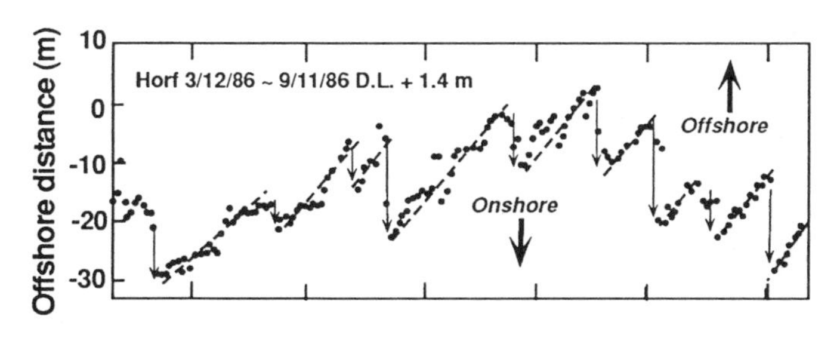

(b) On-offshore changes of shoreline position.

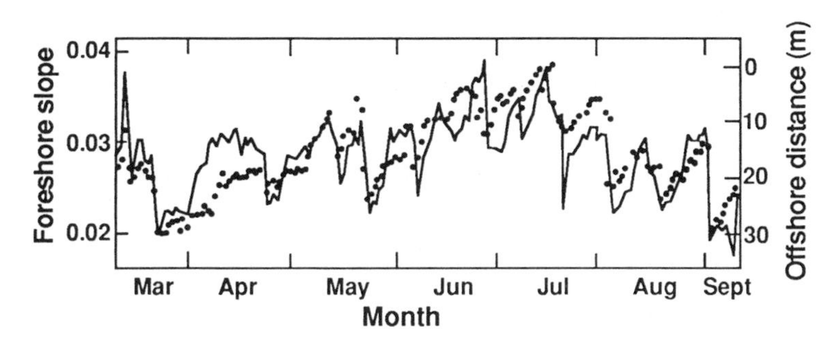

(c) Changes of foreshore beach slope.

Fig. 9. Effects of varying wave energy flux (a) on (b) shoreline position and (c) foreshore beach slope. (Katoh and Yanagishima, 1988)

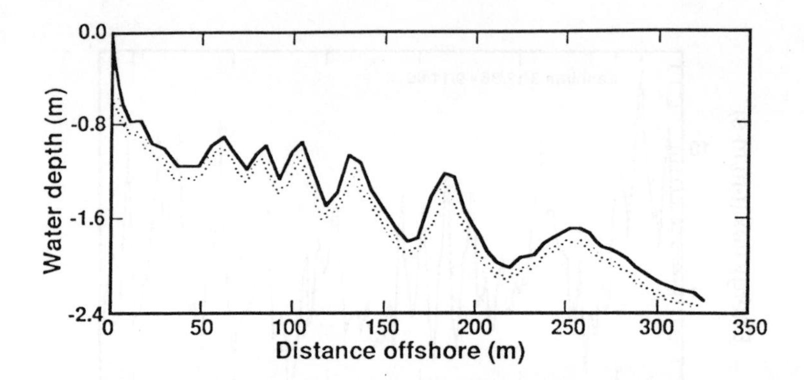

(a) Multiple-barred profile from Chesapeake Bay. (Dolan and Dean (1985)

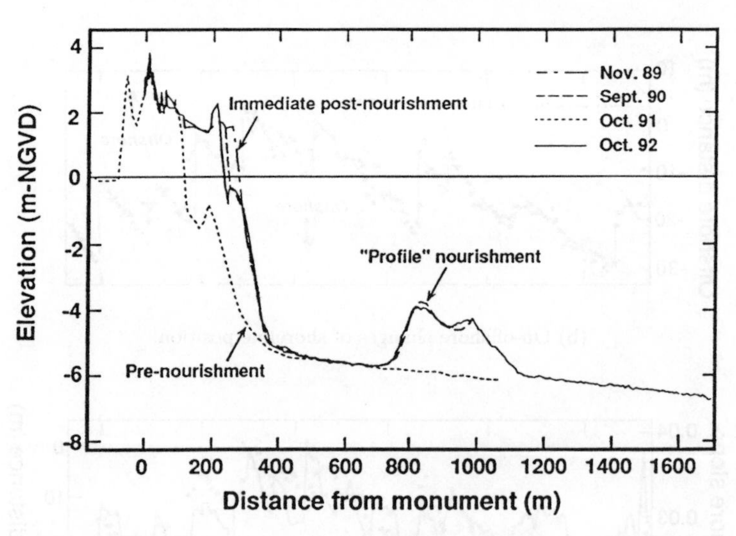

(b) Profiles from monitoring of a beach and profile nourishment project at Perdido Key, Florida.

Fig. 10. Examples of two offshore per profiles.

events, a characteristic profile fronting a shore parallel structure is one with a trough at its base as shown in Fig. 12, from Kriebel (1987) for a profile affected by Hurricane Elena in Pinellas County, Florida, in September 1985.

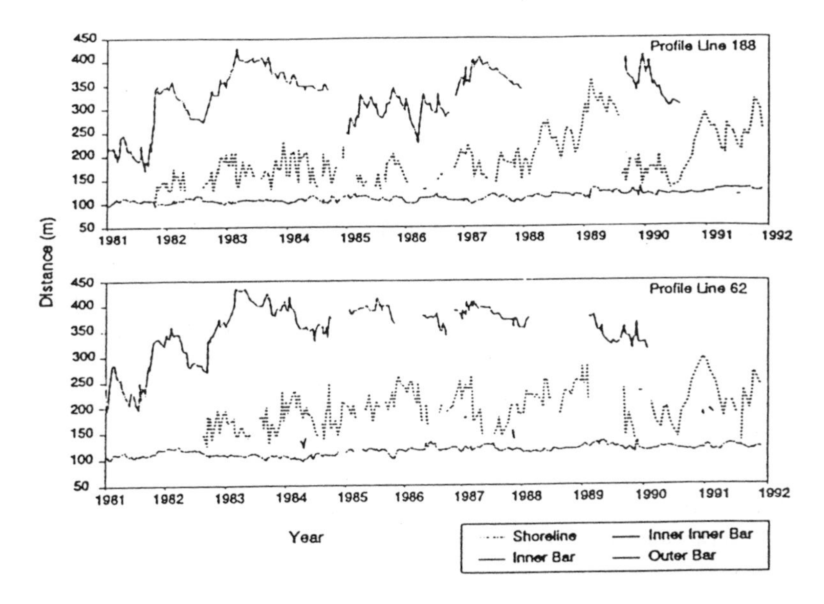

Fig. 11. Variation in shoreline and bar crest positions. Duck, North Carolina. (Lee and Birkemeier, 1993)

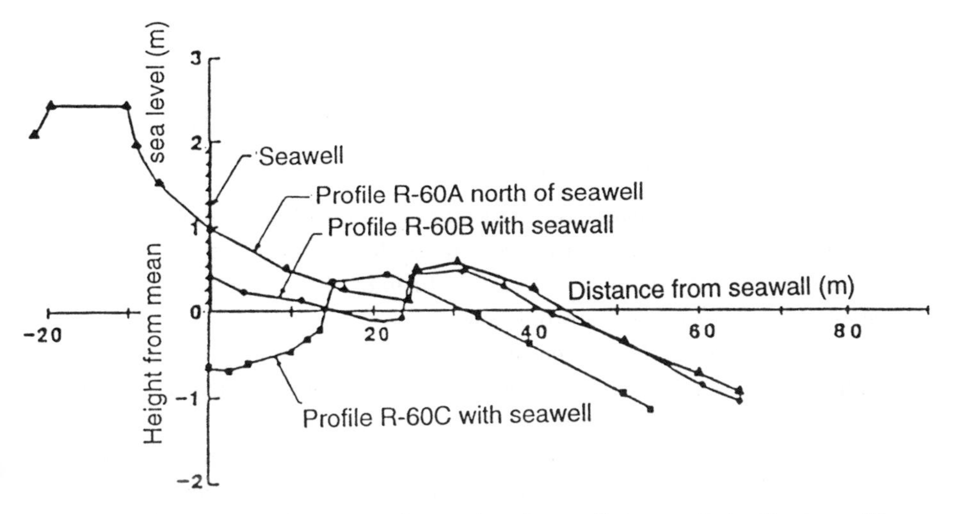

Fig. 12. Comparison of response of natural and seawalled seawalls to Hurricane Elena, September 1985. (Kriebel *et al.*, 1986)

This trough is due to large cross-shore transport gradients immediately seaward of the structure. Although the *hydrodynamic* cause of this scour is not well known, it has been suggested that it is due to a standing wave system with an antinode at the structure. A second more heuristic explanation is that the scour is the result of the seawall preventing removal of the sand from behind the seawall and the transport system removes the sand from as near as possible to the area from which removal would normally occur. Barnett and Wang (1988) have reported on a model study to evaluate the interaction of a seawall with the profile and have found that the additional volume represented by the scour trough is approximately 62% of that which would have been removed landward of the seawall if it had not been present. During mild wave activity, it appears that the profile recovers nearly as it would if the seawall had not been present. The reader is referred to the comprehensive review by Kraus (1987) for additional information on shore parallel structures and their effects on the shoreline.

The interaction of shore perpendicular structures with longshore sediment transport induces cross-shore sediment transport components. Figure 2h shows the profile steepening and flattening on the updrift and downdrift sides respectively of a groin. In this case, seaward transport on the updrift side is assisted by a wave driven seaward flowing current induced by the presence of the groin. Three-dimensional models would require a quantitative knowledge of the mechanics of this transport in order to represent adequately the effects of such structures.

2.4. *Methods of measuring and representing beach profiles*

a. *Methods of profile measurements*

Changes in beach and nearshore profiles are a result of cross-shore and longshore sediment transport. If the longshore gradients in the longshore component can be assumed small, it is possible to infer the volumetric cross-shore transport from two successive profile surveys.

Historically, profiles have been measured using land survey techniques out to wading depths and a boat mounted fathometer for greater depths. Analysis of data from a fathometer requires a water level as a datum. Areas of high tidal range and waves can introduce significant errors into the data base. Accuracy of this type of profile data varies, but is believed to be on the order of ± 0.15 m. More recently, sea sleds or wheel mounted platforms have been used, extending the land surveying approach to depths not otherwise possible, thus reducing the error to approximately ±0.02 m. The reader is referred to Clausner *et al.* (1986) for a review of four profile survey methods and their accuracies.

b. *Descriptive approaches to equilibrium beach profile representation*

The empirical orthogonal function (EOF) method provides a descriptive and quantitative approach to represent both the average beach profile and the predominant variations about the average. The procedure requires as input, a data set consisting of either repetitive beach profiles at one location or a number of profiles from different locations at one time. The case of a number of repetitive profiles at one location is more usual and will serve as the basis for discussion here.

The EOF method decomposes the profile variation into a mean profile and the minimum number of deviations about the mean to describe all measurements most efficiently. An analogy is the vibration of a string suspended between two points. In this case, the natural modes (shapes) can be obtained from the differential equation and the boundary conditions. In the case of beach profile variations, the natural shapes of the various components cannot be obtained *a priori*, i.e., based on physics; however, the EOF method allows the modes to be identified approximately from the data set. Figure 13, from Aubrey (1979), illustrates this procedure graphically.

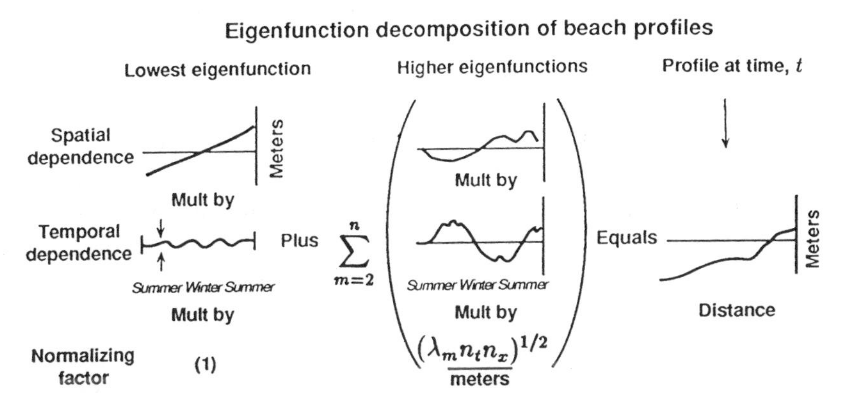

Fig. 13. Schematic of decomposition of beach profile data using empirical orthogonal function approach. (Aubrey, 1979)

3. Engineering Aspects of Cross-Shore Sediment Transport

3.1. *Introduction*

The preceding section has discussed the natural characteristics of beach profiles in equilibrium, the effects that cause disequilibrium and the associated profile changes. This section addresses engineering aspects of profiles and cross-shore sediment transport including the models that have been developed for representing cross-shore transport and profile evolution.

3.2. *Quantitative description of equilibrium beach profiles*

Various models have been proposed for representing equilibrium beach profiles (EBP). Some of these models are based on examination of the geometric characteristics of profiles in nature and some attempt to represent in a gross manner the forces active in shaping the profile. One approach utilized is to recognize the presence of the constructive forces and to hypothesize the dominance of various destructive forces. This approach can lead to simple algebraic forms for the profiles which then can be tested against profile data. Dean (1977) has examined the forms of the EBPs that would result if the dominant destructive forces were

1. wave energy dissipation per unit water volume,
2. wave energy dissipation per unit surface area, and
3. uniform average longshore shear stress across the surf zone.

It was found for all three of these destructive forces, that by using linear wave theory, the EBP could be represented by the following simple algebraic form

$$h(y) = A(D)y^m \tag{10}$$

in which $A(D)$ signifies that the so-called sediment scale parameter, A, depends on the sediment size, D. This form with an exponent, m, equal to two-thirds had been found earlier by Bruun (1954) based on an examination of beach profiles in Denmark and in Monterey Bay, California. Dean (1977) found the value of the exponent m to be two-thirds for the case of wave energy dissipation per unit volume as the dominant force and 0.4 for the other two cases of destructive forces. Comparison with approximately 500 profiles from the east coast and Gulf shorelines of the United States showed that, although there was a reasonably wide spread of the exponents, m, a value of two-thirds provided the best fit to all of the data. This allows the appealing interpretation that the wave energy dissipation per unit water volume causes destabilization of the sediment particle through the turbulence associated with the breaking waves. Thus, dynamic equilibrium results when the level of destabilizing and (unidentified) constructive forces are balanced. The sediment scale parameter, $A(D)$, and the equilibrium wave energy dissipation per unit volume, $\mathcal{D}_*$, are related by

$$A = \left[\frac{24}{5}\frac{\mathcal{D}_*}{\rho g\sqrt{g}\kappa^2}\right]^{2/3} . \tag{11}$$

Moore (1982) and Dean (1987b) have provided representations for the sediment scale parameter, A, as a function of sediment size, D, and fall velocity, w_f, as shown in Fig. 14.

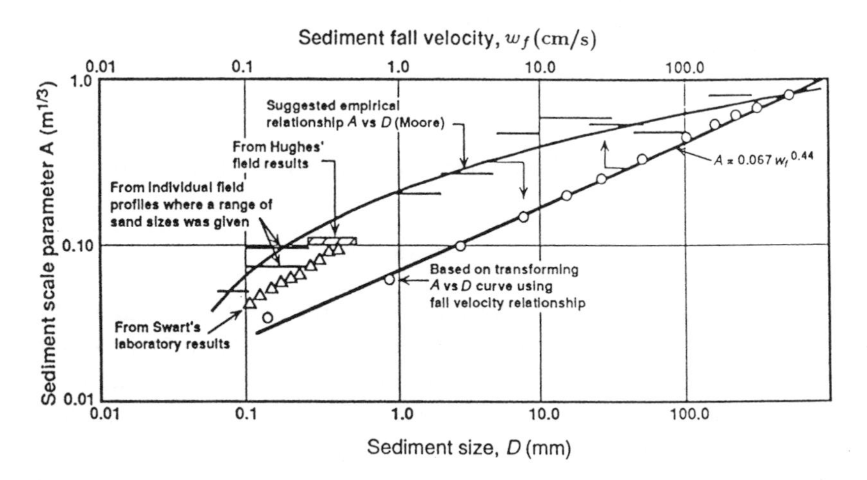

Fig. 14. Variation of sediment scale parameter, A, with sediment size, D, and fall velocity, w_f. (Dean, 1987b)

There are two inherent disadvantages of Eq. (10) with an exponent of two-thirds. The slope of the beach profile at the water line ($y = 0$) is infinite and the form is monotonic, i.e., it cannot represent bars. It has been shown that the first shortcoming can be overcome by recognizing that gravity is also a significant destabilizing force when the profile becomes steep. In this case the form is

$$y = \frac{h}{m} + \left(\frac{h}{A}\right)^{3/2} \quad (12)$$

which unfortunately is significantly more cumbersome to apply. Larson (1988) and Larson and Kraus (1989) have shown that an EBP of the form of Eq. (12) results by replacing the simple breaking wave model leading to Eq. (10) (with an exponent of two-thirds) by the more complex breaking model of Dally *et al.* (1985).

Bodge (1992) and Komar and McDougal (1994) have proposed slightly different forms of an equilibrium profile based on an *exponential* form. Bodge proposed

$$h(y) = h_o(1 - e^{-ky}) \quad (13)$$

in which h_o is the asymptotic depth at a great offshore distance, and k is a decay constant. The form suggested by Komar and McKougal is quite similar

$$h(y) = \frac{m_o}{k}(1 - e^{-ky}) \quad (14)$$

in which m_o/k is the equilibrium depth and m_o can be shown to be the beach face slope

$$\frac{\partial h}{\partial y} = m_o, y = 0\ . \quad (15)$$

Bodge fitted his recommended form to the averages of the ten data sets provided by Dean (1977) and found that the majority (60% to 71%) were fitted better by the exponential fit compared to the Ay^m relationship and that 80% to 86% of the data sets were fit better than the $Ay^{2/3}$ expression. The exponential forms have two free constants which are determined to provide the best fit and thus should agree better in general than for the case in which m is constrained to the 2/3 value. Komar fitted his form to a single Nile Delta profile. Since the exponential profile forms require determination of the two free parameters from the individual profile being represented, they can be applied in a diagnostic manner but not prognostically.

Vellinga (1983, 1986) has recommended an EBP based on a series of small and large scale model tests and comparisons with field data. This EBP was the basis for an updated method for dune erosion prediction in the Netherlands. The purpose of the small scale model tests was to develop profile modeling relationships (including distorted models) and to recommend a method for calculating dune erosion during extreme storms. The large scale model tests were conducted to validate the small scale model results. The principal findings from this study are reviewed in the following paragraphs.

The sediment fall parameter, $H_b/(w_f T)$, was found to be a valid modeling parameter, in which H_b is the breaking wave height, w_f is the sediment fall velocity, and T is the wave period. This parameter preserves the ratio of sediment fall time to the wave period. It was also found that, with the exception of wave run-up, distorted models could be used to investigate dune erosion. The modeling relationships determined were

$$\begin{aligned} n_l/n_d &= (n_d/n_w^2)^{0.28} \\ n_t &= (n_d)^{0.5} \\ n_H &= n_t^2 = n_d \end{aligned} \tag{16}$$

in which n_l and n_d are the horizontal and vertical scale ratios, respectively, n_w is the scale ratio of the sediment fall velocities, and n_t and n_H are the scale ratios of time and wave height.

Based on the model tests, the equilibrium beach profile was determined to be

$$\left(\frac{7.6}{H_{o_s}}\right) h = 0.47 \left[\left(\frac{7.6}{H_{o_s}}\right)^{1.28} \left(\frac{w_f}{0.0268}\right)^{0.56} y + 18\right]^{0.5} - 2.0 \tag{17}$$

in which all units are metric and H_{o_s} is the deep water significant wave height corresponding to the breaking wave height as calculated using linear wave theory.

As shown in Fig. 15, the profile equilibrated out to a depth equal to approximately 0.75 H_{o_s}. This finding is relevant to the depth of closure addressed later.

It is of interest that the equilibrium beach profile developed by Vellinga is somewhat similar to that represented by Eq. (10) with an exponent of two-thirds. One difference is that Vellinga's equation depends on both the sediment characteristics *and* the wave height, whereas Eq. (10) is a function of only the sediment characteristics. It can be shown that the Vellinga equation predicts that an increase in wave height causes a milder beach face slope which is in

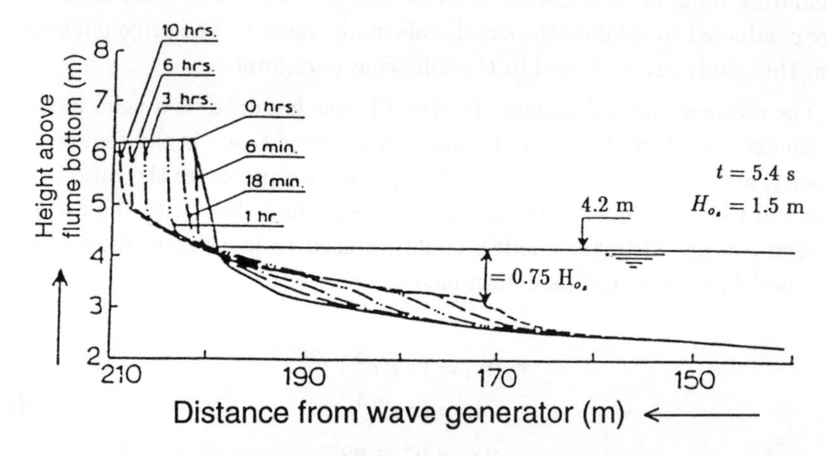

Fig. 15. Erosional profile evolution. Large wave tank results. (Vellinga, 1983)

accord with nature. Differentiating Eq. (17), to determine the profile slope,

$$\frac{\partial h}{\partial y} = \frac{0.235\left(\frac{7.6}{H_{o_s}}\right)^{0.28}\left(\frac{w_f}{0.0268}\right)^{0.56}}{\left[\left(\frac{7.6}{H_{o_s}}\right)^{1.28}\left(\frac{w_f}{0.0268}\right)^{0.56} y + 18\right]^{0.5}} \tag{18}$$

it is readily seen for increasing wave height, the profile slope decreases for small offshore distances, y. The beach face slope is determined by setting $y = 0$ in the equation below

$$\frac{\partial h}{\partial y}|_{y=0} = 0.055\left(\frac{7.6}{H_{o_s}}\right)^{0.28}\left(\frac{w_f}{0.0268}\right)^{0.56} \tag{19}$$

which depends on wave height and sediment fall velocity in the correct qualitative manner.

Figure 16 presents a comparison of profiles based on Eq. (10) (with $m = 2/3$) and (15) for two different wave heights. It can be seen that the two are in reasonable agreement for the smaller wave heights but deviate for the larger wave heights. For a fixed seaward distance, the Vellinga model predicts increasing water depths with increasing wave heights which is counter to observations in nature.

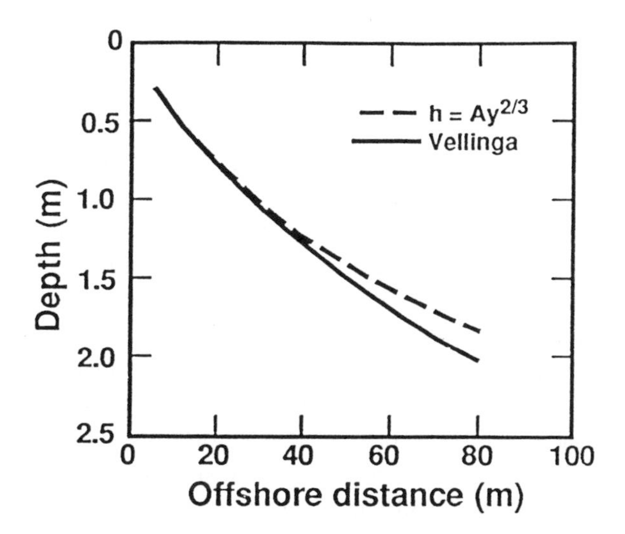

(a) Comparison of $h = Ay^{2/3}$ and the Vellinga model. $w_f = 0.03$ m/s, $H_{os} = 1.0$ m.

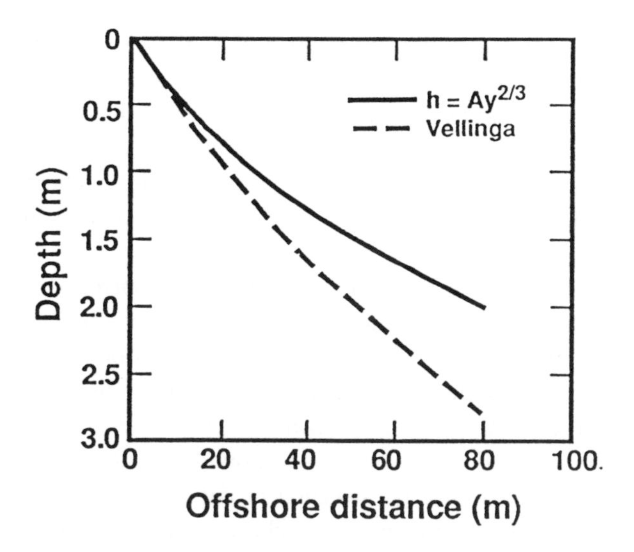

(b) Comparison of $h = Ay^{2/3}$ and the Vellinga model. $w_f = 0.03$ m/s, $H_{os} = 7.6$ m.

Fig. 16. Comparison of equilibrium beach profiles by Vellinga's equation and $h = Ay^{2/3}$.

3.3. *Limits of cross-shore sand transport in the onshore and offshore directions*

For purposes of discussion, it is useful to consider the long- term and short-term limits of cross-shore sediment transport. During short-term erosional events, elevated water levels and high waves are usually present and the seaward limit of interest is that to which significant quantities of sand-sized sediments are transported and deposited. The landward limit is the maximum point of uprush or, in cases in which overwash occurs, may be the depositional limit on the landward side of the barrier island.

It is important to note that sediment particles are in motion to considerably greater depths than those to which significant profile readjustment occurs. This readjustment occurs most rapidly in the shallow portions of the profile and, during erosion, the transport and deposition from these areas cause the leading edge of the deposition to advance into deeper water. This is illustrated in Fig. 15 from Vellinga (1983) in which it is seen that with progressively increasing time, the evolving profile advances into deeper and deeper water. It is also evident from Fig. 15 that the rate of profile evolution is decreasing consistent with an approach to equilibrium. In the case in which no overtopping of a barrier island occurs, the depth of limiting motion is not that to which the sediment particles are disturbed but rather the seaward limit to which the depositional front has advanced. Vellinga recommends that this depth be $0.75\ H_{o_s}$ in which H_{o_s} is the deep water significant wave height computed from the breaking wave height using linear water wave theory.

In the case in which overwash does not occur, the limit of effective transport is commonly taken as the breaking depth, h_b, based on the significant wave height.

If overwash occurs, the landward limit may be controlled by the extent to which the individual uprush and overwash event is competent to transport sediment. Often this distance is determined by the loss of transporting power due to percolation into the beach or by water impounded by the overwash event itself. In the latter case, the landward depositional front will advance at more or less the angle of repose into the impounded water.

The seaward limit of effective seasonal profile fluctuation is a useful engineering concept and is referred to as the "closure depth". Based on laboratory and field data, Hallermeier (1978, 1981) developed the first rational approach to the determination of the closure depth. He defined two depths, the shallower of which delineates the limit of intense bed activity and the deepest seaward of which there is expected to be little sand transport due to waves. The shallower of the two appears to be of the greatest engineering relevance. Based on correlations with the Shields parameter, the shallower depth, h_c, was recommended as

$$\frac{\rho {v_b}^2}{(\rho_S - \rho) g h_c} \approx 0.03 \, . \tag{20}$$

This result was developed into a more meaningful form for application using linear wave theory and transferred to field conditions by rationalizing that the seaward closure depth would be associated with wave conditions that are relatively rare. Hallermeier chose the effective significant wave height, H_e, and the effective wave period T_e, as those that are exceeded only twelve hours per year or only 0.14% of the time. The resulting approximate equation for the depth of closure h_c was determined to be

$$h_c = 2.28 \; H_e - 68.5 \left(\frac{{H_e}^2}{g{T_e}^2} \right) \tag{21}$$

in which H_e can be determined from the annual mean significant wave height, $\overline{H}$, and the standard deviation in wave height, σ_H, as

$$H_e = \overline{H} + 5.6 \sigma_H \tag{22}$$

Birkemeier (1985) evaluated Hallermeier's relationship using high quality field measurements and recommended slightly different constants in the equation proposed by Hallermeier

$$h_c = 1.75 \; H_e - 57.9 \left(\frac{{H_e}^2}{g{T_e}^2} \right) \tag{23}$$

and also found that the following simplified approximation provided nearly as good a fit to the data

$$h_c = 1.57 \; H_e \, . \tag{24}$$

3.4. *Characteristics and occurrence of offshore bars*

Offshore bars are long depositional submerged features along much of the world's shoreline and may occur singly or multiply. Early attempts to correlate bars with causative factors centered on the deep water wave steepness,

H_o/L_o, concluding that an offshore bar would occur if the steepness exceeded 0.04. More recently, it has been suggested that bars could be caused by

1. wave breaking,
2. long (infragravity) wave reflection from the beach,
3. edge waves, and
4. interference of forced and free wave components of the first harmonic of the dominant wave.

Based on field observations, Dean (1973) hypothesized that the sediment was suspended during the crest phase position and that if the fall time were less or greater than one-half wave period, the net transport would be landward or seaward respectively resulting in bar formation in the latter case. This mechanism would be consistent with the wave breaking cause. Further rationalizing that the suspension height would be proportional to the wave height resulted in identification of the so-called fall velocity parameter, $\frac{H_b}{w_f T}$. Examination of small scale laboratory data for which the deep water reference wave height H_o values were available, led to the following relationship for bar formation

$$\frac{H_o}{w_f T} \geq 0.85 \,. \tag{25}$$

Later, Kriebel *et al.* (1986) examined only prototype and large scale laboratory data and found a constant of approximately 2.8 rather than 0.85 as in Eq. (25). Larson and Kraus (1989) and Kraus *et al.* (1991) examined only large wave tank data and proposed the following two relationships for bar formation

$$\frac{H_o}{L_o} \geq 115 \left(\frac{\pi w_f}{gT} \right)^{3/2} \tag{26}$$

and

$$\frac{H_o}{L_o} \leq 0.0070 \left(\frac{H_o}{w_f T} \right)^3 \tag{27}$$

each of which provides a somewhat better fit to the data. Dalrymple (1992) has shown that each of the latter two equations can be represented in terms of a single "Profile Parameter", P, where

$$P = \frac{gH_o^2}{w_f^3 T} \tag{28}$$

and that the criterion for bar formation is that P exceeds about 10,000.

3.5. *Static models for profile response*

One class of profile evolution models is that of static or *geometric models.* In this class, an equilibrium profile is established and the profile responds to the forcing function, usually an increased water level and wave height by satisfying the conservation equation and the landward and seaward limits of profile mobilization. Because the models are static, a sediment transport relationship is not required; however, the implied volumetric sediment transport distribution across the surf zone can be determined through the profile changes and the sediment conservation equation (to be introduced later). The conditions under which these simple models are more appropriate than the dynamic- and process-based models depends on the *time scale* of the process. If the time scale is very long, the geometric models are much more appropriate. If the time scale of the forcing function is shorter than the time scale of the process, then it is necessary to employ a dynamic model. Bruun (1962) proposed the following relationship for shoreline response to sea level rise, S

$$R = S\frac{W_*}{h_* + B} \tag{29}$$

in which R is the shoreline recession $(-\Delta y)$, W_* and $(h_* + B)$ are the width and vertical extent of the active profile. The basis for this equation is seen in Fig. 17 in which the two components of the response are

1. a retreat of the shoreline, $-\Delta y$, which produces a sediment "yield" $-\Delta y(h_* + B)$ and
2. an increase in elevation of the equilibrium profile by an amount of the sea level rise, S, which causes a sediment "demand" equal to $S\ W_*$. Equating the demand and the yield results in Eq. (29).

Equation (29) is known as the "Bruun Rule". It is noted that the Bruun Rule does not depend on the particular profile shape.

Dean and Maurmeyer (1983) later extended Bruun's result to apply to the case of a barrier island in the form

$$R = S\frac{L_* + W + L_L}{h_* - h_L} \tag{30}$$

and the various terms are explained in Fig. 18. Edelman (1972) modified the Bruun Rule to make it more appropriate for larger values of increased water levels and thus applicable for large storm surges. It was assumed that the

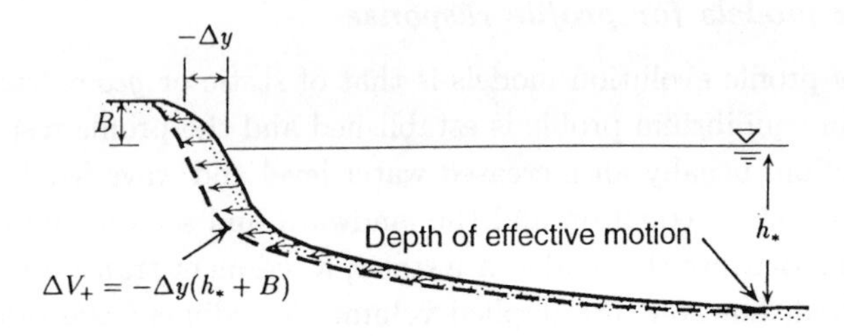

a) Volume of sand "generated" by horizontal retreat, R, $(= -\Delta y)$ of equilibrium profile over vertical distance $(h_* + B)$.

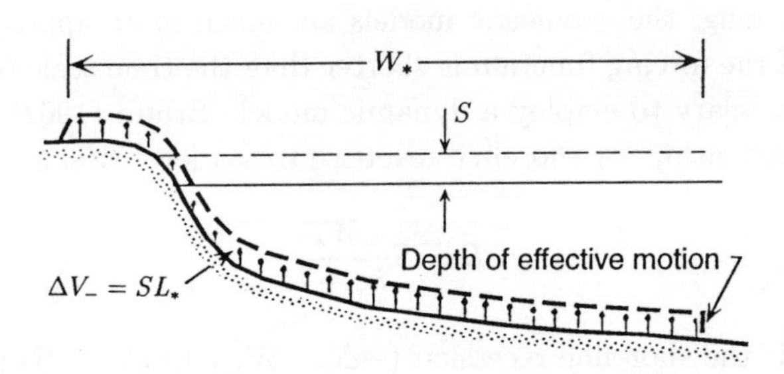

b) Volume of sand required to maintain an equilibrium profile of active width, W_*, due to a rise, S, in mean water level.

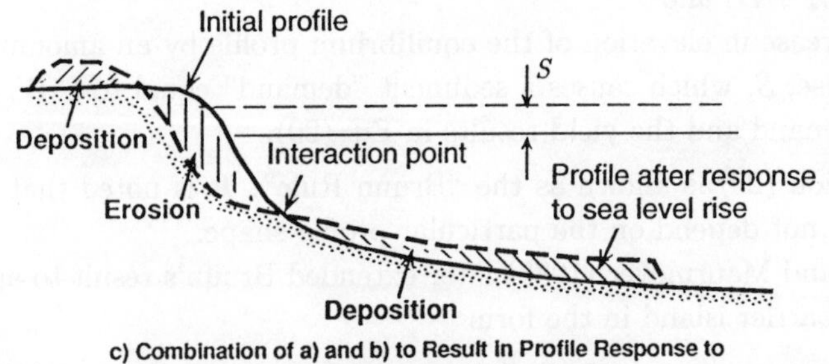

c) Combination of a) and b) to result in profile response to sea level rise to maintain equilibrium profile.

Fig. 17. Components of sand volume balance due to sea level rise and associated profile retreat according to the Bruun Rule.

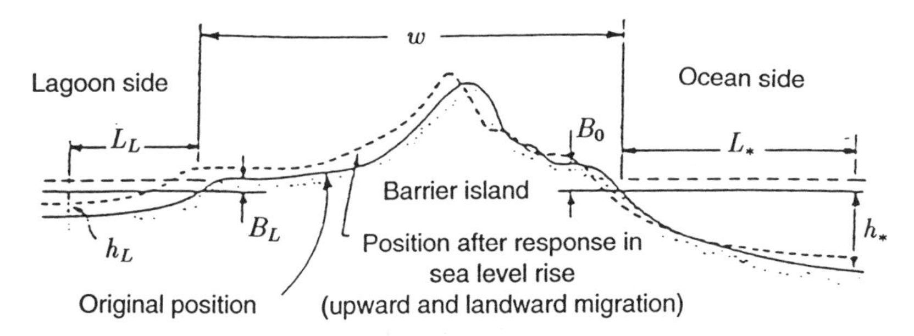

Fig. 18. The Bruun Rule generalized for the case of a barrier island that maintains its form relative to the adjacent ocean and lagoon. (Dean and Maurmeyer, 1983)

profile maintained pace with the elevated water level and thus at each time, the following equation is valid

$$\frac{\partial R}{\partial t} = \frac{\partial S}{\partial t}\left[\frac{W_*}{h_* + B(t)}\right] \tag{31}$$

where now $B(t)$ represents the instantaneous total height of the active profile *above* the current water level. Noting from Fig. 19 that $B(t) = B(0) - S(t)$, and substituting in Eq. (31) and integrating,

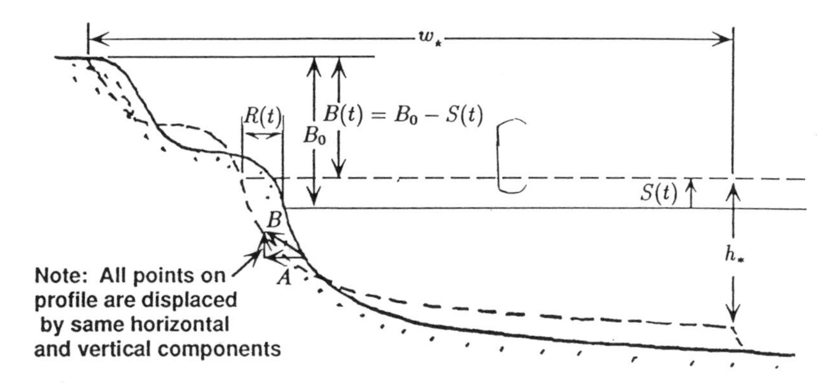

Fig. 19. Elements of the Edelman model.

$$R(t) = W_* \ln\left[\frac{h_* + B(0)}{h_* + B(0) - S(t)}\right] . \tag{32}$$

Using the small argument approximation for the natural logarithm

$$\ln(1+z) \approx z - \frac{z^2}{2} + \frac{z^3}{3} + \ldots \tag{33}$$

it is readily shown that to the first approximation, Edelman's equation (Eq. (32)) is the same as the Bruun Rule (Eq. (29)).

3.6. *Dynamic models for profile response*

An analytical or numerical model for the representation of profile evolution requires a transport equation and an equation for the conservation of sediment. The conservation equation balances the differences between inflows and outflows as predicted by the transport equation. In addition, boundary conditions must be employed at the landward and seaward ends of the active region. These boundary conditions can usually be expressed in terms of a maximum limiting slope such that if the slope is exceeded, adjustment of the profile will occur. This condition is sometimes referred to as "avalanching". The location of the seaward and landward boundaries are usually taken at the limits of wave breaking and wave run-up respectively.

Two types of representations of the physical domain for numerical modeling can be considered as shown in Fig. 20. In the first type shown in Fig. 20a, the cells are finite increments of the distance variable, y. Thus the distance is the independent variable and the depth varies with time. In the second type shown in Fig. 20b, the computational cells are formed by finite increments of the depth range, h. In this case, the independent variable is h and y varies with time for each h value. There is an inherent advantage of the first type since the presence of bars can be represented with no difficulties.

a. *Conservation equation*

The conservation equation is very straightforward and for the computational cell type in Fig. 20a is given by

$$\frac{\partial h}{\partial t} = \frac{\partial q_y}{\partial y} \tag{34}$$

in which y and t are the independent variables. If depth and time are regarded as the independent variables as in Fig. 20b, the conservation equation is

$$\frac{\partial y(h)}{\partial t} = -\frac{\partial q_y}{\partial h} \tag{35}$$

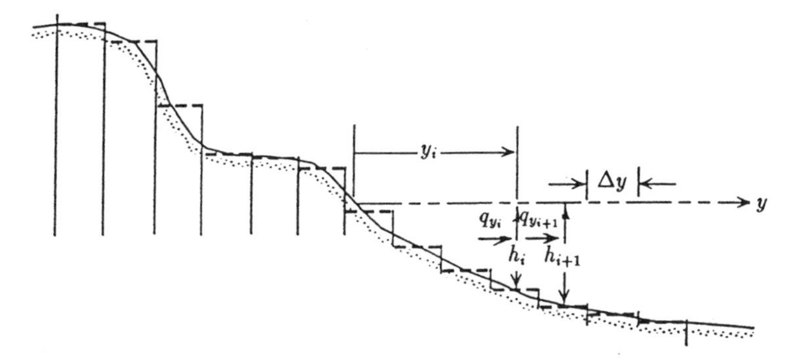

(a) Grid with y and t as the independent variables and h dependent.

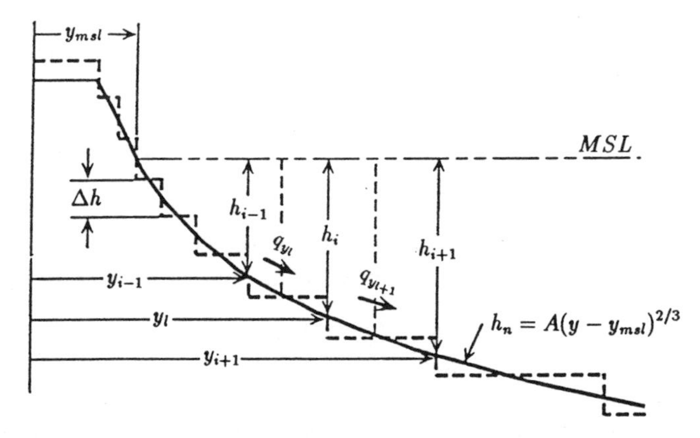

(b) Grid with h and t as the independent variables and y dependent. (Modified from Kriebel and Dean, 1985).

Fig. 20. Two types of grids employed in numerical modeling of cross-shore sediment transport and profile evolution.

in which it is noted that each y value is associated with a depth value (h).

b. *Transport relationships*

Sediment transport relationships can be categorized as "closed loop" models which converge to a target profile and "open loop" models which are not *a priori* constrained to the final (equilibrium) profile. Transport relationships of the "closed loop" type will be reviewed first.

Closed Loop Transport Relationships. One of the first closed loop transport relationships proposed was that of Kriebel (1982) and Kriebel and Dean (1985). Recalling that the EBP represented by Eq. (10) is consistent with uniform wave energy dissipation per unit water volume, $\mathcal{D}_*$, this transport relationship is

$$q_y = K'(\mathcal{D} - \mathcal{D}_*) \tag{36}$$

such that at equilibrium, $\mathcal{D} = \mathcal{D}_*$, which is consistent with Eq. (10) with exponent 2/3. An implicit procedure was used to solve the equations of conservation and transport. Kriebel (1986) has calibrated the model against the large wave tank data of Saville (1957) and the Hurricane Eloise data summarized by Chiu (1977) and has determined the value of the coefficient $K' = 8.7 \times 10^{-6}\ m^4/N$. An example of a numerical solution is shown in Fig. 21. It is noted that a transport equation incorporating the effects of a profile with finite slope at the water line can also readily be employed

$$q_y = K'\left[\left(\frac{\mathcal{D}_*}{m_o} + \frac{5}{16}\rho g \kappa^2 \sqrt{gh}\right)\frac{\partial h}{\partial y} - \mathcal{D}_*\right] \tag{37}$$

Of interest for this transport model is the role of the term $K'\frac{\mathcal{D}_*}{m_o}$. Combining this portion of the transport equation with the conservation equation yields

$$\frac{\partial h}{\partial t} = \left[k'\frac{\mathcal{D}_*}{m_o}\right]\frac{\partial^2 h}{\partial y^2} \tag{38}$$

which is recognized as the heat conduction or diffusion equation. Thus this term, which was previously ascribed to gravity, tends to smooth out irregularities.

The model by Larson (1988) and Larson and Kraus (1989) is called SBEACH. The profile computational domain is subdivided into four regions, each with its own sediment transport relationship. The characteristics of each of these regions are shown in Fig. 22. The transport relationship above the mean water line (Zone IV) varies linearly and thus predicts a change in the position of this feature without change in form. The relationship in Zone III is of the same form as Eq. (37) with the important exception that this equation is used to establish the *magnitude* of the transport. The transport direction (i.e. onshore or offshore) is determined by the following relationship

$$\frac{H_o}{L_o} \gtrless 0.0007(H_o/w_f T)^3\ . \tag{39}$$

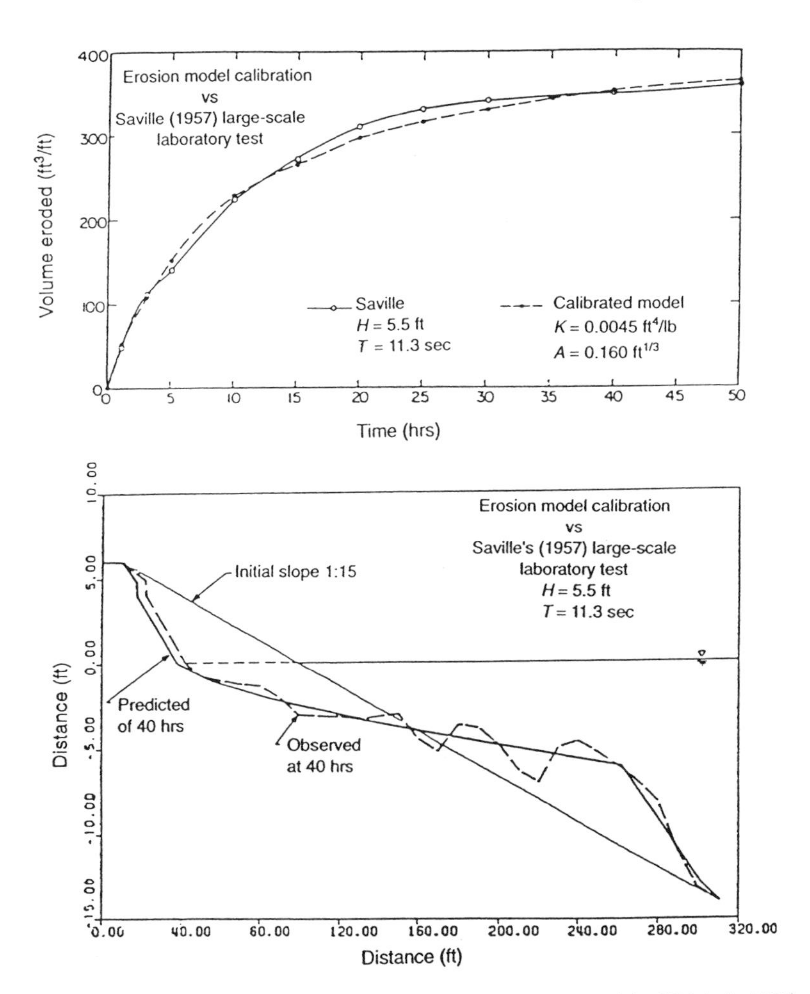

Fig. 21. Example of numerical solution by the Kriebel–Dean model. (Kriebel, 1986)

If the value on the LHS exceeds the RHS, the transport is directed onshore over the entire active profile and vice versa if the LHS is less than the RHS. Zone II is quite short and plays the role of linking Zones II and III. The authors of this model note that this region can be omitted. Zone I is interesting in that it provides an empirical relationship that always results in a bar formation if the transport is seaward. It is noted that Eq. (39) does not include any of the profile characteristics and thus even if the profile is very steep, such as occurs

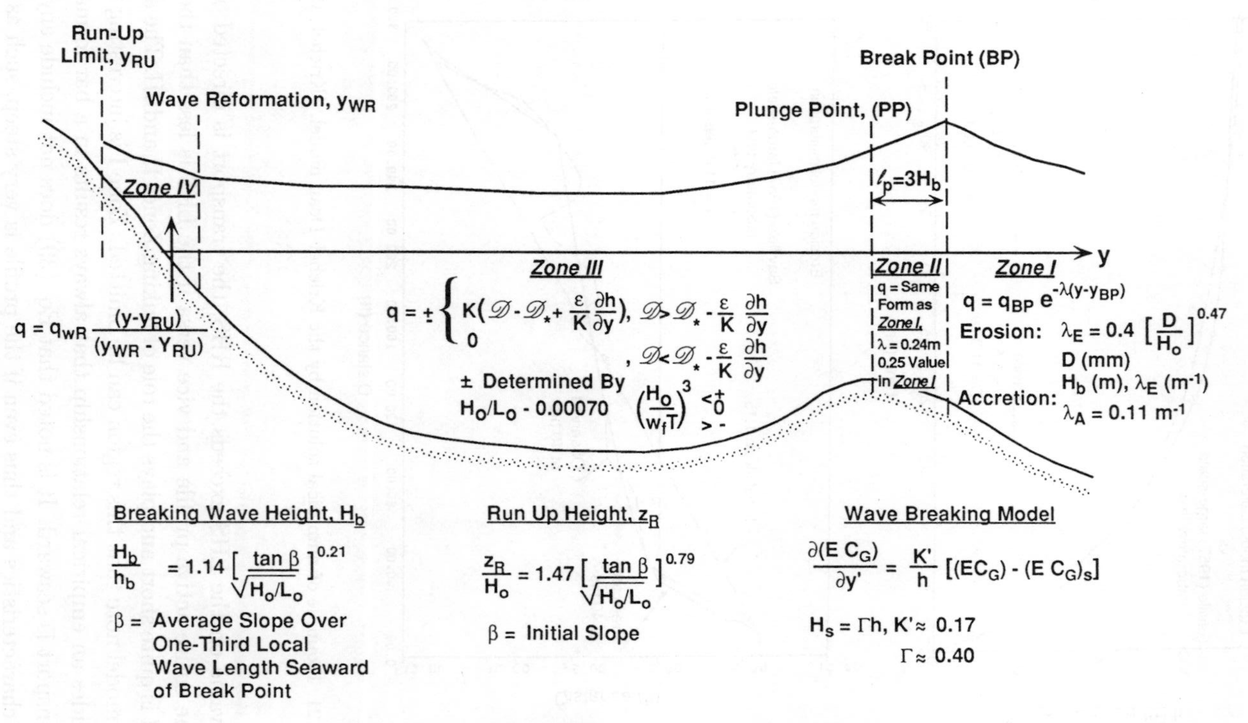

Fig. 22. Summary of SBEACH transport relationships.

after beach nourishment and the wave and sediment characteristics indicate onshore transport, the transport direction utilized in the model is onshore, a seemingly severe shortcoming of this transport representation.

With regard to bar formation, it can be shown from the conservation equation

$$\frac{\partial h}{\partial t} = \frac{\partial q_y}{\partial y} , \tag{40}$$

that a bar will be present if the following is satisfied

$$\int_0^t \frac{\partial^2 q_y}{\partial y^2} dt < -\frac{\partial h(t=0)}{\partial y} . \tag{41}$$

Thus, the formation of a bar depends on the initial profile slope and requires a strong negative curvature in the cross-shore sediment transport.

Two analytical or semi-analytical closed loop models have been documented. Kobayashi (1987) has combined the conservation equation and a form of the transport relationship employed by Moore (1982) and Kriebel and Dean (1985) into one equation representing the profile response. The landward and seaward limits of the domain are treated as moving boundaries and the eroding dune face is represented as a vertical face. An analytical solution is determined for a particular initial condition. It is found that the problem is mathematically similar to a diffusion problem in which the convergence to the steady state problem first occurs rapidly and later approaches the steady state solution asymptotically. The model results are found to approximate the numerical solutions of the Kriebel–Dean model.

Kriebel *et al.* (1991) and Kriebel and Dean (1993) have developed and illustrated a convolution method of representing beach profile evolution. The method is based on the observation from numerical solutions that for a constant forcing function, i.e., water level, a particular contour $y(t)$ tends to approach the equilibrium contour, y_∞, asymptotically in approximate accordance with the following relationship

$$y(t) = y_\infty(1 - e^{-t/T_S}) . \tag{42}$$

The solution is obtained in the form of a convolution integral of the response function in Eq. (42) and the water level history up to the time, t,

$$y(t) = \frac{y_\infty}{T_S} \int_0^t f(\tau) e^{-(t-\tau)/T_S} d\tau \tag{43}$$

in which $f(\tau)$ is the time varying water level and y_∞ is the shoreline displacement associated with the peak water level. It can be verified that integration of Eq. (43) with $f(\tau) = 1$ yields Eq. (42). The method was found to yield very similar results to those obtained through numerical modeling.

Open Loop Transport Relationships. Dally and Dean (1984) developed a numerical model for profile evolution in which the cross-shore transport was considered to be due only to suspended sediment, i.e., no bed load transport was included. Both intermittent and steady components of suspended transport were included in the model. For the intermittent transport it was assumed that the sediment was suspended at the crest phase position and was advected with the sum of the first order wave particle velocity and the steady second order mean velocities as the particle descended to the bottom. Those particles which were in constant suspension were advected by the steady flows which are due to the transfer of momentum and the return flow of the mass transport. The distribution of the second order mean velocities over depth was very similar to that in Eq. (6). The suspended sediment concentration was considered to be exponential as given by

$$C(Z) = C_A \exp\left[\frac{-15(z - z_A)}{h\sqrt{\tau/\rho}}\right]. \tag{44}$$

Examples of the results from this model will be presented later.

Ohnaka and Watanabe (1990) employ the following transport relationship

$$Q_w = [A_w(\tau_b - \tau_{cr}) + A_{wb}\tau_t]\frac{F_D V_b}{\rho g} \tag{45}$$

$$Q_c = [A_c(\tau_b - \tau_{cr}) + A_{cb}\tau_t]\frac{V}{\rho g} \tag{46}$$

for transport driven by waves and currents respectively. In the above equations, A_w, A_c, A_{wb} and A_{cb} are nondimensional coefficients. The term τ_t is the turbulent induced shear stress given by

$$\tau_t = \rho^{1/3}(nf_D E)^{2/3} \tag{47}$$

in which $nf_D E$ is the breaker-induced wave energy dissipation per unit area and time.

A second widely used transport relationship in open loop models is based on the Bagnold Model (Bagnold, 1963) of sediment transport which was originally developed for riverine flows and has been adapted for coastal purposes by Bailard and Inman (1981), Bailard (1981, 1982) and Bowen (1980). The transport model considers both suspended and bedload transport q_s and q_b respectively, given by Bailard (1982) as

$$q_s = \frac{\varepsilon_s f}{8g(s-1)(1-p)w_f}\left[\langle |V_t|^3 V_t\rangle - \frac{\varepsilon_S}{w_f}\tan\beta\langle |V_t|^5\rangle\right] \tag{48}$$

and

$$q_b = \frac{\varepsilon_b f}{8g(s-1)(1-p)\tan\phi}\left[\langle |V_t|^2 V_t\rangle - \frac{\tan\beta}{\tan\phi}\langle |V_t|^3\right] \tag{49}$$

in which the ε values are efficiency factors, f is the Darcy–Weisbach shear stress coefficient, w_f is the sediment fall velocity, and β the local bed slope. It is noted that Bailard (1981) also presents considerably more complicated forms of the Bagnold formulation for specific application to coastal conditions.

4. Evaluation of Cross-Shore Sediment Transport Models

4.1. *Introduction*

Dally and Dean (1984) have proposed the following five criteria for a "good" beach profile evolution model.

1. Generate both normal and storm profile types depending on wave conditions and sediment characteristics.
2. Predict the proper shape of these profiles, i.e.,
   (a) The normal profile should be monotonic and concave upwards; and (b) the bar(s) of the storm profile should have the proper spacing and shape.
3. Correctly predict the rate of evolution.
4. Respond to changes in water level due to tides, storm surges, or long-term fluctuations.
5. Approach an equilibrium if all the relevant parameters are held constant.

These criteria with the additional capabilities of representing overwash and the interaction with engineered structures still appear appropriate and will be used in evaluating the various models. Because most models are very complex and the programming is extremely detailed, the evaluation will generally be based on published results. Moreover, in some cases, only a qualitative evaluation will be possible.

4.2. *Model evaluation*

Analytical and numerical models will be reviewed in the chronological order in which they were published. As was done for the description for the various models, closed and open loop types will be reviewed separately.

a. *Closed loop models*

Model of Kriebel and Dean (1985)

Results from this model were presented in Fig. 21. This model satisfies all the criteria except those relating to the offshore bars, i.e., Criteria 1 and 2. Kriebel (1990) has added the capabilities of representing overwash and interaction with seawalls.

SBEACH Model Larson (1988) and Larson and Kraus (1989, 1990)

This model has been compared with laboratory and field data as shown in Figs. 23, 24, and 25. Figure 23 presents results for the two sets of available large scale wave tank data. The upper panel shows a comparison from the Saville (1957) data and the comparisons include calculations at various times and the measured profile at forty hours. The lower panel is for data from the large Japanese tank and presents calculations at various times and the measured profile at thirty and a half hours. The calculations are in quite good agreement with the data. The waves used in these tests were monochromatic which tend to favor well developed and concentrated bars. Figure 23 is a comparison from Duck, North Carolina, and compares the evolution of a profile over a three-day period in which the wave heights were on the order of 1.5 m. The initial profile included a bar located approximately 40 m from shore. During the period of interest, the bar migrated seaward approximately 65 m. The simulations provide a reasonable qualitative representation of the evolution. The bar becomes more subdued at its initial location, but in contrast to measurements is still present at the final time. Also, the bar had started to form slightly seaward of the measured location. The calculations showed substantially greater erosion at the shoreline than measured. This example demonstrates the extreme difficulty in simulating an actual event in nature. The effect of the January 1992 storm on a recently completed and

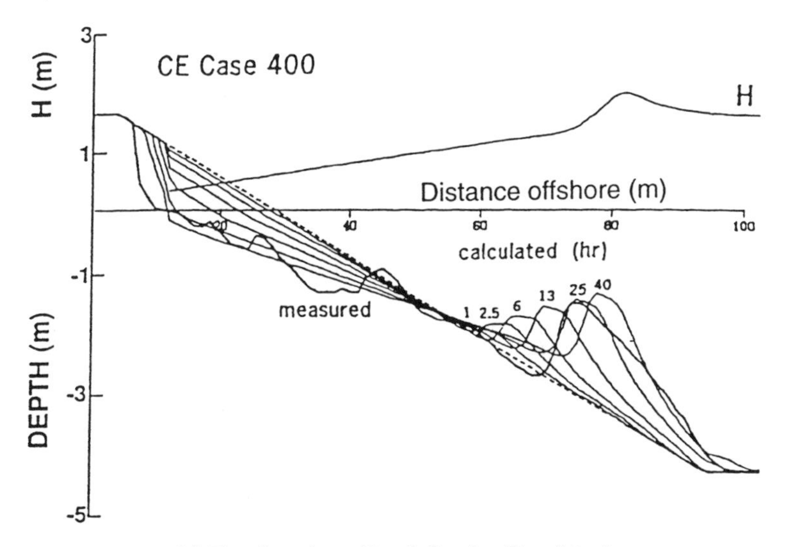

(a) Test from large Beach Erosion Board tank.

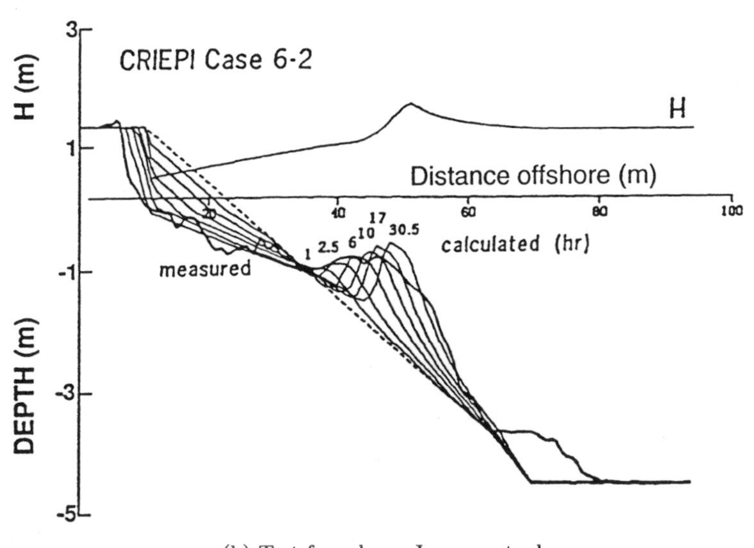

(b) Test from large Japanese tank.

Fig. 23. Comparison of SBEACH against two tests from large scale wave tanks. (Larson and Kraus, 1989)

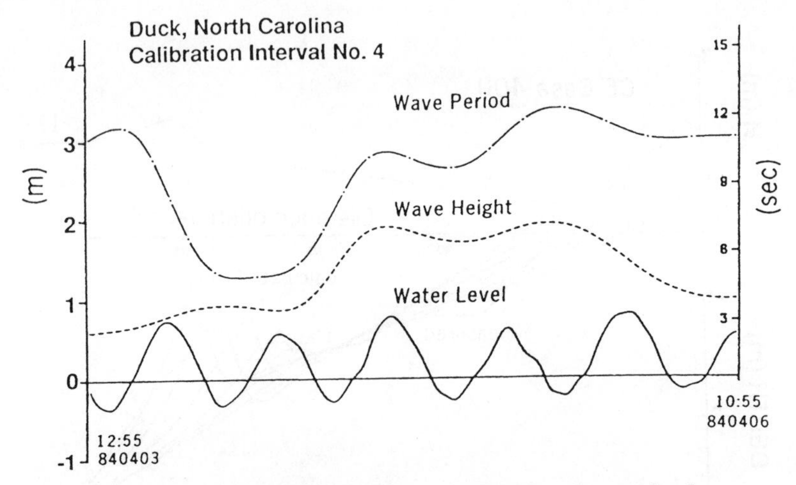

(a) Input measured wave height, wave period, and water level.

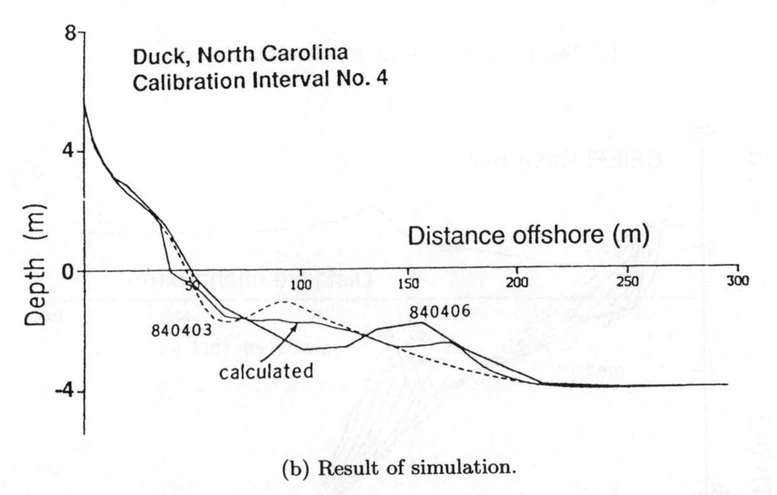

(b) Result of simulation.

Fig. 24. Results of test of SBEACH against profile evolution data from Duck, North Carolina. (Larson and Kraus, 1989)

documented beach nourishment project at Ocean City, Maryland provided an additional opportunity to test SBEACH. The model, which at that time did not include overwash was modified to include this effect. The results of applying the model with and without overwash are shown in Fig. 25. It is seen that by including the effect of overwash, there is a slightly improved fit at the 74th Street profile (Figs. 25a and b). In both cases, the model predicts the presence of a well-developed bar at a depth of about 3 m, a feature which is totally absent in the measured profiles. This difference is probably the result of the model development being based on monochromatic wave tank tests which, as noted before, tend to accentuate bar formation. In summary, SBEACH appears to perform well with the possible exception that the condition applied to determine the *direction* of sediment transport does not include the beach profile characteristics (slope). The beach slope and the gravity induced flows will contribute significantly to the transport direction in many applications of interest. SBEACH does include capabilities of representing overwash and interaction with seawalls.

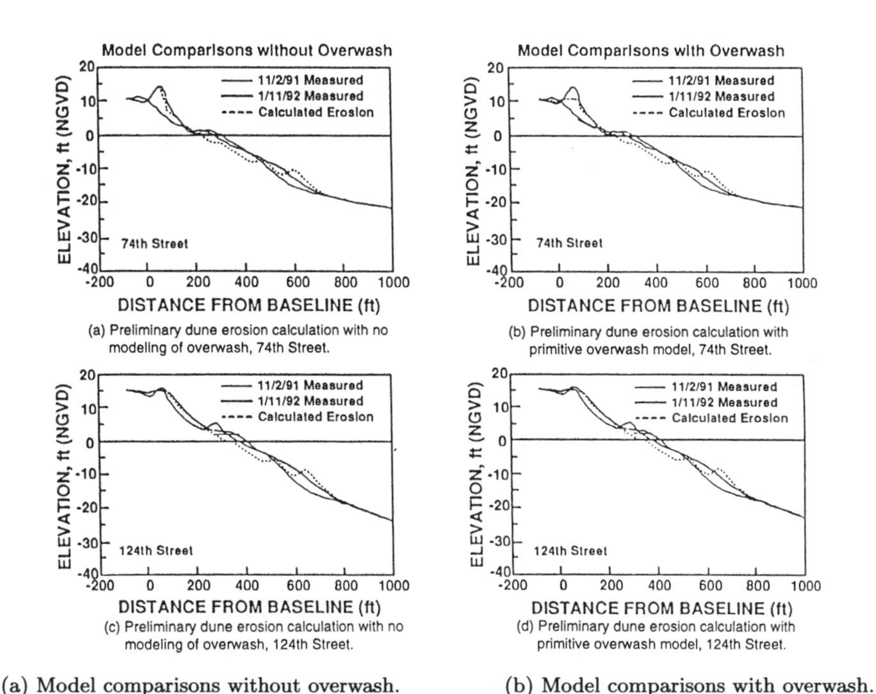

(a) Model comparisons without overwash. (b) Model comparisons with overwash.

Fig. 25. Test of SBEACH against profile evolution resulting from the January 1992 storm, Ocean City, Maryland. Model comparisons shown with and without overwash. (From Kraus *et al.*, 1993)

b. *Open loop models*

Model of Dally and Dean (1984)

This model includes representation of only suspended sediment transport. Example model results for various sand sizes and wave heights are presented in Fig. 26a. A comparison of calculated and measured large wave tank profiles is presented in Fig. 26b. It was found that this model satisfied all of the original criteria discussed earlier with the exception of approaching a stable profile if the model were allowed to run for a very long time. The model does not include capabilities of overwash or seawall interaction.

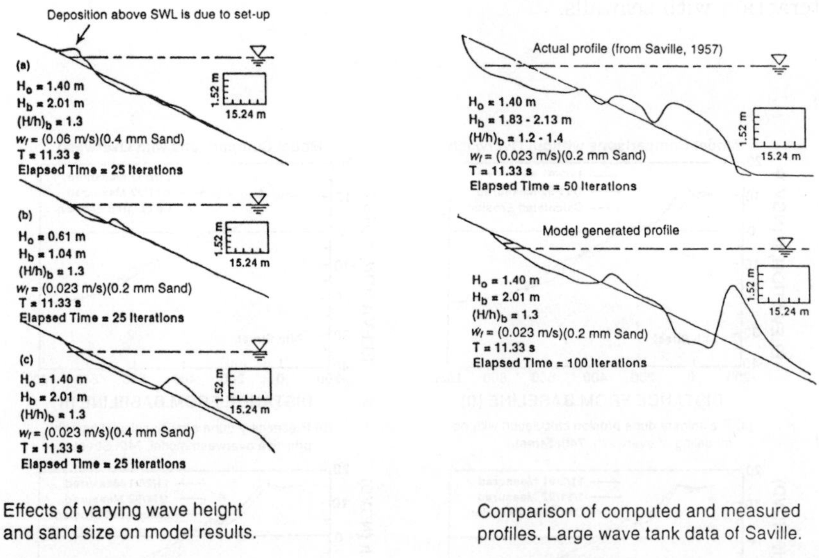

Fig. 26. Example output to illustrate characteristics of the Dally-Dean model. (Dally and Dean, 1984)

Bailard Model (1981) and Watanabe Model (1987)

The Bailard energetics transport relationship has been applied by a number of investigators and incorporated into several profile evolution models. One of the difficulties appears to be the need to reliably determine the velocity moments appearing in the expressions for the suspended and bed load transport components (Eqs. (48) and (49)). To test the model while circumventing the possible inaccuracies in estimating these quantities, both Bailard (1982) and Bowen (1980) have estimated the transport based on velocity measurements. Figure 27, (Bailard, 1982) compares the predicted and measured volume changes at Torrey Pines, California. A somewhat similar approach by Bowen (1980) was inconclusive. One test that can be applied to the Bailard model is whether it would predict a reasonable equilibrium beach profile for the case of zero net sediment transport. Figure 28 (Bailard, 1981) presents a relationship between the normalized equilibrium beach slope and the dimensionless depth for various values of the parameter $\pi H/(w_f T)$. Considering as an example the following conditions: $H = 1$ m, $h = 2$ m, $T = 8$ s, $w_f = 0.03$ m/s, an unrealistic slope of greater than 60° is found.

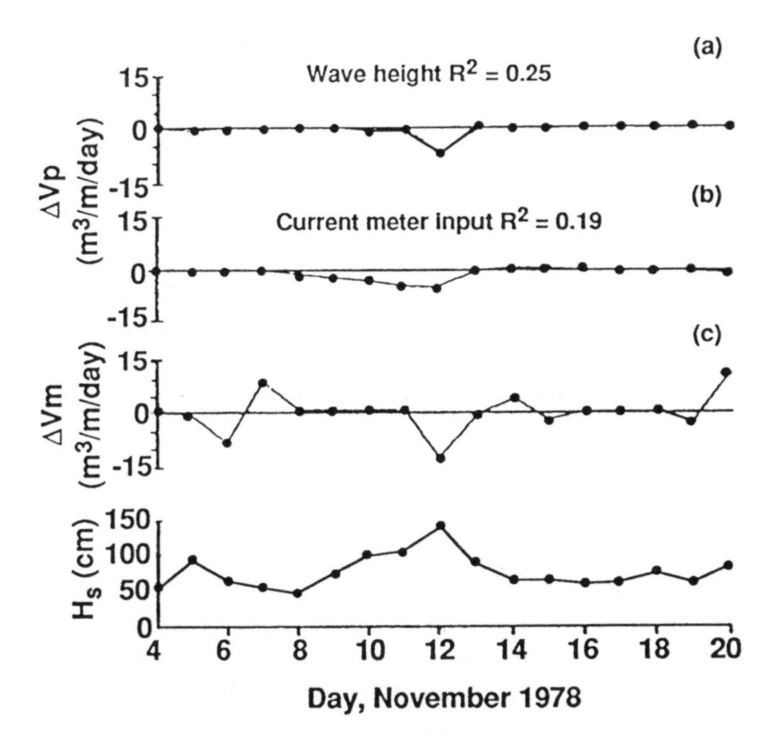

Fig. 27. Comparison of measured volume changes (c) with predicted based on: (a) wave height input and (b) current meter input. Test of Bailard model against Torrey Pines, California data. (Bailard, 1982)

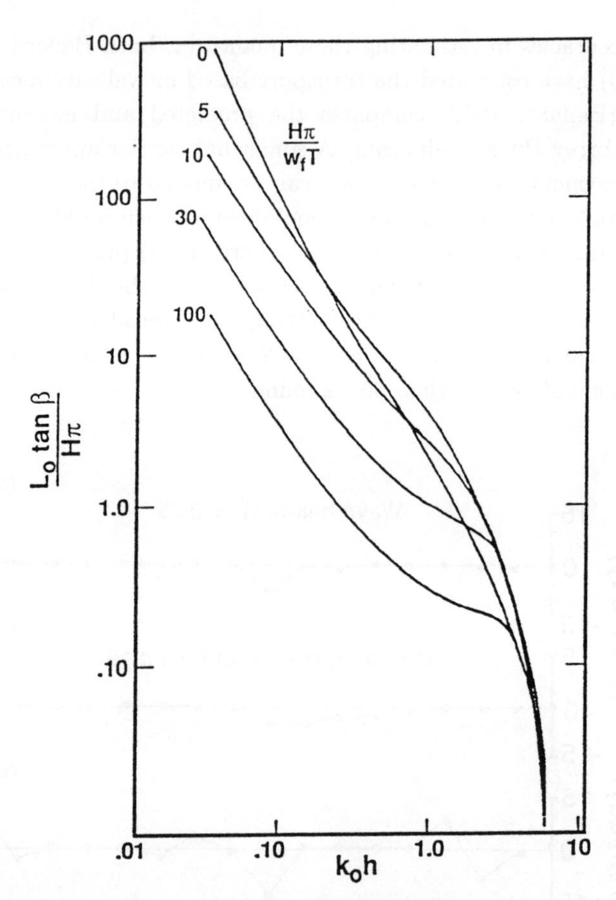

Fig. 28. Variation of equilibrium beach slope with fall velocity parameter and wave number. All in nondimensional forms. Normal wave incidence. (Bailard, 1981)

Roelvink and Stive (1989) have conducted the most thorough and well documented evaluation of the Bailard model. Two laboratory experiments were conducted with a beach commencing with a uniform slope of 1:40 and a second test commencing with an initial bar. The experiments included active long wave absorption at the wavemaker but did not attempt to generate the appropriate long wave system associated with the Jonswap spectrum used in the tests. The measured hydrodynamics were compared with those computed from a model by Battjes and Janssen (1978) as calibrated by Battjes and Stive (1985). As shown by the example in Fig. 29, the computed third and fourth

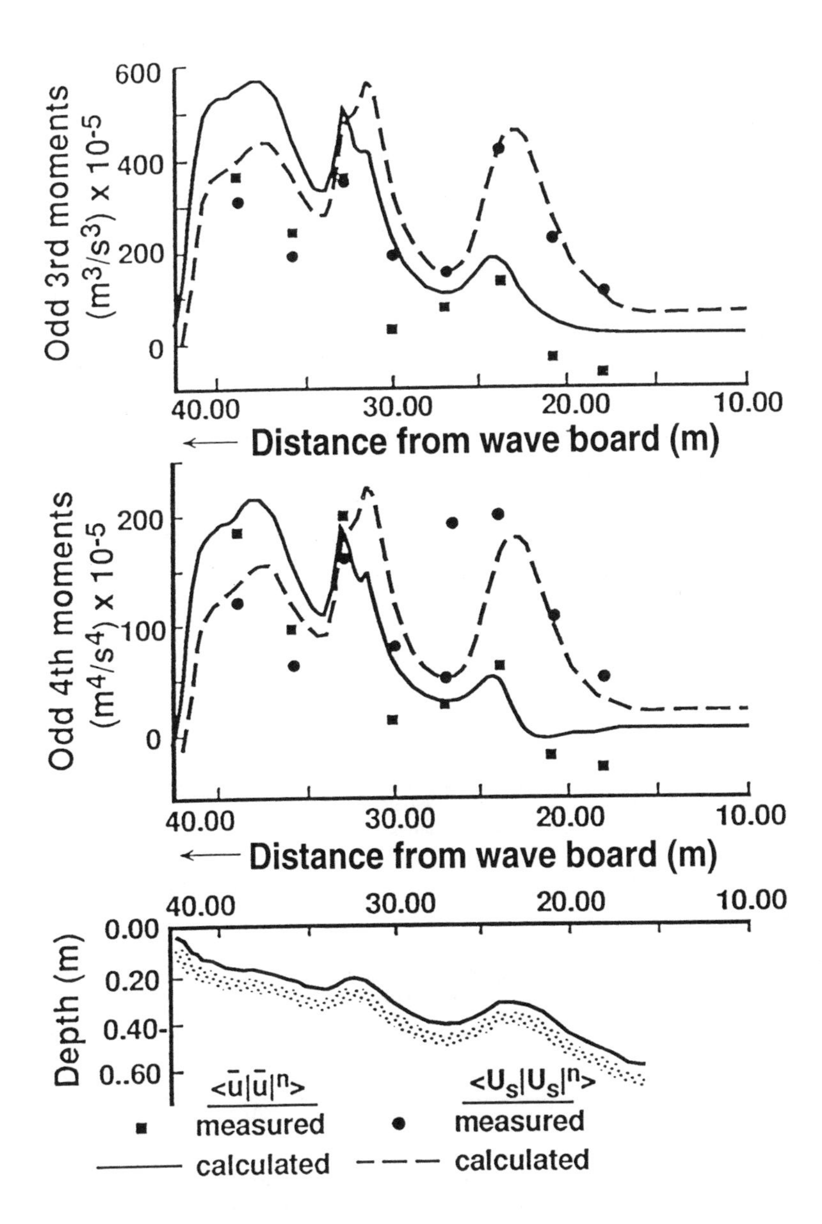

Fig. 29. Comparison of measured and computed third and fourth velocity moments. (Roelvink and Stive, 1989)

moments are in reasonable agreement with those measured. It was found desirable to modify the Bailard model by including a contribution of the wave breaking induced turbulence, termed a "stirring" effect. The modifications due to this term are shown in Fig. 30 in which the measured total transport is compared with the calculated total transport with and without the additional term. It was also found necessary to include a constant factor of 2 to the thus modified transport formulation. The augmented Bailard model was applied to calculate the profile evolution, and the comparisons with measurements for the initially planar beach are shown in Fig. 31 for conditions of all terms included in the calculations and sequentially omitting one term to demonstrate its relative importance. In this comparison, "asymmetry" refers to the nonlinear wave effect in which the forward velocities are of greater magnitude and of shorter duration than the seaward velocities; "no lag" refers to the effect of the wave setup occurring some distance landward of the breaking point as contrasted to considering that the momentum is transferred directly to wave setup: the other terms are self explanatory. It is of interest in these comparisons that if we adopt the following definition for the presence of a bar,

$$\frac{\partial^2 h}{\partial y^2} > 0 , \tag{49}$$

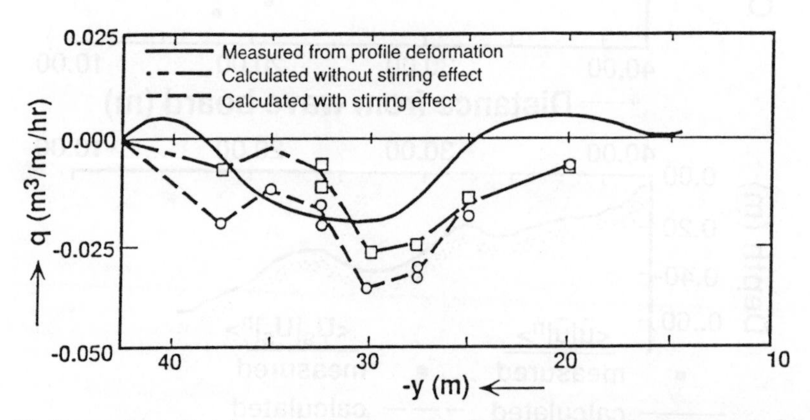

Fig. 30. Comparison of total measured and computed transport distributions across the surf zone with and without the additional effect due to turbulence. (Roelvink and Stive, 1989)

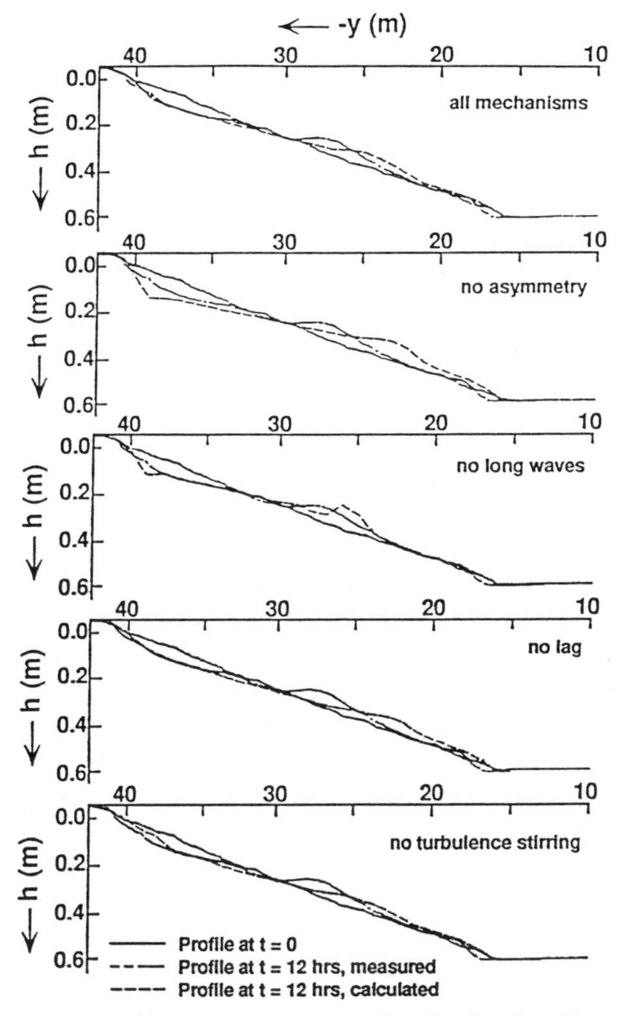

Fig. 31. Comparison of measured and predicted profiles. Predicted profiles presented with various terms omitted. (From Roelvink and Stive, 1989)

immediately landward of the feature, although the measurements clearly show a bar present, only one of the calculated profiles, that with "no long waves", has a bar. The other calculated profiles contain a feature which can be described as a "terrace". The profile resulting from running the model for an additional twelve hours is shown in Fig. 32 which shows that a feature consistent with

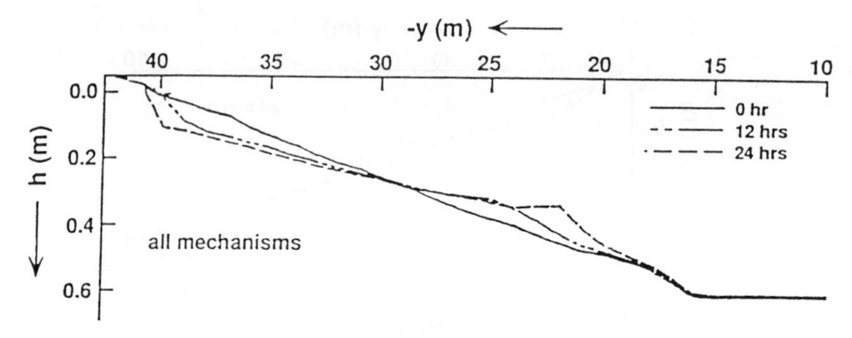

Fig. 32. Effect of running the numerical model for an additional twelve hours. No wave tank data were available for comparison for the additional twelve hours. (Roelvink and Stive, 1989)

the definition of a bar is now present and that this feature continues to move seaward and the shoreline continues to recede. No measurements were available after the initial twelve hours. Only limited information was presented for the second case in which the initial profile contained a bar feature. The measured and computed profiles for this case after twelve hours of run time are presented in Fig. 33. In this case, the model provides a poor representation of the evolved profile. In fact, it appears that the mean square deviation between the final calculated and measured profiles is greater than between the measured initial and final profiles. Thus, a better agreement would have been obtained by setting all coefficients in the transport equation to zero. Finally, the authors state that the model does not approach a stable form for long run times.

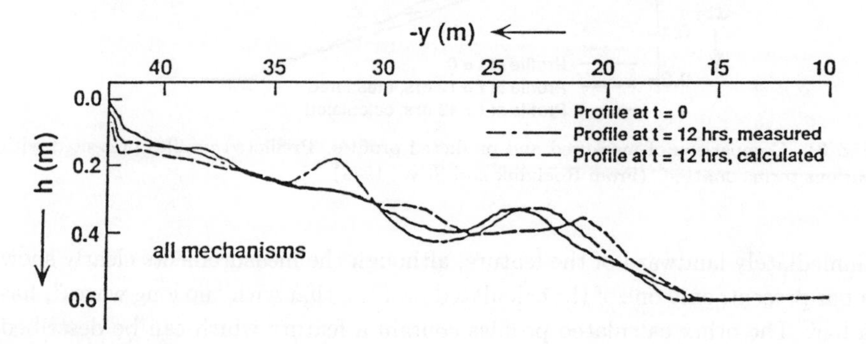

Fig. 33. Comparison of measured and computed profiles for case of a barred initial profile. (Roelvink and Stive, 1989)

Wise *et al.* (1991) have evaluated the Bailard transport model along with the depth integrated hydrodynamic model of Kobayashi *et al.* (1989) using the laboratory measurements of Roelvink and Stive (1989). This paper is especially well detailed and provides the reader with a step by step comparison of the measured and predicted hydrodynamics. The comparisons for cross-shore variations in measured and predicted wave heights, undertow and velocity variance in the low frequency range were presented as were the velocity moments required for application of Eqs. (48) and (49). No comparisons for profile evolution were presented; however, the measured and computed average sediment transport rates were compared over twelve hours of testing and showed reasonable qualitative agreement.

Southgate and Nairn (1993) and Nairn and Southgate (1993) have provided a thorough description of their hydrodynamic and sediment transport models respectively. The Bailard transport relationship is employed; several applications to engineering problems of interest are illustrated.

Wang *et al.* (1992) describe an encouraging application of the Watanabe transport relationship for profile evolution modeling.

Hedegaard *et al.* (1992) have reported on a comprehensive testing of six European models of the "open-loop" kind. The individual models were developed by the Danish Hydraulics Institute (DHI), Delft Hydraulics Laboratory (DH), Hydraulics Research Station (HR), University of Liverpool (UL) and the Laboratoire National d'Hydraulique (LNH). This paper has six coauthors, one from each of the agencies for which a model was tested. Transport relationships included those of Bailard and Watanabe. Variables highlighted in the paper include wave transformation, velocity distributions and profile evolution. The testing was based on both an intercomparison of results from the various models and comparison of the predicted profile evolution with the test results from the large wave tank in Hannover. Tests included both regular and irregular waves. It is not clear whether the comparisons presented included improvements due to adjustments to the models. It is evident that in order to maintain the stability of the profiles, it was necessary in at least some cases to smooth either the transport or the profile although the methods and frequencies of smoothening were not quantified. The results from the regular and irregular wave tests are described in the following paragraphs.

The regular test results were carried out with an incident wave height and period of 1.5 m (in a water depth of 5 m) and 6 s. The median sediment size was 0.33 mm and the initial beach slope was 1:20. Figure 34, which presents

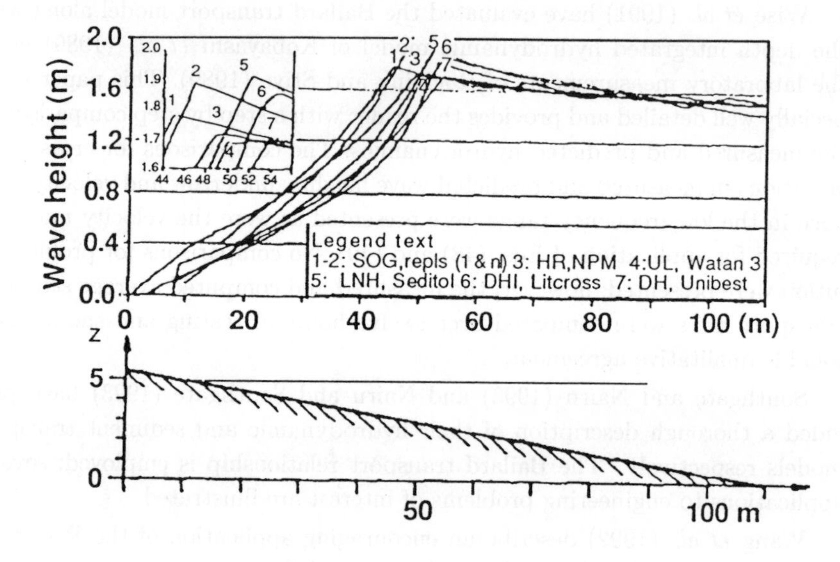

Fig. 34. Intercomparison of calculated shoaling results. (Hedegaard *et al.*, 1992)

the intercomparison of wave shoaling results for monochromatic waves on a beach of uniform slope, shows that outside the breaking zone, the models were in reasonable agreement and that the range of differences within the breaking zone is within approximately ±0.2 m. Figure 35 compares the calculated velocity distributions over depth at five locations across the profile. Within the breaking zone, the maximum shoreward and seaward velocities ranged from approximately 0.1 to 1 m/s, a factor of five. Outside the breaking zone, the maximum velocities ranged from 0.3 to 0.6 m/s, a factor of 2.0. It is evident that velocity profiles exhibiting a strong landward velocity near the upper portions of the water column are responding to the transfer of surface momentum within the breaking zone and those profiles with predominant seaward velocities are outside the breaking zone and are primarily a return flow due to the mass transport. It is of interest that one of the models also predicts the average vertical velocity inside the breaking zone and that the predicted maximum vertical velocity near the break point was approximately 0.1 m/s which exceeds the sediment fall velocity. The calculated initial cross-shore transport rates on a planar beach are compared in Fig. 36. It is shown that substantial differences exist and the need is evident in some models for smoothing or other

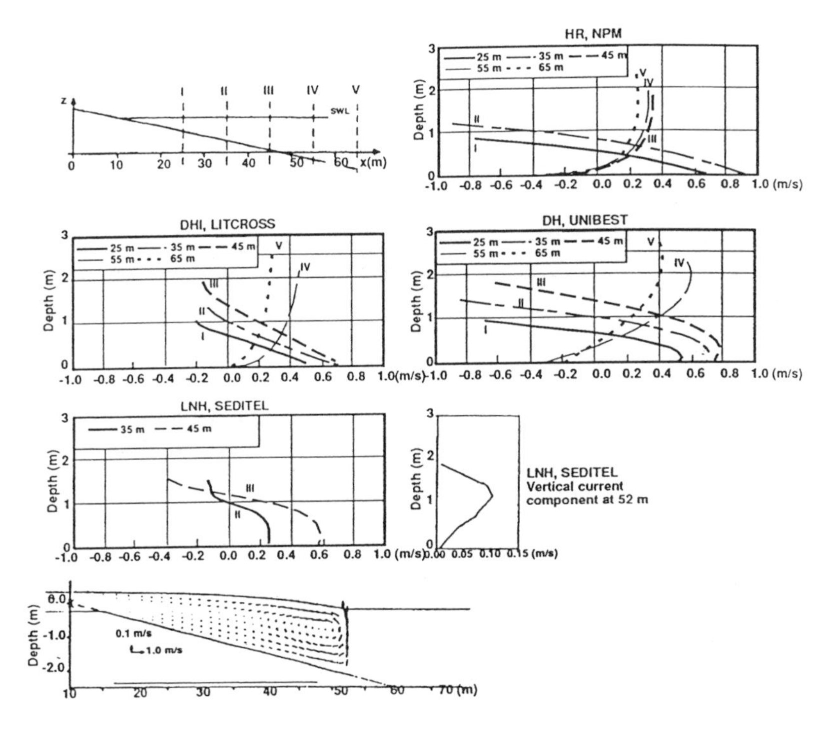

Fig. 35. Intercomparison of calculated velocity distributions at five locations across the profile. (Hedegaard *et al.*, 1992)

stabilization in order to prevent the profile from becoming unstable. It is of interest that the magnitudes of the extreme initial transport vary over a factor of approximately 18. The calculated and measured profiles are compared in Fig. 37 after 4.3 hours of evolution. All of the models correctly predict the formation of bars and the location and height of the bar is reasonably represented by most models.

The irregular wave tests were conducted with a Jonswap spectrum with a significant wave height, $H_s = 1.5$ m (at a distance of 5 m in front of the wavemaker in a water depth of 5 m), a peak wave period of 6 s and well sorted sand with a median diameter of 0.22 m. The measured profile evolution, root mean square wave heights, and spectra at various locations and times are presented in Fig. 38. It is immediately evident that with irregular waves, no bar was produced, a somewhat surprising result with the smaller sediment size

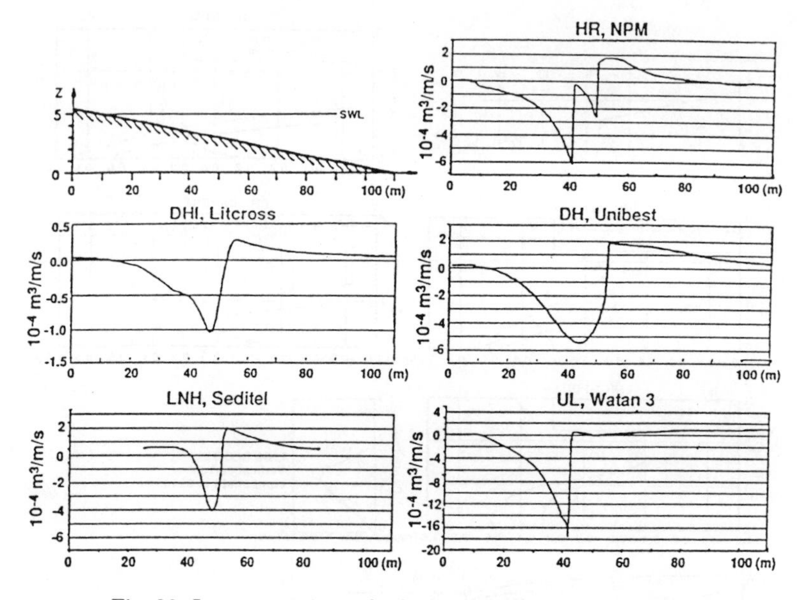

Fig. 36. Intercomparison of calculated initial transport rates.

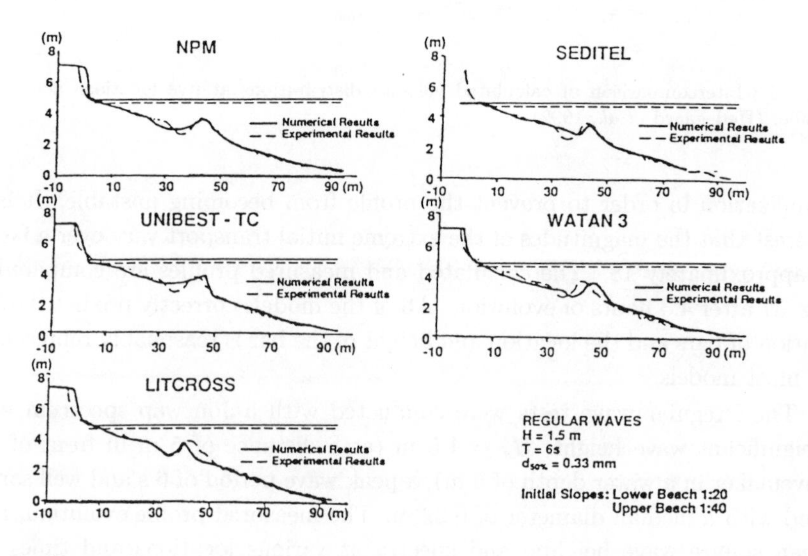

Fig. 37. Measured and predicted profiles for regular wave tests. (Hedegaard *et al.*, 1992)

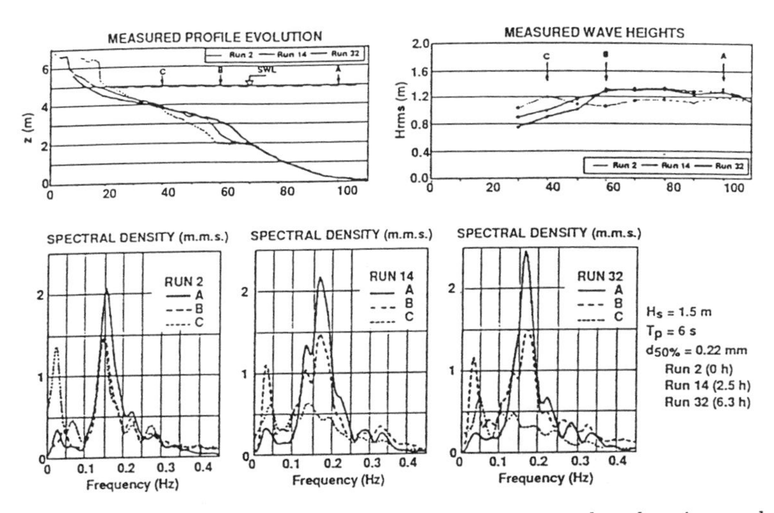

Fig. 38. Measured profile evolution, wave heights and spectra at three locations and times. (Hedegaard *et al.*, 1992)

which, taken alone, should increase the tendency for bar formation. This result supports the hypothesis that bars are the result of wave breaking repeatedly at a fixed point as predicted by the models for regular waves. Measured and calculated transport rates by the five models are presented in Fig. 39. The measured bathymetry was used as input to these calculations. The maximum measured transport rates occur during the first measurement period with a peak of 12×10^{-4} $m^3/m/s$. The corresponding peak calculated values range from 3×10^{-4} $m^3/m/s$ to 8×10^{-4} $m^3/m/s$. It is also of interest that the calculated transport distributions are much more irregular than the measured distributions. Rather than a decrease in the peak transport rate with time as was the case for the measurements, three of the four models predicted the peak transport to occur at later times. One of the models appears to become more unstable with time. The final comparison is for measured and calculated profiles and is shown in Fig. 40. Consistent with the measurements, none of the models predicted bar formation, although results of two of the three models shown exhibit features significantly different from those that occurred in the measurements. In particular, the "HR, NPM" model has an unrealistic foreshore and the "UL, WATAN3" has a very irregular profile and a feature close to the shoreline that appears to approximate a subdued bar. It is not

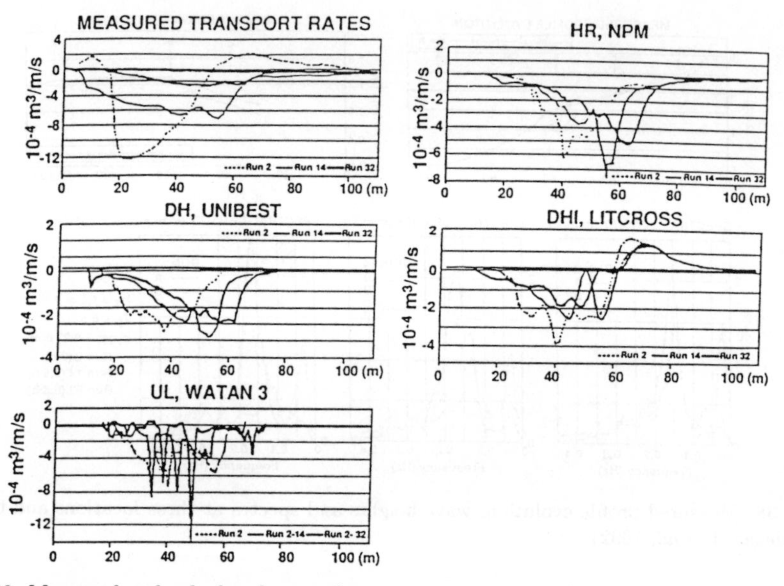

Fig. 39. Measured and calculated cross-shore transport rates at three times. Irregular wave tests. (Hedegaard *et al.*, 1992)

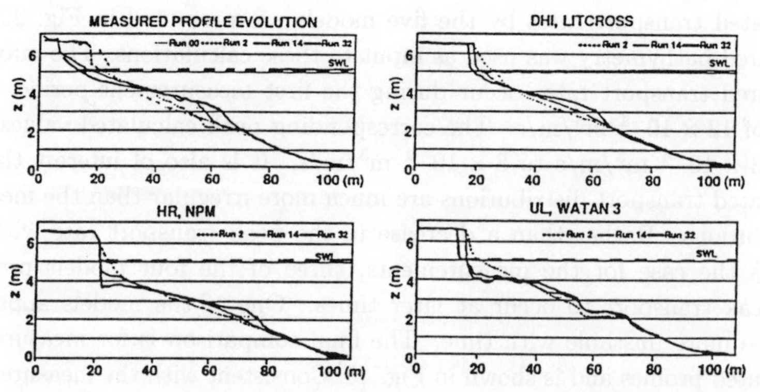

Fig. 40. Measured and calculated profile evolution at three times. (Hedegaard *et al.*, 1992)

clear whether or not these models would remain stable if allowed to run for a long time.

5. Summary and Conclusions

The forces, mechanisms and formulations for profile evolution and cross-shore sediment transport have been reviewed and evaluated on the basis of physics and comparisons of numerical model predictions with available laboratory tests of beach profile evolution. The following summary statements result from this review.

Geometric or "static" models consider the profile to remain in equilibrium with the conditions causing profile modification. Thus, there is not a need to specify cross-shore sediment transport relationships explicitly; however, the volumetric transport can be deduced from the profile changes and the sediment conservation equation. Geometric models are more appropriate for those changes characterized by time scales that are longer than the process under consideration, for example, shoreline response to sea level change.

Seasonal shoreline fluctuations are a manifestation of significant cross-shore sediment transport and can amount to 30 m or so. The causes of these changes are not well understood or predictable but may be the result of greater wave intensity and perhaps greater wave steepness. Because of the engineering significance of seasonal shoreline changes, they should be investigated and may offer a basis for evaluating cross-shore sediment transport models.

Several factors have been proposed as causative agents for the formation of offshore bars, including undertow, wave reflection, infragravity waves, edge waves, and interference between forced and free harmonics of the fundamental wave. Model test results available from large wave tank facilities, field studies, theoretical considerations and results from the more complete open loop numerical models support undertow and waves breaking repeatedly at the same point as the dominant mechanism. Infragravity waves, which result from the nonlinear interaction of the dominant components in the spectrum but are not responsible for bar formation, are probably important in the inner portion of the surf zone where the primary wave system decreases in magnitude due to breaking. There is only one closed loop model that includes the generation of bars.

On a decadal scale, overwash is a significant process by which barrier islands keep pace with relative sealevel rise. At least two closed loop models account for the effects of overwash. Greater efforts should be made to ensure that overwash is appropriately represented in closed loop models and, on a parallel track, to enhance our capabilities to incorporate overwash in open loop models.

It has not been possible to discern from the literature that an open loop model exists that converges to a stable profile for long run times without significant smoothing. Also it is not clear whether or not *some* comparisons of these models against particular laboratory data sets included calibration specific to that application or, if carried out, the degree of calibration. Thus, it appears that at the present time open loop models should be considered in the research rather than the applications arena. Progress has been impeded by the very great hydrodynamic and sediment transport complexities in the surf zone. Although the identity of the relevant "forces" are understood, their individual large values at equilibrium results in a delicate balance which, when disturbed, is difficult to predict with the altered hydrodynamics and/or profile geometry. In particular, open loop models do not incorporate a strong feedback mechanism that will assure equilibration under wide ranges of conditions.

Both bed load and suspended load transport components are potentially significant over the range of sediment characteristics and wave conditions of engineering interest. It would appear that suspended load transport would be dominant for energetic wave conditions, finer sand, profiles significantly out of equilibrium, near the bar region and in the shallower portion of the profile where the surface induced turbulence due to breaking penetrates to the bottom. Although models including only the suspended component have yielded reasonable results and predicted all known effects of wave and sediment characteristics at least qualitatively correctly, under less energetic conditions bed load could be important.

The major manifestation of the interaction of a shore parallel structure such as a seawall with erosive storm conditions is the scour trough that occurs immediately seaward of the structure. Wave reflection has been proposed as a possible cause of this scour. An alternate heuristic explanation is that the seaward transport processes remove the sand from as near to the location that they would if the structure had not been present. A seawall interaction has been included in at least two closed loop models. More effort should be directed to developing a rational understanding of the processes which cause scour and to incorporating these processes in closed and open loop models.

Although results based on the application of the Bagnold/Bailard sediment transport model are encouraging, it appears appropriate from physical considerations to account for the mobilizing effects of energy dissipation due to wave breaking as was done by Roelvink and Stive.

In future evaluations of model performance it would be useful to present quantitative measures of predictive skill such as the mean square error between predicted and measured profile changes.

References

Aubrey, D. G. (1979). Seasonal patterns of onshore/offshore sediment movement. *J. Geophysical Research* **84** (C10):6347–6354.

Bagnold, R. A. (1940). Beach formation by waves: some model experiments in a wave tank. *J. Institution of Civil Engineers* **15**:27–52.

Bagnold, R. A. (1963). Mechanics of marine sedimentation, in *The Sea: Ideas and Observations.* Vol. 3. Interscience.

Bailard, J. A. (1981). An energetics total load sediment transport model for a plane sloping beach. *J. Geophysical Research* **86** (C11):10938–10954.

Bailard, J. A. (1982). Modeling on-offshore sediment transport in the surf zone. *Proc. 18th Int. Conf. Coastal Eng. ASCE.* 1419–1438.

Bailard, J. A. and D. L. Inman (1981). An energetics bedload model for a plane sloping beach: Local transport. *J. Geophysical Research.* **86** (C3):2035–2043.

Barnett, M. and H. Wang (1988). Effects of a vertical seawall on profile response. *Proc. 21st Int. Conf. Coastal Eng. ASCE.* Chpt. 111, 1493–1507.

Battjes, J. A. and J. P. F. M. Janssen (1978). Energy loss and set-up due to breaking of random waves. *Proc. 16th Int. Conf. Coastal Eng. ASCE.* Chpt. 32, 569–587.

Battjes, J. A. and M. J. F. Stive (1985). Calibration and verification of a dissipation model for random breaking waves. *J. Geophysical Research* **90** (C5):9159–9167.

Birkemeier, W. A. (1985). Field data on seaward limit of profile change. *J. Wtrwy., Port, Coast. and Oc. Eng. ASCE.* **111** (3):598–602.

Bodge, K. R. (1992). Representing equilibrium beach profiles with an exponential expression, *J. Coastal Research* **8** (1):47–55.

Bowen, A. J. (1980). Simple models of nearshore sedimentation: Beach profiles and longshore bars. *The Coastline of Canada*, ed. S. B. McCann, Geological Survey of Canada 1–11.

Bruun, P. (1954). Coast erosion and the development of beach profiles. *U.S. Army Beach Erosion Board Technical Memorandum* No. 44.

Bruun, P. (1962). Sea level rise as a cause of shore erosion. *J. Wtrwy., Port, Coast. and Oc. Eng.* **118**:117–130.

Chiu, T. Y. (1977). Beach and dune response to Hurricane Eloise of September 1975. *Proc. Coastal Sediments 1977, ASCE.* 116–134.

Clausner, J. E., W. A. Birkemeier, and G. R. Clark (1986). Field comparison of four nearshore survey systems. Miscellaneous Paper CERC-86-6. Coastal Engineering Research Center, U.S. Army Waterways Experiment Station.

Dally, W. R. and R. G. Dean (1984). Suspended sediment transport and beach profile development. *J. Wtrwy., Port, Coast. and Oc. Eng. ASCE.* **110** (1):15–33.

Dally, W. R., R. G. Dean, and R. A. Dalrymple (1985). Wave height variation across beaches of arbitrary profile. *J. Geophysical Research* **90** (C6):11917–11927.

Dalrymple, R. A. (1992). Prediction of storm/normal beach profiles. *J. Wtrwy., Port, Coast. and Oc. Eng. ASCE.* **118**, (2):193–200.

Dean, R. G. (1973). Heuristic models of sand transport in the surf zone. *Proc. Conf. Eng. Dynamics in the Surf Zone*, Sydney, Australia.

Dean, R. G. (1977). Equilibrium beach profiles: U.S. Atlantic and Gulf coasts, Ocean Engineering Report No. 12. Department of Civil Engineering, University of Delaware, Delaware.

Dean, R. G. (1987a). Additional sediment input into the nearshore region. *Shore and Beach* **55** (3–4):76–81.

Dean, R. G. (1987b), Coastal sediment processes: toward engineering solutions. *Coastal Sediments 1987, Specialty Conf. Advances in Understanding of Coastal Sediment Processes, ASCE*, New Orleans, Louisiana. Vol. 1, 1–24.

Dean, R. G. and E. M. Maurmeyer (1983). Models for beach profile response. *CRC Handbook on Beach Erosion and Coastal Processes*, ed. P. D. Komar, Chpt. 7, 151–166.

Dette, H. H. and K. Uliczka (1987). Prototype investigations on time-dependent dune recession and beach erosion, *Proc. Conf. Coastal Sediments* 1987, ASCE. 1430–1444.

Edelman, T. (1972). Dune erosion during storm conditions, *Proc. 13th Int. Conf. Coastal Eng. ASCE*. 1305–1312.

Hallermeier, R. J. (1978). Uses for a calculated limit depth to beach erosion. *Proc. 16th Int. Conf. Coastal Eng. ASCE*, Hamburg. 1493–1512.

Hallermeier, R. J. (1981). A profile zonation for seasonal sand beaches from wave climate. *Coastal Eng.* **4**:253–277.

Hedegaard, I. B., J. A. Roelvink, H. Southgate, P. Penchon, J. Nicholson, and L. Hamm (1992). Intercomparison of coastal profile models. *Proc. 23rd Int. Conf. Coastal Eng. ASCE*. Chpt. 162, 2108–2121.

Katoh, K. and S. Yanagishima (1988). Predictive model for daily changes of shoreline. *Proc. 21st Int. Conf. Coastal Eng. ASCE*, Hamburg. Chpt. 93, 1493–1512.

Kobayashi, N. (1987). Analytical solutions for dune erosion by storms. *J. Wtrwy., Port, Coast. and Oc. Eng.* **113** (4):401–418.

Kobayashi, N., G. S. DeSilva, and K. D. Watson (1989). Wave transformation and swash oscillation on gentle and steep slopes. *J. Geophysical Research* **94** (C1):951–966.

Komar, P. D. and W. G. McDougal (1994). An analysis of exponential beach profiles. *J. Coastal Research*, **10** (1):59–69.

Kraus, N. C. (1987). The effects of seawalls on a beach: A literature review, *Proc. ASCE Specialty Conf. Coastal Sediments 1987*. 945–960.

Kraus, N. C., M. Larson, and D. L. Kriebel (1991). Evaluation of beach erosion and accretion predictors. *Proc. Conf. Coastal Sediments 1991, ASCE*. 572–587.

Kraus, N. C. and R. A. Wise (1993). Simulation of January 4, 1992 storm erosion at Ocean City, Maryland. *Shore and Beach* **61** (1):34–40.

Kriebel, D. L. (1982). Beach and dune response to hurricanes. M.Sc. Thesis. Civil Engineering Department, University of Delaware. 349.

Kriebel, D. L. (1986). Verification study of a dune erosion model. *Shore and Beach.* **54** (3):13–20.

Kriebel, D. L. (1987). Beach recovery following Hurricane Elena. *Proc. Conf. Coastal Sediments 1987, ASCE.* 990–1005.

Kriebel, D. L. and R. G. Dean (1985). Numerical simulation of time-dependent beach and dune erosion, *Coastal Eng.* **9**:221–245.

Kriebel, D. L., W. R. Dally, and R. G. Dean (1986). Undistorted Froude model for surf zone sediment transport. *Proc. 20th Int. Conf. Coastal Eng., ASCE.* 1296–1310.

Kriebel, D. L., N. C. Kraus, and M. Larson (1991). Engineering methods for predicting beach profile response. *Proc. Specialty Conf. Coastal Sediments 1991, ASCE.* 557–571.

Kriebel, D. L. and R. G. Dean (1993). Convolution method for time-dependent beach-profile response, *J. Wtrwy., Port, Coast. and Oc. Eng. ASCE.* **119** (2):204–227.

Larson, M. (1988). Quantification of beach profile change. Report No. 1008, Department of Water Resources and Engineering, University of Lund, Lund, Sweden.

Larson, M. and N. C. Kraus (1989). SBEACH: Numerical model for simulating storm-induced beach change, Report 1: Empirical foundation and model development. U.S. Army Coastal Engineering Research Center, U.S. Army Waterways Experiment Station, Technical Report CERC-89-9.

Larson, M. and N. C. Kraus (1990). SBEACH: Numerical model for simulating storm-induced beach change, Report 2: Numerical formulation and model tests. U.S. Army Coastal Engineering Research Center, Waterways Experiment Station, Technical Report CERC-89-9.

Lee, G.-H. and W. A. Birkemeier (1993). Beach and nearshore survey data: 1985-1991 CERC Field Research Facility. U.S. Army Corps of Engineers, Waterways Experiment Station, Technical Report CERC-93-3.

Longuet-Higgins, M. S. (1953). Mass transport in water waves, *Philosophical Trans. Royal Society of London*, Ser. A **245**:535–581.

Longuet-Higgins, M. S. and R. W. Stewart (1964). Radiation stresses in water waves; a physical discussion with applications. *Deep Sea Research* **2**:529–540.

Moore, B. D. (1982). Beach profile evolution in response to changes to water level and wave height. M.Sc. Thesis, Department of Civil Engineering, University of Delaware.

Nairn, R. B. and H. N. Southgate (1993). Deterministic profile modeling of nearshore processes. Part 2: Sediment transport and beach profile development. *Coastal Engineering* **19** (1,2):57-96.

Ohnaka, S. and A. Watanabe (1990). Modeling of wave-current interaction and beach change. *Proc. 22nd Int. Conf. Coastal Eng., ASCE*, Chpt. 185, 2443–2456.

Phillips, O. M. (1977). *The Dynamics of the Upper Ocean.* Cambridge University Press.

Roelvink, J. A. and M. J. F. Stive (1989). Bar-generating cross-shore flow mechanisms on a beach, *J. Geophysical Research* **94** (C4):4785–4800.

Saville, T. (1957). Scale effects in two-dimensional beach studies. *Trans. 7th General Meeting Int. Association of Hydraulic Research* **1**:A3.1–A3.10.

Southgate, H. N. (1991). Beach profile modeling: Flume data comparisons and sensitivity tests. *Proc. Conf. Coastal Sediments 1991, ASCE.* 1829–1841.

Southgate, H. N. and R. B. Nairn (1993). Deterministic profile modeling of nearshore processes. Part 1: Waves and currents, *Coastal Eng.* **19** (1,2):27–56.

Stive, M. J. F. and J. A. Battjes (1984). A model for offshore sediment transport. *Proc. 19th Int. Conf. Coastal Eng. ASCE.* Chpt. 97, 1420–1436.

Uliczka, K. and R. B. Nairn (1991). Cross-shore sediment transport modeling and comparison with tests at prototype scale. *Proc. Conf. Coastal Sediments 1991, ASCE.* 462–476.

Vellinga, P. (1983). Predictive computational model for beach and dune erosion during storm surges. *Proc. ASCE Specialty Conf. Coastal Structures 1983.* 806–819.

Vellinga, P. (1986). Beach and dune erosion during storm surges. Dissertation, Delft Technical University, Delft Hydraulics Laboratory Communications No. 372, 185 pages.

Wang, H. G. Miao and L.-H. Lin (1992). A time-dependent nearshore morphological response model. *Proc. 23rd Int. Conf. Coastal Eng. ASCE.* 2513–2527.

Watanabe, A. (1987). 3-dimensional numerical model of beach evolution, *Proc. Conf. Coastal Sediments 1987, ASCE.* 802–817.

Wise, R. A., N. Kobayashi, and A. Wurjanto (1991). Cross-shore sediment transport under irregular waves in surf zones, *Proc. Conf. Coastal Sediments 1991, ASCE.* 658–673.

CONCEPTUAL DESIGN OF RUBBLE MOUND BREAKWATERS

JENTSJE W. VAN DER MEER

This paper first gives an overall view of physical processes involved with rubble mound structures and a classification of these structures. Governing parameters are described as hydraulic parameters (related to waves, wave run-up and run-down, overtopping, transmission and reflection) and as structural parameters (related to waves, rock, cross-section and response of the structure). The description of hydraulic response is divided into:

- wave run-up and run-down,
- wave overtopping,
- wave transmission,
- wave reflection.

The main part of the paper describes the structural response (stability) which is divided into:

- rock armor layers,
- armor layers with concrete units,
- underlayers, filters, toe protection and head,
- low-crested structures,
- berm breakwaters.

The description of hydraulic and structural response is given in design formulas or graphs which can be used for a conceptual design of rubble mound structures.

1. Introduction

The design tools given in this paper on rubble mound structures are based on tests of schematized structures. Structures in prototype may differ (substantially) from the test-sections. Results based on these design tools can therefore only be used in a conceptual design. The confidence bands given for most formulas support the fact that reality may differ from the mean curve. It is advised to perform physical model investigations for detailed design of all important rubble mound structures.

1.1. *Processes involved with rubble mound structures*

The processes involved with stability of rubble mound structures under wave (possibly combined with current) attack are given in a basic scheme in Fig. 1.

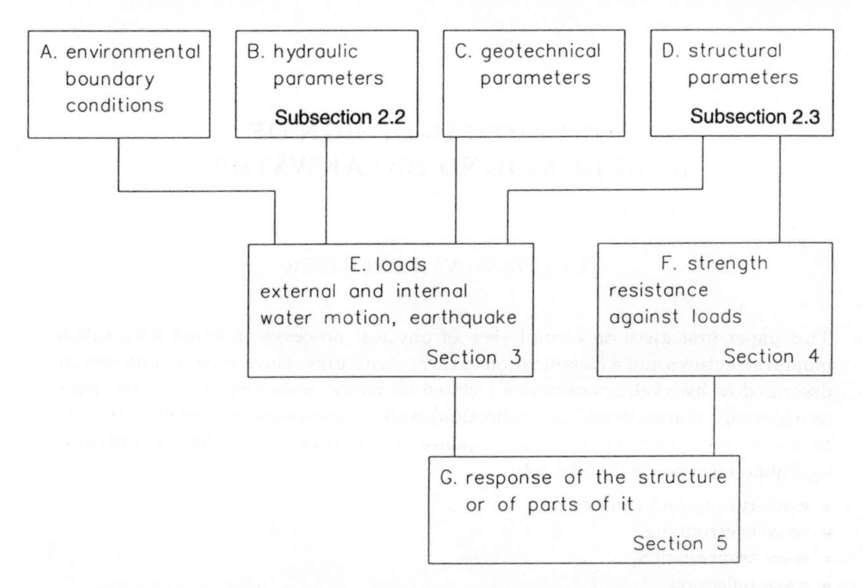

Fig. 1. Basic scheme of assessment of rubble mound structure response.

The environmental conditions (wave, current, and geotechnical characteristics) lead to a number of parameters which describe the boundary conditions at or in front of the structure (A in Fig. 1). These parameters are not affected by the structure itself, and generally, the designer of a structure has no influence on them. Wave height, wave height distribution, wave breaking, wave period, spectral shape, wave angle, currents, foreshore geometry, water depth, set-up and water levels are the main hydraulic environmental parameters. These environmental parameters are not described in this paper. A specific geotechnical environmental condition is an earthquake.

Governing parameters can be divided into parameters related to hydraulics (B in Fig. 1), related to geotechnics (C) and to the structure (D). Hydraulic parameters are related to the description of the wave action on the structure (hydraulic response). These hydraulic parameters are described in Subsecs. 2.1 and 2.2. The main hydraulic responses are wave run-up, run-down, wave overtopping, wave transmission and reflection. These are covered in Sec. 3.

Geotechnical parameters are related to, for instance, liquefaction, dynamic gradients, and excessive pore pressures. They are not described in this paper.

The structure can be described by a large number of structural parameters (D) and some important ones are the slope of the structure, the mass and mass density of the rock, rock or grain shape, surface smoothness, cohesion, porosity, permeability, shear and bulk moduli, and the dimensions and cross-section of the structure. The structural parameters related to hydraulic stability are described in Subsecs. 2.3.

The loads on the structure or on structural elements are given by the environmental, hydraulic, geotechnical, and structural parameters together (E in Fig. 1). These loads can be divided into those due to external water motion on the slope, loads generated by internal water motion in the structure, and earthquakes. The external water motion is affected amongst others things by the deformation of the wave (breaking or not breaking), the run-up and run-down, transmission, overtopping and reflection. These topics are described in Subsec. 2.2. The internal water motion describes the penetration or dissipation of water into the structure, the variation of pore pressures and the variation of the freatic line. These topics are not treated in this paper.

Almost all structural parameters might have some or large influence on the loads. Size, shape, and grading of armor stones have influence on the roughness of the slope, and therefore, on run-up and run-down. Filter size and grading, together with the above-mentioned characteristics of the armor stones, have an influence on the permeability of the structure, and hence, on the internal water motion.

The resistance against the loads (waves, earthquakes) can be called the strength of the structure (F in Fig. 1). Structural parameters are essential in the formulation of the strength of the structure. Most of them also influence the loads, as described above.

Finally the comparison of the strength with the loads leads to a description of the response of the structure or elements of the structure (G in Fig. 1), the description of the so-called failure mechanisms. The failure mechanism may be treated in a deterministic or probabilistic way.

Hydraulic structural responses are stability of armor layers, filter layers, crest and rear, toe berms, and stability of crest walls and dynamically stable slopes. These structural responses are described in Sec. 4. Geotechnical responses or interactions are slip failure, settlement, liquefaction, dynamic response, internal erosion, and impacts. They are not described in this paper.

Figure 1 can also be used in order to describe the various ways of physical and numerical modeling of the stability of coastal and shoreline structures. A black box method is used if the environmental parameters (A in Fig. 1) and the hydraulic (B) and structural (D) parameters are modeled physically, and the responses (G) are given in graphs or formulas. A description of water motion (E) and strength (F) is not considered.

A grey box method is used if parts of the loads (E) are described by theoretical formulations or numerical models which are related to the strength (F) of the structure by means of a failure criterion or reliability function. The theoretical derivation of a stability formula might be the simplest example of this.

Finally, a white box is used if all relevant loads and failure criterions can be described by theoretical/physical formulations or numerical models, without empirical constants. It is obvious that it will take a long time and a lengthy research effort before coastal and shoreline structures can be designed by means of a white box.

The colors black, grey, and white, used for the methods described above do not suggest a preference. Each method can be useful in a design procedure.

This paper will deal with physical processes and design tools, which means that design tools should be described so that

- they are easily applicable;
- the range of application should be as wide as possible;
- research data from various investigations should, wherever possible, be combined and compared, rather than giving the data of different investigations separately.

1.2. *Classification of rubble mound structures*

Rubble mound structures can be classified by the use of the $H/\Delta D$ parameter, where H = wave height, Δ = relative mass density, and D = characteristic diameter of structure, armor unit (rock or concrete), stone, shingle, or sand. Small values of $H/\Delta D$ give structures as caissons or structures with large armor units. Large ones imply gravel beaches and sand beaches.

Only two types of structures have to be distinguished if the response of the various structures is concerned. These types can be classified into statically stable and dynamically stable.

Statically stable structures are those where no or minor damage is allowed under design conditions. Damage is defined as displacement of armor units. The mass of individual units must be large enough to withstand the wave forces during design conditions. Caissons and traditionally designed breakwaters belong to the group of statically stable structures. The design is based on an optimum solution between design conditions, allowable damage, and cost of construction and maintenance. Static stability is characterised by the design parameter damage, and can roughly be classified by $H/\Delta D = 1\text{–}4$.

Dynamically stable structures are structures where profile development is concerned. Units (stones, gravel, or sand) are displaced by wave action until a profile is reached where the transport capacity along the profile is reduced to a very low level. Material around the still-water level is continuously moving during each run-up and run-down of the waves, but when the net transport capacity has become zero, the profile has reached an equilibrium. Dynamic stability is characterised by the design parameter profile, and can roughly be classified by $H/\Delta D > 6$.

The structures concerned in this paper are rock armored breakwaters and slopes and berm type breakwaters. The structures are roughly classified by $H/\Delta D$ =1–10.

An overview of the types of structures with different $H/\Delta D$ values is shown in Fig. 2, which gives the following rough classification:

1. $H/\Delta D < 1$ Caissons or seawalls.
   No damage is allowed for these fixed structures. The diameter, D, can be the height or width of the structure.
2. $H/\Delta D = 1\text{–}4$ Stable breakwaters.
   Generally uniform slopes are applied with heavy artificial armor units or natural rock. Only little damage (displacement) is allowed under severe design conditions. The diameter is a characteristic diameter of the unit, such as the nominal diameter.
3. $H/\Delta D = 3\text{–}6$ S-shaped and berm breakwaters.
   These structures are characterised by more or less steep slopes above and below the still-water level with a more gentle slope in between which reduces the wave forces on the armor units. Berm breakwaters are designed with a very steep seaward slope and a horizontal berm just above the still-water level. The first storms develop a more gentle profile which is stable further on. The profile changes to be expected are important.
4. $H/\Delta D = 6\text{–}20$ Rock slopes/beaches.
   The diameter of the rock is relatively small and cannot withstand severe wave attack without displacement of material. The profile which is being developed under different wave boundary conditions is the design parameter.

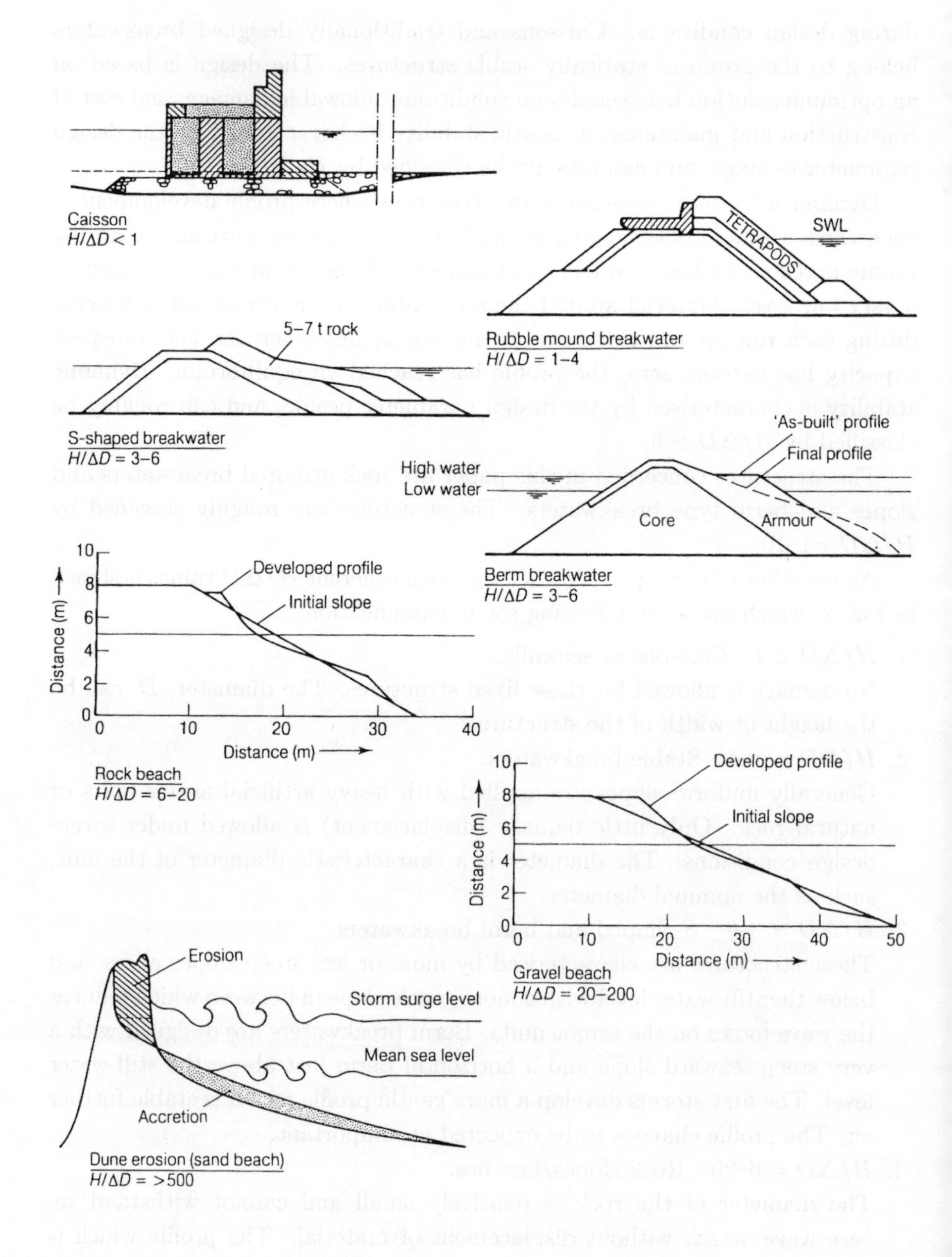

Fig. 2. Type of structure as a function of $H/\Delta D$.

5. $H/\Delta D = 15$–500 Gravel beaches.
   Grain sizes, roughly between ten centimeters and four millimeters, can be classified as gravel. Gravel beaches will change continuously under varying wave conditions and water levels (tide). Again the development of the profile is one of the design parameters.
6. $H/\Delta D > 500$ Sand beaches (during storm surges).
   Also material with very small diameters can withstand severe wave attack. The Dutch coast is partly protected by sand dunes. The dune erosion and profile development during storm surges is one of the main design parameters. Extensive basic research has been performed on this topic (Vellinga, 1986 and Steetzel, 1993).

2. Governing Parameters

The wave boundary conditions can mainly be described by the wave height, period, or steepness and the surf similarity parameter.

The main hydraulic responses of rubble mound structures to wave conditions are wave run-up and run-down, overtopping, transmission, and reflection. The governing parameters related to these hydraulic responses are illustrated in Fig. 3, and are discussed in this section. The hydraulic responses itself are described in Sec. 3.

Finally, a large number of structural parameters can be defined, related to waves, rock, the cross-section and the response of the structure. The structural parameters are treated in this section, the structural responses in Sec. 4.

2.1. *Wave parameters*

Wave conditions are given principally by the incident wave height at the toe of the structure, H, usually as the significant wave height, H_s (average of the highest 1/3 of the waves) or $H_{m0}(4\sqrt{m_0}$, based on the spectrum); the mean or peak wave periods, T_m or T_p (based on statistical or spectral analysis); the angle of wave attack, β, and the water depth, h.

The wave height distribution at deep water can be described by the Rayleigh distribution and in that case one characteristic value, for instance, the significant wave height, describes the whole distribution. In shallow and depth limited water, the highest waves break and in most cases the wave height distributions can no longer be described by the Rayleigh distribution. In those situations, the actual wave height distribution may be important to consider,

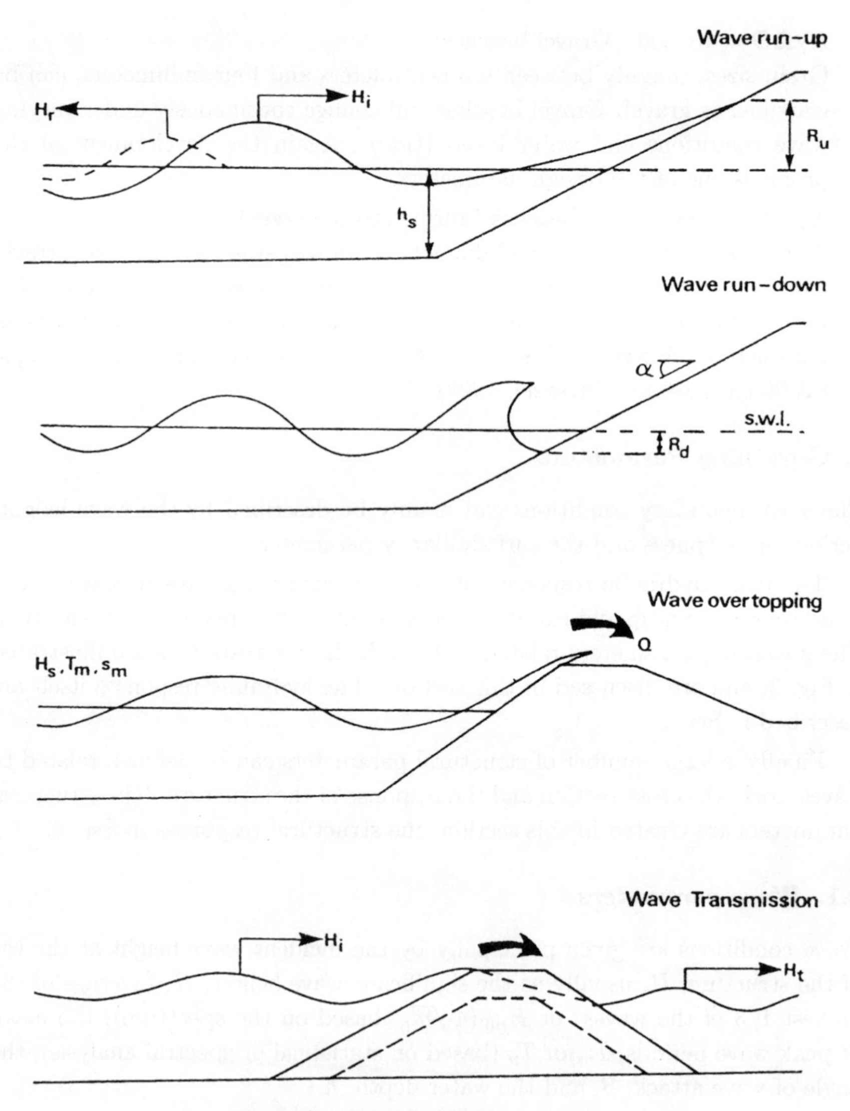

Fig. 3. Governing hydraulic parameters.

or another characteristic value other than the significant wave height. Characteristic values often used are the 2% wave height, $H_{2\%}$, and the $H_{1/10}$, being the average of the highest ten percent of the waves. For a Rayleigh distribution, the values $H_{2\%} = 1.4\ H_s$ and $H_{1/10} = 1.27\ H_s$ hold.

The influence of the wave period is often described as a wave length and related to the wave height, resulting in a wave steepness. The wave steepness, s, can be defined by using the deep water wave length, $L = gT^2/2\pi$:

$$s = \frac{2\pi H}{gT^2} \tag{1}$$

If the situation considered is not really in deep water (in most cases), the wave length at deep water and at the structure differ and therefore, a fictitious wave steepness is obtained. In fact, the wave steepness as defined in Eq. (1) describes a dimensionless wave period. Use of H_s and T_m or T_p in Eq. 1 gives a subscript to s, respectively s_{om} and s_{op}.

The most useful parameter describing wave action on a slope, and some of its effects, is the surf similarity or breaker parameter, ξ, also termed the Iribarren number Ir:

$$\xi = \tan\alpha/\sqrt{s} \tag{2}$$

The surf similarity parameter has often been used to describe the form of wave breaking on a beach or structure (see Fig. 4). It should be noted that

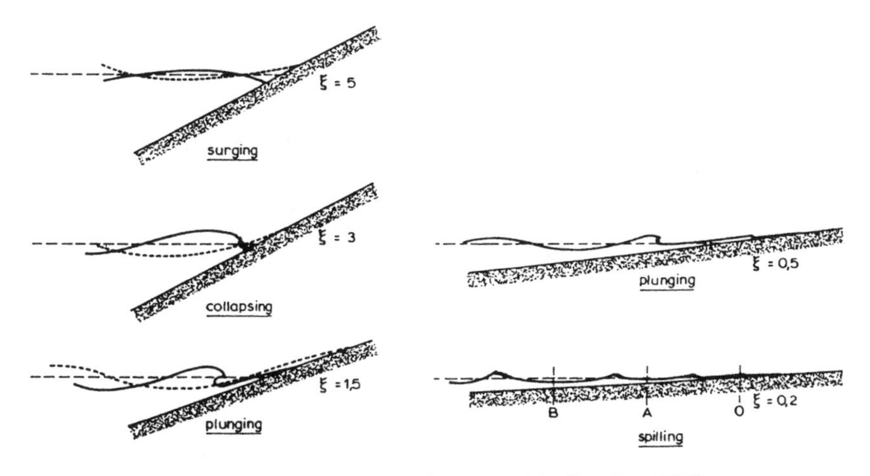

Fig. 4. Breaker types as a function of ξ. (Battjes, 1974)

different versions of this parameter are defined within this paper, reflecting the approaches of different researchers. In this section, ξ_m and ξ_p are used when s is described by s_{om} or s_{op}.

2.2. *Hydraulic parameters*

2.2.1. *Run-up and run-down*

Wave action on a rubble mound structure will cause the water surface to oscillate over a vertical range generally greater than the incident wave height. The extreme levels reached in each wave, termed run-up and run-down, R_u and R_d respectively, and defined relative to the static water level, constitute important design parameters (see Fig. 3). The design run-up level will be used to determine the level of the structure crest, the upper limit of protection or other structural elements, or as an indicator of possible overtopping or wave transmission. The run-down level is often taken to determine the lower extent of main armor protection and/or a possible level for a toe berm.

Run-up and run-down are often given in a dimensionless form, R_{ux}/H_s and R_{dx}/H_s, where the subscript x describes the level considered (for instance 2%) or significant (s).

2.2.2. *Overtopping*

If extreme run-up levels exceed the crest level, the structure will overtop. This may occur for relatively few waves under the design event, and a low overtopping rate may often be accepted without severe consequences for the structure or the area protected by it. Sea walls and breakwaters are often designed on the basis that some (small) overtopping discharge is to be expected under extreme wave conditions. The main design problem therefore reduces to one of dimensioning the cross-section geometry such that the mean overtopping discharge, q, under design conditions remains below acceptable limits.

The most simple dimensionless parameter, Q, for the mean overtopping discharge, q, can be defined by

$$Q = \frac{q}{\sqrt{gH_s^3}} \tag{3}$$

Sometimes the wave steepness and slope angle also have influence on the overtopping and in that case, the definition of dimensionless overtopping discharge as in Eq. (3) may be extended by including s_{om} or s_{op} and/or $\cot\alpha$. Various definitions can be found in Owen (1980), Bradbury *et al.* (1988), and De Waal and Van der Meer (1992).

2.2.3. *Wave transmission*

Breakwaters with relatively low crest levels may be overtopped with sufficient severity to excite wave action behind. Where a breakwater is constructed of relatively permeable construction, long wave periods may lead to transmission of wave energy through the structure. In some cases, the two different responses will be combined.

The quantification of wave transmission is important in the design of low-crested breakwaters intended to protect beaches or shorelines, and in the design of harbor breakwaters where long wave periods transmitted through the breakwater could cause movement of ships or other floating bodies.

The severity of wave transmission is described by the coefficient of transmission, C_t, defined in terms of the incident and transmitted wave heights, H_i and H_t respectively, or the total incident and transmitted wave energies, E_i and E_t:

$$C_t = \frac{H_t}{H_i} = \sqrt{\frac{E_t}{E_i}} \tag{4}$$

2.2.4. *Wave reflections*

Wave reflections are of importance on the open coast, and at commercial and small boat harbours. The interaction of incident and reflected waves often lead to a confused sea in front of the structure, with occasional steep and unstable waves of considerable hazard to small boats. Reflected waves can also propagate into areas of a harbor previously sheltered from wave action. They will lead to increased peak orbital velocities, increasing the likelihood of movement of beach material. Under oblique waves, reflection will increase littoral currents and hence local sediment transport.

All coastal structures reflect some proportion of the incident wave energy. This is often described by a reflection coefficient, C_r, defined in terms of the incident and reflected wave heights, H_i and H_r respectively, or the total incident and reflected wave energies, E_i and E_r:

$$C_r = \frac{H_r}{H_i} = \sqrt{\frac{E_r}{E_i}} \tag{5}$$

When considering random waves, values of C_r may be defined using the significant incident and reflected wave heights as representative of the incident and reflected energies.

2.3. *Structural parameters*

Structural parameters can be divided into four categories which will be treated in this section, and are related to:

- waves;
- rock;
- the cross-section; and
- the response of the structure.

2.3.1. *Structural parameters related to waves*

The most important parameter which gives a relationship between the structure and the wave conditions has been used in Subsec. 1.2. In general, the $H/\Delta D$ gives a good classification. For the design of rubble mound structures, this parameter should be defined in more detail.

The wave height is usually the significant wave height, H_s, defined either by the average of the highest one third of the waves or by $4\sqrt{m_0}$. For deep water, both definitions give more or less the same wave height. For shallow water conditions substantial differences may be present.

The relative buoyant density is described by

$$\Delta = \frac{\rho_r}{\rho_w} - 1 \tag{6}$$

where,

ρ_r = mass density of the rock (saturated surface dry mass density),
ρ_w = mass density of water.

The diameter used is related to the average mass of the rock and is called the nominal diameter:

$$D_{n50} = \left(\frac{M_{50}}{\rho_r}\right)^{1/3} \tag{7}$$

where,

D_{n50} = nominal diameter,
M_{50} = median mass of unit given by 50% on mass distribution curve.

The parameter $H/\Delta D$ changes to $H_s/\Delta D_{n50}$.

Another important structural parameter is the surf similarity parameter, which relates the slope angle to the wave period or wave steepness, and which gives a classification of breaker types. The surf similarity parameter $\xi(\xi_m, \xi_p$, with $T_m, T_p)$ is defined in Subsec. 2.1.

For dynamically stable structures with profile development, a surf similarity parameter cannot be defined as the slope is not straight. Furthermore, dynamically stable structures are described by a large range of $H_s/\Delta D_{n50}$ values. In that case, it is possible to also relate the wave period to the nominal diameter and to make a combined wave height – period parameter. This parameter is defined by

$$H_oT_o = \frac{H_s}{\Delta D_{n50}} * T_m\sqrt{\frac{g}{D_{n50}}} \tag{8}$$

The relationship between $H_s/\Delta D_{n50}$ and H_oT_o is listed in Table 1.

Table 1. Relationship between $H_s/\Delta D_{n50}$ and H_oT_o.

Structure	$H_s/\Delta D_{n50}$	H_oT_o
Statically stable breakwaters	1–4	< 100
Rock slopes and beaches	6–20	200–1500
Gravel beaches	15–500	1000–200,000
Sand beaches	> 500	$> 200,000$

Another parameter which relates both wave height and period (or wave steepness) to the nominal diameter was introduced by Ahrens (1987). In the Shore Protection Manual (SPM, 1984), $H_s/\Delta D_{n50}$ is often called the stability number, N_s. Ahrens included the local wave steepness in a modified stability number, N_s^*, defined by

$$N_s^* = N_s s_p^{-1/3} = \frac{H_s}{\Delta D_{n50}} s_p^{-1/3} \tag{9}$$

In this equation, s_p is the local wave steepness and not the deep water wave steepness. The local wave steepness is calculated using the local wave length from the Airy theory, where the deep water wave steepness is calculated by Eq. (1). This modified stability number, N_s^*, has a close relationship with H_oT_o

Table 2. Wave height-period parameters.

$$\frac{H_s}{\Delta D_{n50}} = N_s$$

$$\frac{H_s}{\Delta D_{n50}} s_p^{-1/3} = N_s^*$$

$$\frac{H_s}{\Delta D_{n50}} T_m \sqrt{\frac{g}{D_{n50}}} = H_oT_o$$

$$\frac{H_s}{\Delta D_{n50}} s_m^{-0.5} \sqrt{\frac{2\pi H_s}{D_{n50}}} = H_oT_o$$

$$\xi_m = \frac{\tan\alpha}{\sqrt{s_{om}}} = \frac{\tan\alpha}{\sqrt{\frac{2\pi H_s}{gT_m^2}}}$$

defined by Eq. (8). An overview of possible wave height-period parameters is given in Table 2.

2.3.2. *Structural parameters related to rock*

The most important parameter which is related to the rock is the nominal diameter defined by Eq. (7). Related to this is, of course, M_{50}, the 50% value on the mass distribution curve. The grading of the rock can be given by the D_{85}/D_{15}, where D_{85} and D_{15} are the 85% and 15% values of the sieve curves, respectively, or by D_{n85}/D_{n15}, based again on the mass distribution curves. These are the most important parameters as far as stability of armor layers is concerned. Examples of gradings are shown in Table 3, showing the relationship between classes of stone (here simply taken as M_{85}/M_{15} and D_{85}/D_{15}).

Further details of recommended methods of specifying gradings and of suggested gradings are given in the CUR/CIRIA Manual (1991), Subsec. 3.2.2.4.

Table 3. Examples of gradings.

Narrow grading $D_{85}/D_{15} < 1.5$		Wide grading $1.5 < D_{85}/D_{15} < 2.5$		Very wide grading $D_{85}/D_{15} > 2.5$	
Class	D_{85}/D_{15}	Class	D_{85}/D_{15}	Class	D_{85}/D_{15}
15–20t	1.10	1–9t	2.08	50–1000 kg	2.71
10–15t	1.14	1–6t	1.82	20–1000 kg	3.68
5–10t	1.26	100–1000 kg	2.15	10–1000 kg	4.64
3-7t	1.33	100–500 kg	1.71	10–500 kg	3.68
1–3t	1.44	10–80 kg	2.00	10–300 kg	3.10
300–1000 kg	1.49	10–60 kg	1.82	20–300 kg	2.46

2.3.3. *Structural parameters related to the cross-section*

There are many parameters related to the cross-section and most of them are obvious. Figure 5 gives an overview. The parameters are:

Crest freeboard, relative to still water level	R_c
Armor crest freeboard relative to swl	A_c
Difference between crown wall and armor crest	F_c
Armor crest level relative to the seabed	h_c
Structure width	B
Width of armor berm at crest	G_c
Thickness of armor, underlayer, filter	t_a, t_u, t_f
Area porosity	n_a
Angle of structure slope	α
Depth of the toe below still-water level	h_t

The permeability of the structure has an influence on the stability of the armor layer. This depends on the size of filter layers and core and can be given by a notional permeability factor, P. Examples of P are shown in Fig. 6, based on the work of Van der Meer (1988a).

The lower limit of P is an armor layer with a thickness of two diameters on an impermeable core (sand or clay) and with only a thin filter layer. This lower boundary is given by $P = 0.1$. The upper limit of P is given by a homogeneous structure which consists only of armor stones. In that case, $P = 0.6$. Two other values are shown in Fig. 6 and each particular structure should be compared with the given structures in order to make an estimation of the P factor. It should be noted that P is not a measure of porosity!

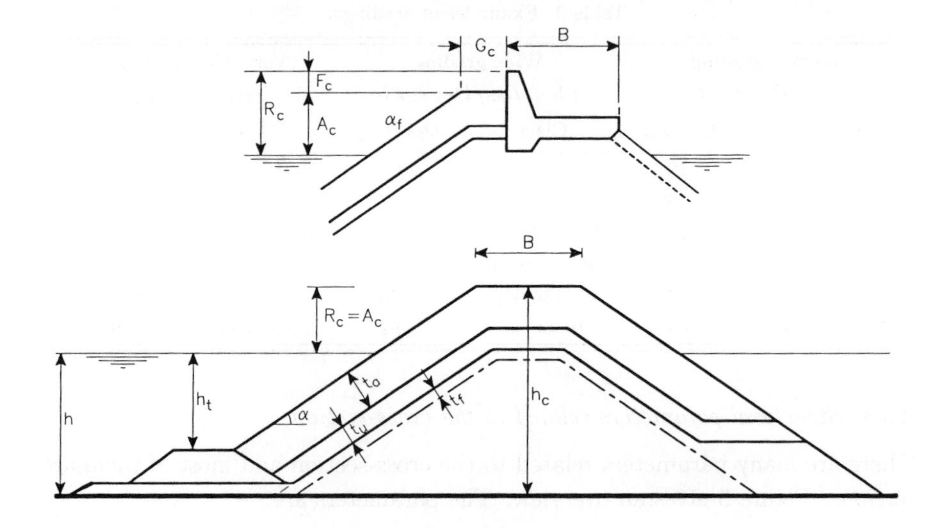

Fig. 5. Governing parameters related to the cross-section.

The estimation of P from Fig. 6 for a particular structure must, more or less, be based on engineering judgement. Although the exact value may not precisely be determined, a variation of P around the estimated value may well give an idea about the importance of the permeability.

The permeability factor P can also be determined by using a numerical (pc)-model that can give the volume of water that penetrates through the armor layer during run-up. Calculations should be done for the structures with $P = 0.5$ and 0.6 (see Fig. 6) and for the actual structure. As $P = 0.1$ gives no penetration at all, a graph can be made with P versus penetrated volume of water. The calculated volume for the actual structure then gives in the graph the P-value. The procedure has been described in more detail in Van der Meer (1988a).

With numerical models developed by Kobayashi and Wurjanto (1989, 1990), Van Gent (1993), or Engering *et al.* (1993), the penetration of water during run-up can be calculated fairly easy.

2.3.4. *Structural parameters related to the response of the structure*

The behaviour of the structure can be described by a few parameters. Statically

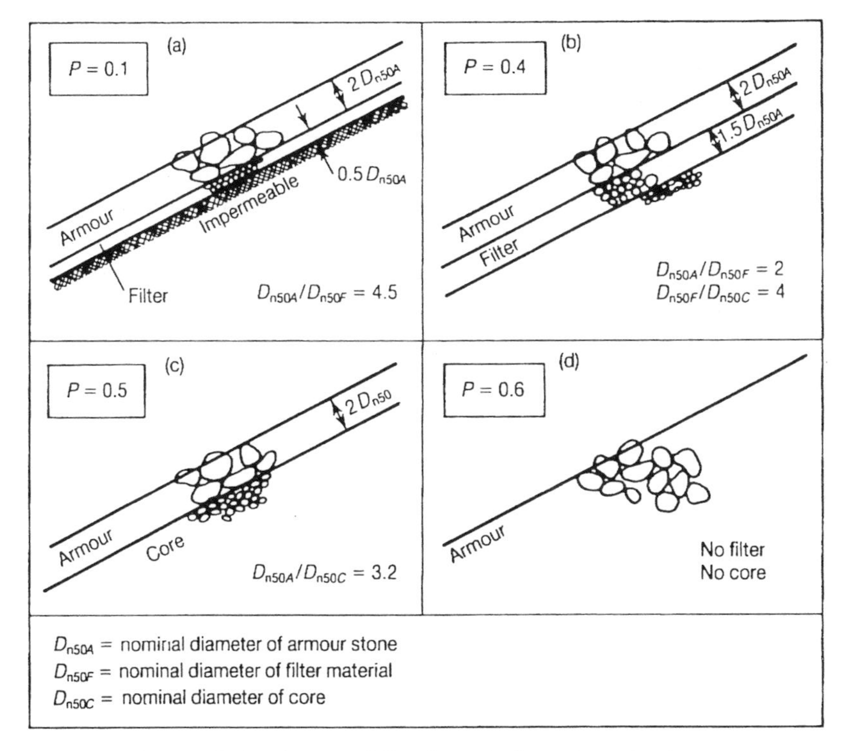

Fig. 6. Notional permeability factor P for various structures.

stable structures are described by the development of damage. This can be the amount of rock that is displaced or the displaced distance of a crown wall. Dynamically stable structures are described by a developed profile.

The damage to the armor layer can be given as a percentage of displaced stones related to a certain area (the whole or a part of the layer). In this case, however, it is difficult to compare various structures as the damage figures are related to different totals for each structure. Another possibility is to describe the damage by the erosion area around still-water level. When this erosion area is related to the size of the stones, a dimensionless damage level is presented which is independent of the size (slope angle and height) of the structure. This damage level is defined by

$$S = \frac{A_e}{D_{n50}^2} \tag{10}$$

where,

S = damage level,

A_e = erosion area around still-water level.

A plot of a structure with damage is shown in Fig. 7, the damage level taking both settlement and displacement into account. A physical description of the damage, S, is the number of squares with a side D_{n50} which fit into the erosion area. Another description of S is the number of cubic stones with a side of D_{n50} eroded within a D_{n50} wide strip of the structure. The actual number of stones eroded within this strip can be more or less than S, depending on the porosity, the grading of the armor stones and the shape of the stones. Generally, the actual number of stones eroded in a D_{n50} wide strip is equal to 0.7 to 1 times the damage S.

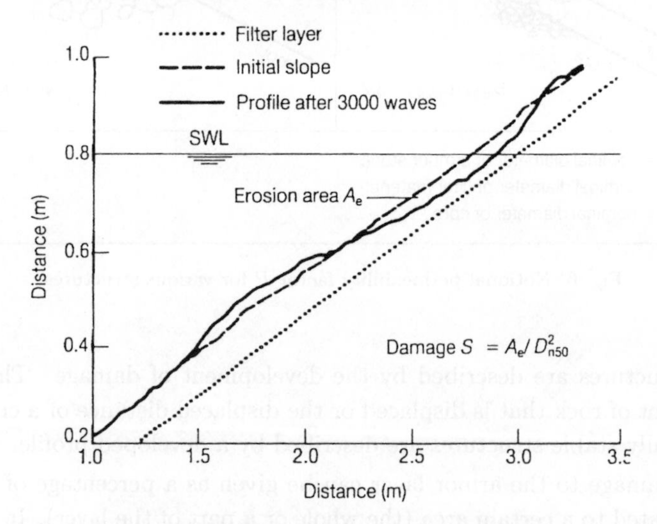

Fig. 7. Damage S based on erosion area A_e.

The limits of S depend mainly on the slope angle of the structure. For a two diameter thick armor layer, the values in Table 4 can be used. The initial damage of S = 2–3 is according to the criterion of the Hudson formula which gives 0–5% damage. Failure is defined as exposure of the filter layer. For S values higher than 15–20, the deformation of the structure results in an S-shaped profile and should be called dynamically stable.

Another definition is suggested for damage to concrete armor units. Such damage can be defined as the relative damage, N_o, which is the actual number of units (displaced, rocking, etc.) related to a width (along the longitudinal axis of the structure) of one nominal diameter, D_n. For cubes, D_n is the side of the cube; for tetrapods, $D_n = 0.65\ D$, where D is the height of the unit; for accropode, $D_n = 0.7D$; and for Dolosse, $D_n = 0.54\ D$ (with a waist ratio of 0.32).

Table 4. Design values of S for a two diameter thick armor layer.

Slope	Initial damage	Intermediate damage	Failure
1:1.5	2	3–5	8
1:2	2	4–6	8
1:3	2	6–9	12
1:4	3	8–12	17
1:6	3	8–12	17

An extension of the subscript in N_o can give the distinction between units displaced out of the layer, units rocking within the layer (only once or more times), etc. In fact the designer can define his own damage description, but the actual number is related to a width of one D_n. The following damage descriptions will be used in this paper:

$$N_{od} = \text{units displaced out of the armor layer (hydraulic damage)},$$
$$N_{or} = \text{rocking units},$$
$$N_{omov} = \text{moving units}, \ N_{omov} = N_{od} + N_{or}\ .$$

The definition of N_{od} is comparable with the definition of S, although S includes displacement and settlement, but does not take into account the porosity of the armor layer. Generally, S is about twice N_{od}.

Dynamically stable structures are those where profile development is accepted. Units (stones, gravel, or sand) are displaced by wave action until a profile is reached where the transport capacity along the profile is reduced to a minimum. Dynamic stability is characterised by the design parameter profile.

An example of a schematised profile is shown in Fig. 8. The initial slope was 1:5 which is relatively gentle and one should note that Fig. 8 is shown on a distorted scale. The profile consists of a beach crest (the highest point of the profile), a curved slope around still-water level (above still-water level steep, below gentle), and a steeper part relatively deep, below still-water level. For gentle slopes (shingle slope > 1:4), a step is found at this deep part. The profile is characterised by a number of lengths, heights, and angles and these were related to the wave boundary conditions and structural parameters (Van der Meer, 1988a).

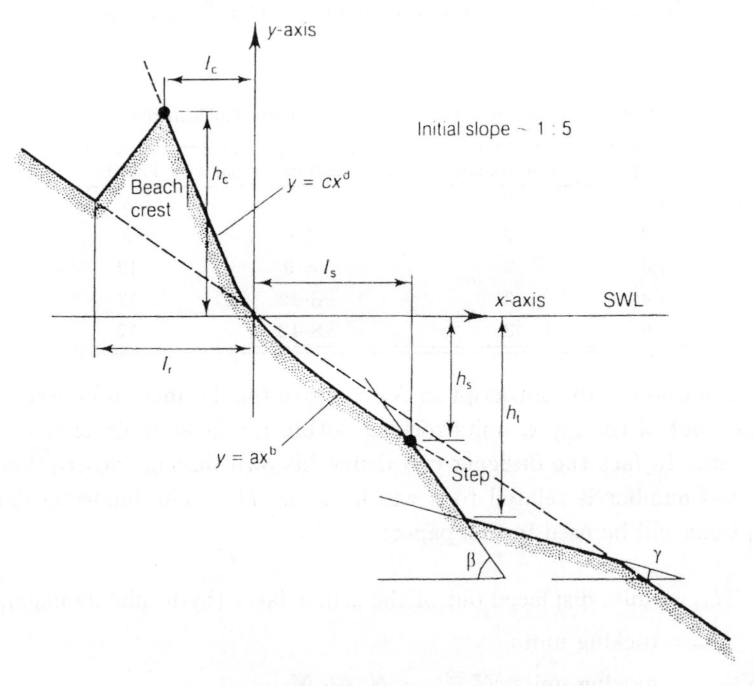

Fig. 8. Schematised profile on a 1:5 initial slope.

3. Hydraulic Response

This section presents methods that may be used for the calculation of the hydraulic response parameters which were also given in Fig. 3:

- run-up and run-down levels;
- overtopping discharges;
- wave transmission; and
- wave reflections.

Where possible, the prediction methods are identified with the limits of their application. These methods are generally available to describe the hydraulic response for only a few simplified cases, either because tests have been conducted for a limited range of wave conditions or because the structure geometry tested often represents a simplification in relation to many practical structures. It will therefore be necessary to estimate the performance for real situations from predictions for related (but dissimilar) structure configurations. Where this is not possible, or the predictions are less reliable than expected, physical model tests should be conducted.

3.1. *Wave run-up and run-down*

Predictions of R_u and R_d may be based on simple empirical equations, supported by model test results, or on numerical models of wave/structure interaction. One-dimensional numerical models of wave run-up have been developed by Kobayshi and Wurjanto (1989, 1990), Van Gent (1993), and Engering *et al.* (1993). A two-dimensional model has been developed by Van der Meer *et al.* (1992). These numerical models will not be discussed here.

All calculation methods require parameters to be defined precisely. Run-up and run-down levels are defined relative to still-water level (see Fig. 3). On some bermed and shallow slopes, run-down levels may not fall below still-water. All run-down levels in this paper are given as positive if below still-water level, and all run-up levels will also be given as positive if above it.

The upward excursion is generally greater than the downward, and the mean water level on the slope is often above still-water level. Again this may be most marked on bermed and shallow slopes. These effects often complicate the definition, calculation, or measurement of run-down parameters.

Much of the field data available on wave run-up and run-down apply to gentle and smooth slopes. Some laboratory measurements have been made on steeper smooth slopes and on porous armored slopes. Prediction methods for smooth slopes may be used directly for armored slopes that are filled or fully grouted with concrete or bitumen. These methods can also be used for rough nonporous slopes with an appropriate reduction factor.

The behaviour of waves on rough porous (rubble mound) slopes is very different from that on nonporous slopes, and the run-up performance is not well predicted by adapting equations for smooth slopes. Different data must be used. This difference is illustrated in Fig. 9, where significant relative run-up, R_{us}/H_s, is plotted for both smooth and rock slopes. The greatest divergence between the performance of the different slope types is seen for $1 < \xi_p < 5$. For ξ_p above 6 or 7, the run-up performance of smooth and porous slopes tends to be very similar values. In that case, the wave motion is surging up and down the slope without breaking, and the roughness and porosity is then less important.

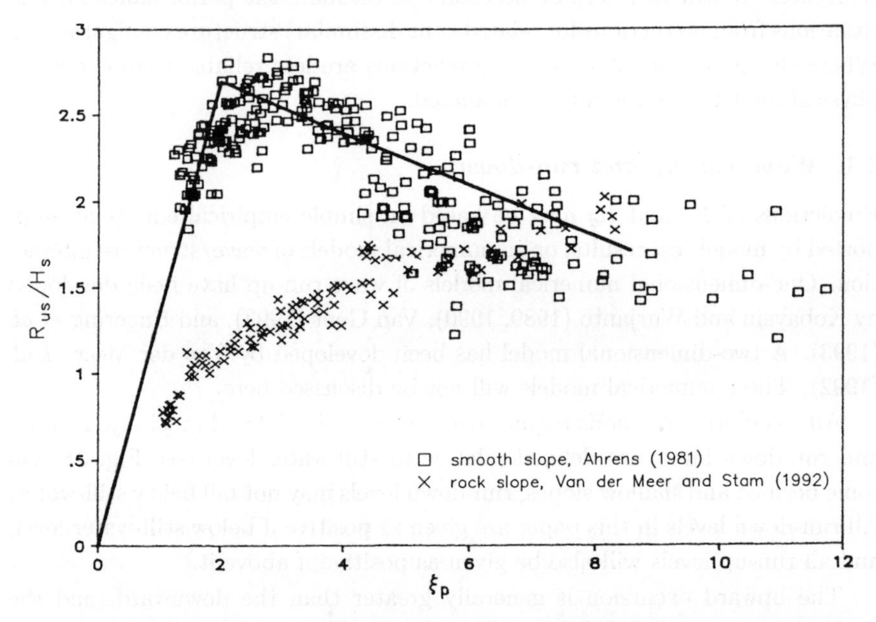

Fig. 9. Comparison of relative significant run-up for smooth and rubble mound slopes.

Delft Hydraulics has recently performed various applied fundamental research studies in physical scale models on wave run-up and overtopping on various structures (De Waal and Van der Meer, 1992). Run-up has extensively been measured on rock slopes. The influence on run-up and overtopping of berms, roughness on the slope (also one layer of rock) and shallow water, has been measured for smooth slopes. Finally, the influence of short-crested waves and oblique (long- and short-crested) waves has been studied on wave run-up and overtopping. All research was commissioned by the Technical Advisory Committee for Water Defenses (TAW) in The Netherlands. De Waal and Van der Meer (1992) give an overall view of the final results, such as design formulas and design graphs. Only the main results will be summarized here, then run-up and run-down on rock slopes will be dealt with.

A general run-up formula can be given with a smooth slope as a basis. This formula for the 2% relative run-up $R_{u2\%}$ is given by

$$\frac{R_{u2\%}}{H_s} = 1.5\gamma\xi_p \text{ with a maximum of } 3.0\ \gamma \tag{11}$$

where H_s = the significant wave height, γ = a total reduction factor for various influences, and ξ_p = the surf similarity parameter based on the peak period. This general formula is shown in Fig. 10.

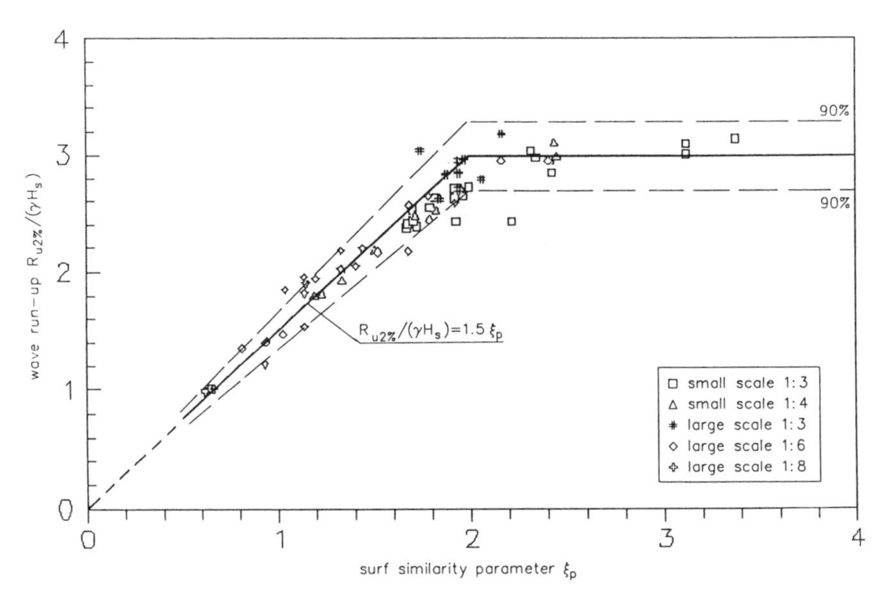

Fig. 10. General formula for wave run-up on a smooth slope, including various effects.

The influence of berms, roughness, shallow water, and oblique wave attack on wave run-up and overtopping can be given as reduction factors γ_b, γ_f, γ_h, and γ_β, respectively. They are defined as the ratio of run-up on a slope considered to that on a smooth impermeable slope under otherwise identical conditions (TAW, 1974). The total reduction factor then becomes

$$\gamma = \gamma_b\gamma_f\gamma_h\gamma_\beta. \tag{12}$$

For the reduction factors one is referred to De Waal and Van der Meer (1992). An overview of Eq. (12) with all test results is given in Fig. 11. In this figure, the vertical axis gives the relative run-up, divided by the total reduction factor. Equation (11) can be considered as the average of the data (except for a small overprediction around $\xi_p = 2$). The reliability of the equation can be described by assuming the factor 1.5 as a stochastic variable with a normal distribution with mean 1.5 and variation coefficient (standard deviation divided by mean) of 0.085. The 90% confidence bands, for instance, can then be calculated by using $1.5 + 1.64^*0.085 = 1.64$ and $1.5 - 1.64^*0.085 = 1.36$ in Eq. (11). These bands are also shown in Fig. 11.

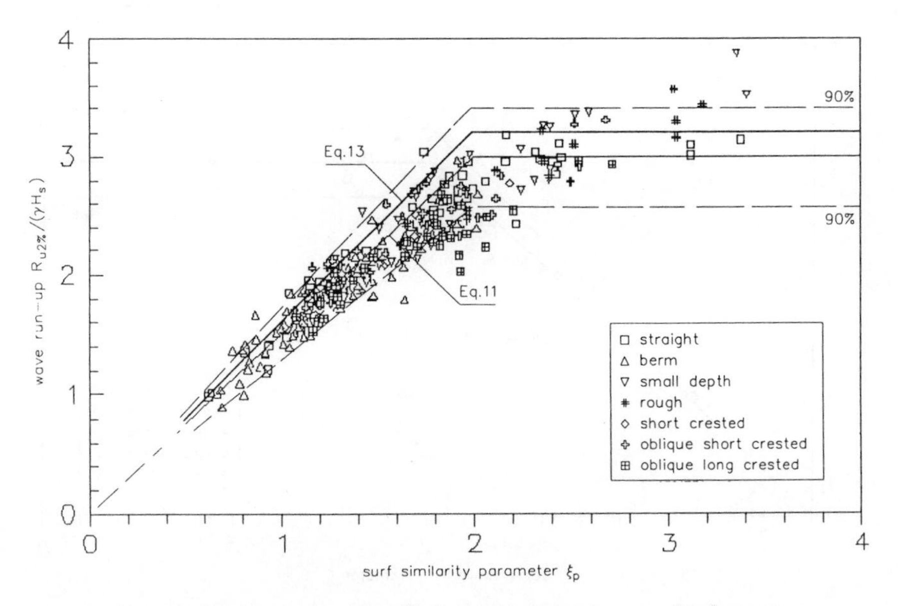

Fig. 11. Summary of test results for various structures and influences.

In a probabilistic design method, one can use Eq. (11) with the given variation coefficient. For a guideline, it is often a practice to include some safety in a formula, say, about one standard deviation. Therefore, the (deterministic) design formula used in The Netherlands is not Eq. (11), but

$$\frac{R_{u2\%}}{H_s} = 1.6\gamma\xi_p \text{ with a maximum of } 3.2\,\gamma. \tag{13}$$

Run-up levels on rubble slopes armored with rock armor or rip-rap have been measured in laboratory tests. In many instances, the rubble core has been reproduced as fairly permeable, except for those particular cases where an impermeable core has been used. Therefore, test results often span a range within which the designer must interpolate.

Analysis of test data from measurements by Van der Meer and Stam (1992) has given prediction formulas for rock slopes with an impermeable core, described by a notional permeability factor $P = 0.1$, and porous mounds of relatively high permeability given by $P = 0.4 - 0.6$. The notional permeability factor P was described in Subsec. 2.3.3, Fig. 6.

Two sets of empirically derived formulas can be given for a run-up on rock slopes. The first set gives the run-up as a function of the surf similarity or breaker parameter. Coefficients for various run-up levels were derived. Secondly, the run-up was described as a Weibull distribution, including all possible run-up levels.

The prediction formulas for run-up versus surf similarity parameters are

$$\frac{R_{ux}}{H_s} = a\xi_m \text{ for } \xi_m < 1.5 \tag{14}$$

$$\frac{R_{ux}}{H_s} = b\xi_m^c \text{ for } \xi_m > 1.5 \; . \tag{15}$$

The run-up for permeable structures ($P > 0.4$) is limited to a maximum:

$$\frac{R_{ux}}{H_s} = d \; . \tag{16}$$

In the above equations, coefficients are given in Table 5. Values for the coefficients a, b, c, and d have been determined for levels of $i = 0.1\%, 1\%, 2\%, 5\%$, 10%, significant, and mean run-up levels.

Table 5. Coefficients in run-up Eqs. (14)–(16) for rock slopes.

Run-up level	a	b	c	d
0.1%	1.12	1.34	0.55	2.58
1%	1.01	1.24	0.48	2.15
2%	0.96	1.17	0.46	1.97
5%	0.86	1.05	0.44	1.68
10%	0.77	0.94	0.42	1.45
sign.	0.72	0.88	0.41	1.35
mean	0.47	0.60	0.34	0.82

Results of the tests and the equations are shown, for example, values of $i =$ 2%, and significant, for each of $P = 0.1$ and $P > 0.4$, in Figs. 12 and 13.

The reliability of Eqs. (14)–(16) can be described by assuming coefficients a, b, and d as stochastic variables with a normal distribution. The variation coefficients for these coefficients are 7% for $P < 0.4$ and 12% for $P \geq 0.4$. Confidence bands can be calculated based on these variation coefficients.

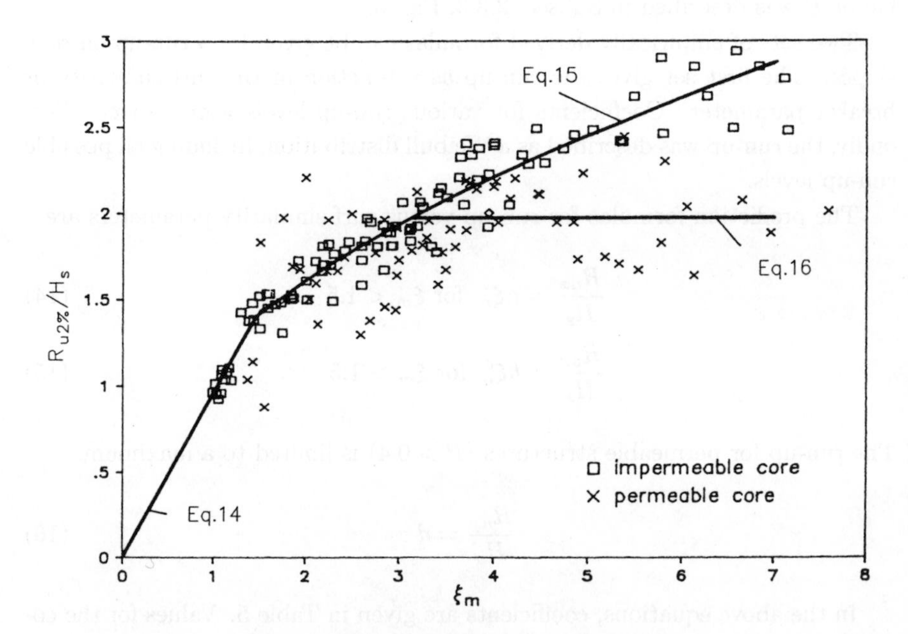

Fig. 12. Relative 2% run-up on rock slopes.

The second method is to describe the run-up as a Weibull distribution:

$$p = P\{R_u > R_{up}\} = e^{-\left(\frac{R_{up}}{b}\right)^c} \tag{17}$$

or

$$R_{up} = b(-\ln p)^{\frac{1}{c}} \tag{18}$$

where,

$p =$ probability (between 0 and 1),

$R_{up} =$ run-up level exceeded by p * 100% of the run-up levels,

$b =$ scale parameter,

$c =$ shape parameter.

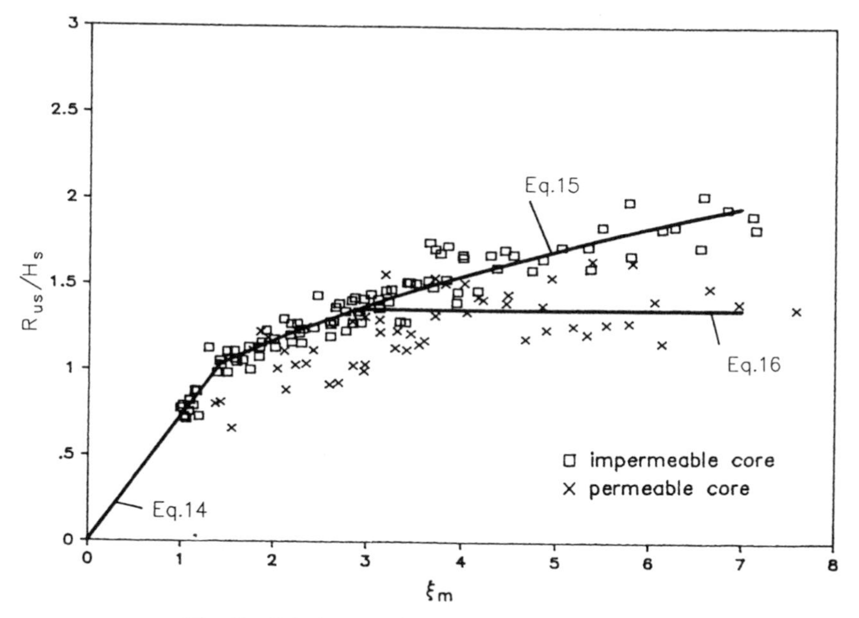

Fig. 13. Relative significant run-up on rock slopes.

The shape parameter c defines the shape of the curve. For $c = 2$, a Rayleigh distribution is obtained. The scale parameter can be described by,

$$\frac{b}{H_s} = 0.4 s_{om}^{-0.25} \cot\alpha^{-0.2}. \tag{19}$$

The shape parameter is described by,

a. for plunging waves:

$$c = 3.0 \xi_m^{-0.75}, \tag{20}$$

b. for surging waves,

$$c = 0.52 P^{-0.3} \xi_m^P \sqrt{\cot\alpha}\ . \tag{21}$$

The transition between Eqs. (20) and (21) is described by a critical value for the surf similarity parameter, ξ_{mc},

$$\xi_{mc} = [5.77 P^{0.3} \sqrt{\tan\alpha}]^{\frac{1}{P+0.75}} \tag{22}$$

For $\xi_m < \xi_{mc}$, Eq. (20) should be used and for $\xi_m > \xi_{mc}$, Eq. (21). The formulas are only applicable for slopes with $\cot\alpha \leq 2$. For steeper slopes the distributions on a 1:2 slope may give a first estimation.

Examples of run-up distributions are shown in Fig. 14. The reliability of Eqs. (20)–(22) can be described by assuming b as a stochastic variable with a normal distribution. The variation coefficient of b is 6% for $P < 0.4$ and 9% for $P \geq 0.4$. Confidence bands can be calculated by means of these variation coefficients.

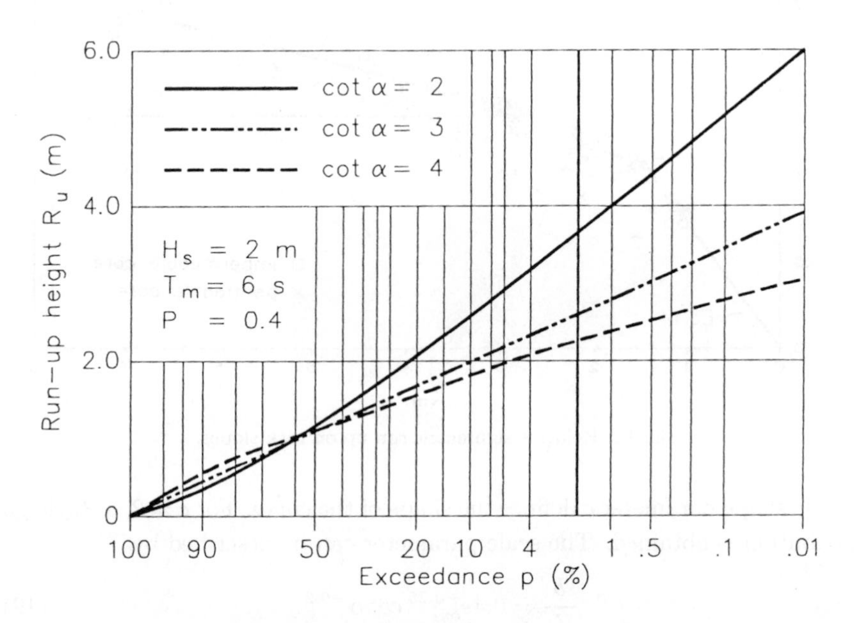

Fig. 14. Run-up distributions on a rock slope.

Run-down levels on porous rubble slopes are also influenced by the permeability of the structure, and by the surf similarity parameter. Analysis of the 2% run-down level on the sections tested by Van der Meer (1988a) has given an equation which includes the effects of structure permeability and wave steepness:

$$\frac{R_{d2\%}}{H_s} = 2.1\sqrt{\tan\alpha} - 1.2P^{0.15} + 1.5e^{-60s_{om}}. \tag{23}$$

Test results are shown in Fig. 15 for an impermeable and a permeable core. The presentation with ξ_m only gives a large scatter. Including the slope angle and the wave steepness separately and also the permeability as in Eq. (23) reduces the scatter considerably.

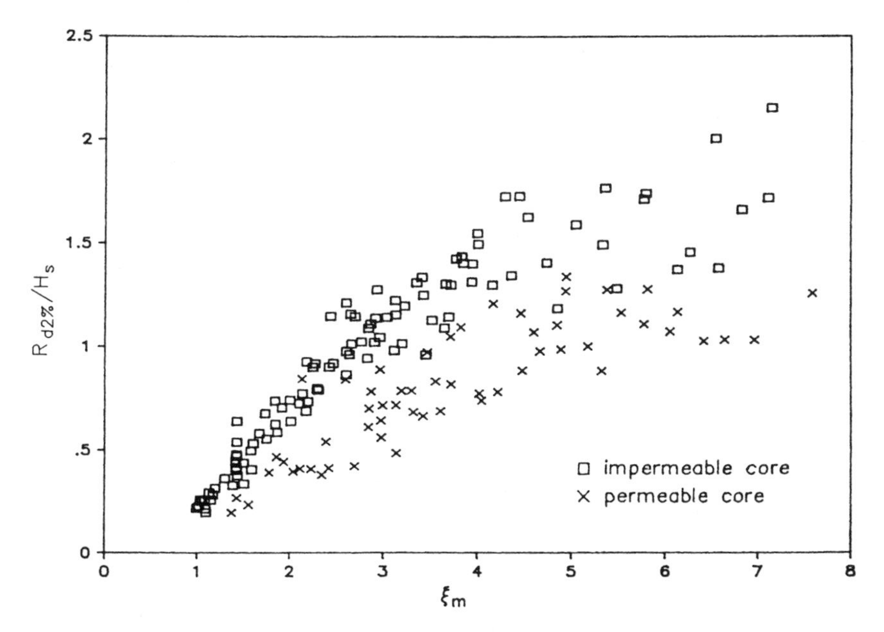

Fig. 15. Run-down $R_{d2\%}/H_s$ on impermeable and permeable rock slopes.

3.2. *Overtopping*

In the design of many sea walls and breakwaters, the controlling hydraulic response is often the wave overtopping discharge. Under random waves this varies greatly from one wave to another. There are some data available to quantify this variation. For many cases it is sufficient to use the mean discharge, q, usually expressed as a discharge per metre run (m^3/s per m).

Suggested critical values of q for various design situations are summarised in Fig. 16. This incorporates recommended limiting values of the mean discharge for the stability of crest and rear armor to types of seawalls and/or the safety of vehicles and people. Most data in Fig. 16 refer to old Japanese data. De Gerloni *et al.* (1991) and Franco (1993) investigated critical overtopping discharges at vertical breakwaters for (model) cars and people (model and prototype). Their results give larger allowable overtopping (see Fig. 16). Dutch results in Delft Hydraulics' large Deltaflume on a prototype grass dike showed that situations with overtopping more than 1–5 l/s per meter were not accessible anymore for people by foot. This situation is also given in Fig. 16.

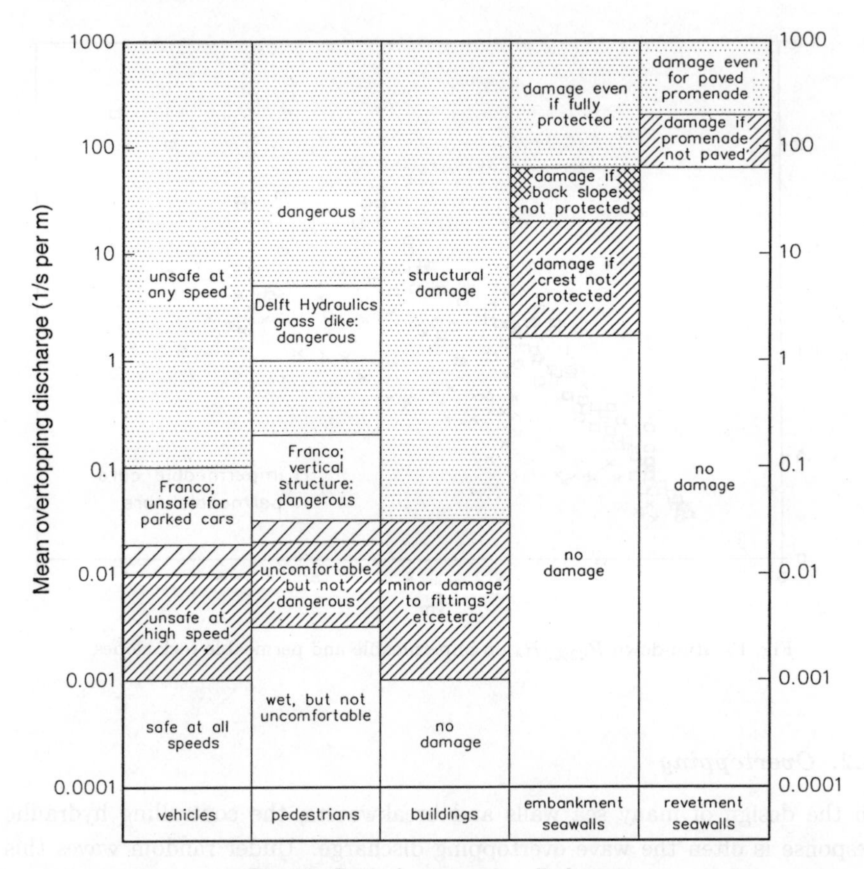

Fig. 16. Critical overtopping discharges.

The calculation of overtopping discharge for a particular structure geometry, water level, and wave condition is based on empirical equations fitted to hydraulic model test results. The data available on overtopping performance is restricted to a few structural geometries. A well-known and wide data set applies to plain and bermed smooth slopes without crown walls (Owen, 1980). More restricted studies have been reported by TAW (1974), Bradbury *et al.* (1988), and Aminthi and Franco (1988). Recently, Delft Hydraulics finished two extensive studies on wave run-up and overtopping (De Waal and Van der Meer, 1992).

Each of these studies have developed dimensionless parameters of the crest freeboard for use in prediction formulas. Different dimensionless groups have been used by each author. The simplest of such parameters is the relative freeboard, R_c/H_s. This simple parameter, however, omits the important effects of the wave period or wave steepness and slope angle.

For plain and bermed smooth slopes, Owen (1980) relates a dimensionless discharge parameter, Q, to a dimensionless freeboard parameter, R, by an exponential equation of the form

$$Q = ae^{-bR/\gamma}. \tag{24}$$

The definitions of Q and R that were given by Owen (1980) are

$$Q = \frac{q}{\sqrt{gH_s^3}}\sqrt{\frac{s_{om}}{2\pi}} \tag{25}$$

$$R = \frac{R_c}{H_s}\sqrt{\frac{s_{om}}{2\pi}} \tag{26}$$

and values for the coefficients a and b were derived from the test results and are given in Table 6.

Table 6. Values of the coefficients a and b in Eq. (24) for straight smooth slopes. (Owen, 1980).

slope	a	b
1:1	0.00794	20.12
1:1.5	0.0102	20.12
1:2	0.0125	22.06
1:3	0.0163	31.9
1:4	0.0192	46.96
1:5	0.025	65.2

De Waal and Van der Meer (1992) used Owen's data on smooth slopes and also data from Führböter *et al.* (1989), besides their own extensive measurements (see Subsec. 3.1). They derived two approaches, one by relating overtopping to wave run-up and the other by treating overtopping separately. Both methods are given here.

The simplest approach for determining wave overtopping is followed when the crest freeboard R_c is related to an expected run-up level on a nonovertopped slope, say, the $R_{u2\%}$. This "shortage in run-up height" can then be described by $(R_{u2\%} - R_c)/H_s$. Equations (11) and (12) can be used to determine $R_{u2\%}$, including all influences of berms, etc. (see Subsec. 3.1). For rock slopes, Eqs. (14) to (20) can be used.

The simplest dimensionless description of overtopping is given by Eq. (3). Figure 17 shows the final results on overtopping and gives all available data, including data of Owen (1980), Führböter *et al.* (1989), and various tests at Delft Hydraulics. The horizontal axis gives the "shortage in run-up height" $(R_{u2\%} - R_c)/H_s$. For the zero value, the crest height is equal to the 2% run-up height. For negative values, the crest height is even higher and overtopping will be (very) small. For a value of 1.5, the crest level is 1.5 H_s lower than the 2% run-up height and overtopping will obviously be large. The vertical axis gives the logarithmic of the mean dimensionless overtopping discharge Q.

Figure 17 gives about 500 data points. The formula that describes more or less the average of the data is given by an exponential function as suggested by Owen (1980):

$$Q = \frac{q}{\sqrt{gH_s^3}} = 8.10^{-5} e^{3.1 \frac{(R_{u2\%} - R_c)}{H_s}} \tag{27}$$

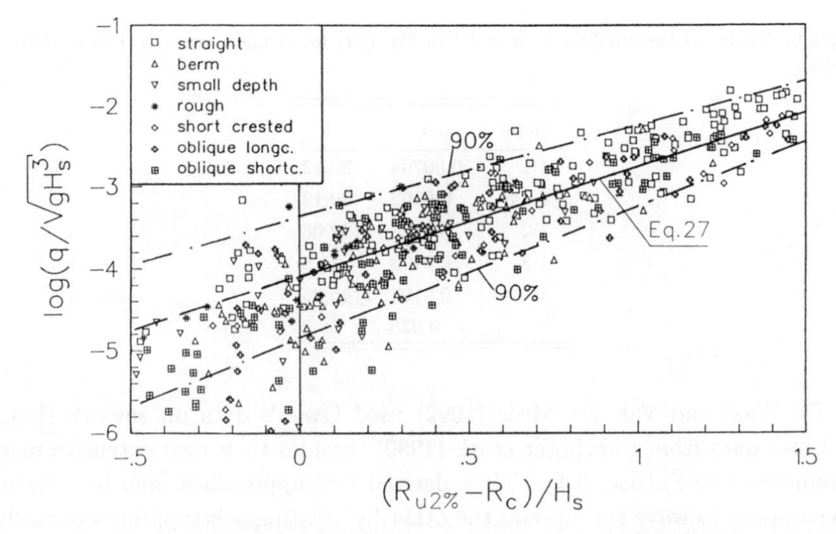

Fig. 17. Wave overtopping on slopes as a function of wave run-up.

The reliability of Eq. (27) can be given by assuming that $\log Q$ (and not Q) has a normal distribution with a variation coefficient $V = \sigma/\mu = 0.11$.

Reliability bands can then be calculated for various practical values of mean overtopping discharges.

The approach described above has a limited application if the overtopping volume becomes fairly large, or if the crest height is much lower than the 2% run-up height, especially when reduction factors for influences of berms, etc., are applied. In those cases above, the method should be applied with care and the method described below is preferred.

From the analysis of De Waal and Van der Meer (1992), it was found that wave overtopping should be divided into situations with plunging and with surging waves. In fact, this corresponds with wave run-up (see Figs. 10 and 11). The dimensionless overtopping discharge is given by Q_b and Q_n, and the dimensionless crest height by R_b and R_n, respectively, where the index b means breaking (plunging, $\xi_p < 2$) waves and n means nonbreaking (surging, $\xi_p > 2$) waves. The definitions are

$$Q_b = \frac{q}{\sqrt{gH_s^3}} \sqrt{\frac{s_{op}}{\tan\alpha}}\,, \tag{28}$$

$$R_b = \frac{R_c}{H_s} \frac{\sqrt{s_{op}}}{\tan\alpha} \frac{1}{\gamma}\,, \tag{29}$$

$$Q_n = \frac{q}{\sqrt{gH_s^3}}\,, \tag{30}$$

$$R_n = \frac{R_c}{H_s} \frac{1}{\gamma}\,. \tag{31}$$

The total reduction factor γ is described in Subsec. 3.1.

The average of all data for breaking waves is shown in Fig. 18 and can be described again by an exponential function:

$$Q_b = 0.06 e^{-5.2 R_b}\,. \tag{32}$$

The reliability of Eq. (32) can be given by assuming the coefficient 5.2 to be a stochastic variable with normal distribution with mean 5.2 and standard deviation 0.55.

A similar graph is shown in Fig. 19 for nonbreaking waves and the average of the data can be described by

$$Q_n = 0.2 e^{-2.6 R_n}\,. \tag{33}$$

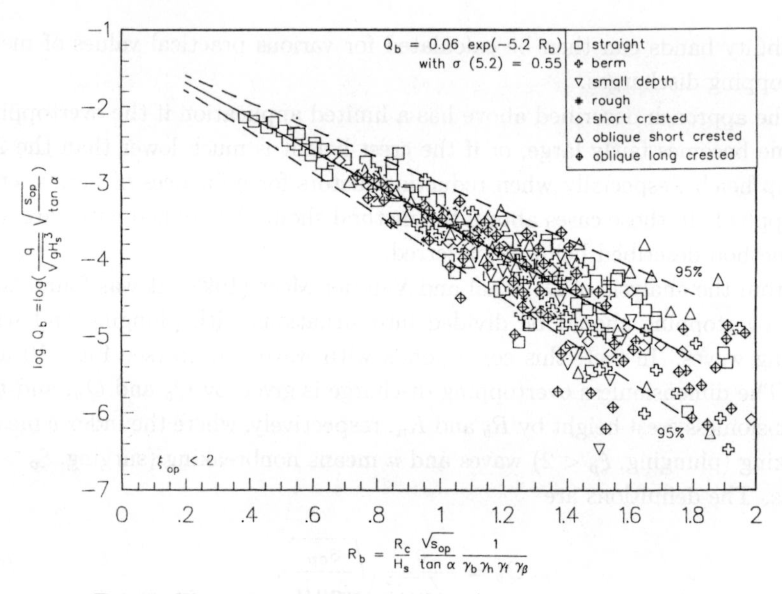

Fig. 18. Wave overtopping for breaking (plunging) waves, $\xi_p < 2$.

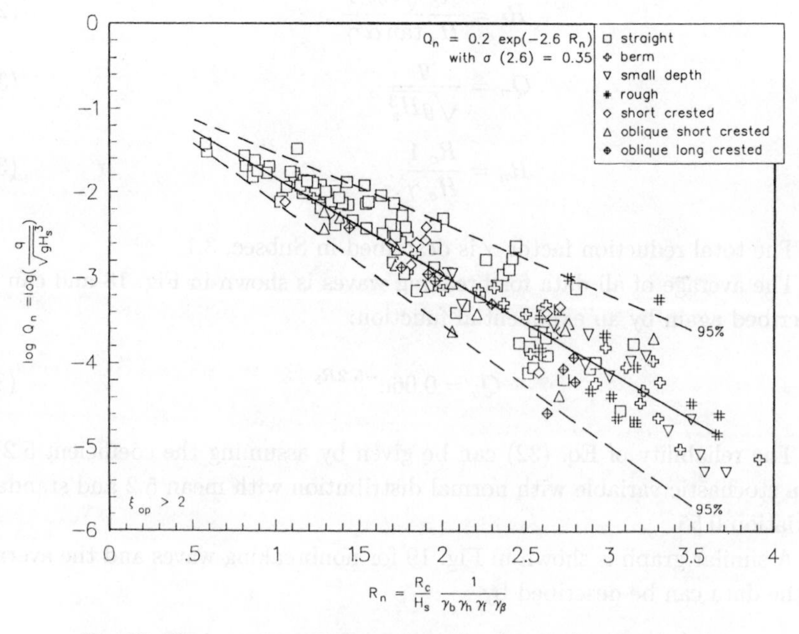

Fig. 19. Wave overtopping for nonbreaking (surging) waves, $\xi_p > 2$.

The reliability can now be described by a standard deviation of 0.35 for the coefficient 2.6.

Surprisingly, there are very little data available describing the overtopping performance of rock armored sea walls without crown walls. However, the results from two tests by Bradbury *et al.* (1988) may be used to give estimates of the influence of wave conditions and relative freeboard. Again the test results have been used to give values of coefficients in an empirical equation. Bradbury *et al.* give the following prediction formula:

$$Q = aR^{-b} \tag{34}$$

with

$$Q = \frac{Q}{\sqrt{gH_s^3}}\sqrt{\frac{s_{om}}{2\pi}}\,, \tag{35}$$

$$R = \left(\frac{R_c}{H_s}\right)^2 \sqrt{\frac{s_{om}}{2\pi}}\,. \tag{36}$$

Values of a and b have been calculated from the results of tests with a rock armored slope at 1:2 with the crest details shown in Fig. 20. For section $A, a = 3.7{*}10^{-10}$ and $b = 2.92$. For section $B, a = 1.3{*}10^{-9}$ and $b = 3.82$.

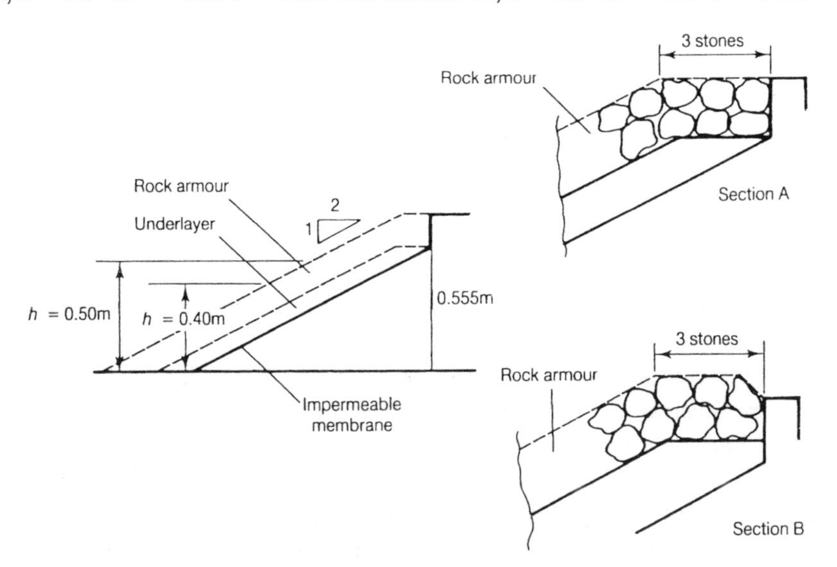

Fig. 20. Overtopped rock structures with low crown wall. (Bradbury *et al.*, 1988)

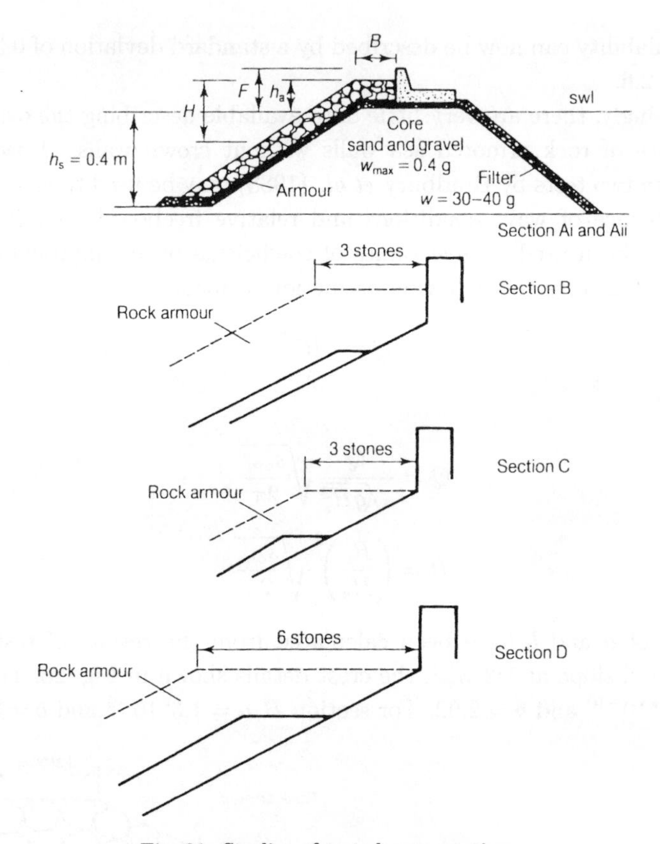

Fig. 21. Studies of tested cross-sections.

More data are available to describe the overtopping performance of rock armored structures with crown walls. Tests have been conducted by Bradbury *et al.* (1988) and Aminthi and Franco (1988), and the results have been used to determine values for coefficients a and b in Eq. (34). The cross-sections studied are illustrated in Fig. 21 and values for the coefficients are given in Table 7. Each of these studies have used different dimensionless freeboard and discharge parameters in empirical equations that are valid for a range of different structural configurations. The data have not been analysed as a single set, and the user may, therefore, need to compare the results given by more than one prediction method.

Table 7. Coefficients a and b in Eq. (34) for overtopping discharges over cross-sections in Fig. 21.

Section	Slope angle	B/H_s	a	b
Ai	1:2.0	1.10	1.7×10^{-8}	2.41
		1.85	1.8×10^{-7}	2.30
		2.60	2.3×10^{-8}	2.68
Aii	1:1.33	1.10	5.0×10^{-8}	3.10
		1.85	6.8×10^{-8}	2.65
		2.60	3.1×10^{-8}	2.69
B	1:2	0.79–1.7	1.6×10^{-9}	3.18
C	1:2	0.79–1.7	5.3×10^{-9}	3.51
D	1:2	1.6–3.3	1.0×10^{-9}	2.82

3.3. *Transmission*

Structures such as breakwaters constructed with low crest levels will transmit wave energy into the area behind the breakwater. The transmission performance of low-crested breakwaters is dependent on the structure geometry, principally, the crest freeboard, crest width and water depth, permeability, and on the wave conditions, principally, the wave height and period.

Hydraulic model test results measured by Seelig (1980), Powell and Allsop (1985), Daemrich and Kahle (1985), Ahrens (1987), and Van der Meer (1988a) have been reanalysed by Van der Meer (1990b) to give a single prediction method. This relates C_t to the relative crest freeboard, R_c/H_s. The data used are plotted in Fig. 22. The prediction equations describing the data may be summarised:

Range of validity — Equation

$$-2.0 < \frac{R_c}{H_s} < -1.13 \qquad C_t = 0.80 \tag{37}$$

$$-1.13 < \frac{R_c}{H_s} < 1.2 \qquad C_t = 0.46 - 0.3\frac{R_c}{H_s} \tag{38}$$

$$1.2 < \frac{R_c}{H_s} < 2.0 \qquad C_t = 0.10 \tag{39}$$

Figure 22 shows all results together with the equations. These equations give a very simplistic description of the data available, but will often be sufficient for a preliminary estimate of performance. The upper and lower bounds

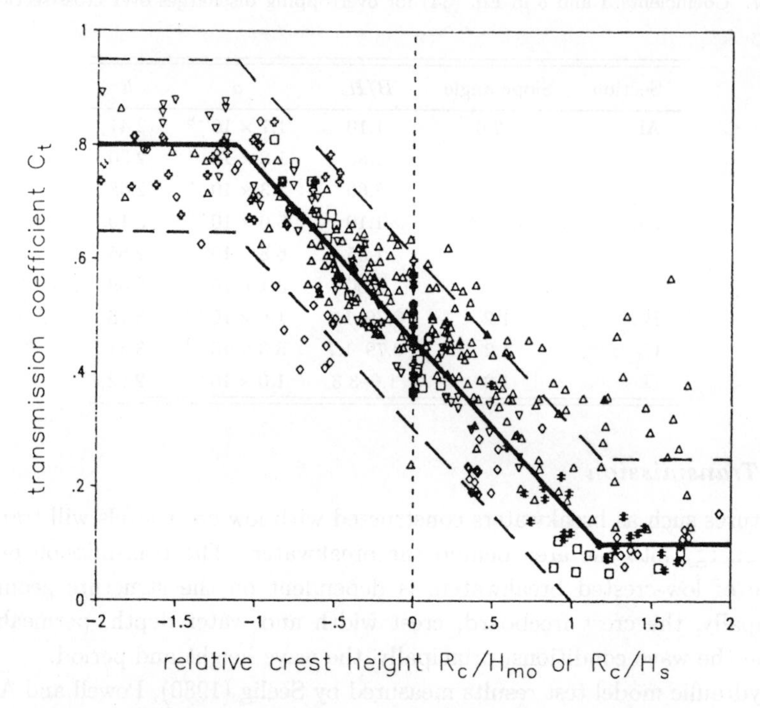

Fig. 22. Wave transmission over and through low-crested structures.

of the data considered are given by lines 0.15 higher (or lower) than the mean lines given above. This corresponds with the 90% confidence bands (the standard deviation was 0.09).

A second analysis on the data was performed by Daemen (1991) and he also performed more tests on wave transmission. A summary has been described by Van der Meer and d'Angremond (1991) and Van der Meer and Daemen (1994) and is given here.

Until now wave transmission has been described in the conventional way as a function of R_c/H_s. It is not clear, however, that the use of this combination of crest freeboard and wave height produces similar results with, on the one hand, constant R_c and variable H_s and, on the other, variable R_c and constant H_s. Moreover, when R_c becomes zero, all influence of the wave height is lost which leads to a large scatter in the graph at $R_c = 0$.

Therefore, it was decided to separate R_c and H_s in the second analysis.

The mass or nominal diameter of the armor layer of a rubble mound structure is determined by the extreme wave attack that can be expected during the lifetime of the structure. There is a direct relationship between the design wave height and the size of armor rock, which is often given as the stability factor $H_s/\Delta D_{n50}$, where Δ is the relative buoyant density. It can be concluded that the nominal diameter of the armor layer characterises the rubble mound structure. It is, therefore, also a good parameter to characterise both the wave height and the crest height in a dimensionless way.

The relative wave height can then be given as H_s/D_{n50}, in accordance with the stability factor, and the relative crest height by R_c/D_{n50}, being the number of rocks that the crest level is above or below still-water level.

Moreover, a separation into H_s/D_{n50} and R_c/D_{n50} enables a distinction between various cases. For example, low H_s/D_{n50} values (smaller than 1 to 2) produce low waves traveling through the crest and high H_s/D_{n50} values (3 to 5) yield situations under extreme wave attack. Finally, D_{n50} can be used to describe other breakwater properties as the crest width B. This yields the parameter B/D_{n50}.

The primary parameters for wave transmission can now be given as

Relative crest height:	R_c/D_{n50}
Relative wave height:	H_s/D_{n50}
Fictitious wave steepness:	s_{op}
And possibly:	B/D_{n50} .

The outcome of the analysis on wave transmission, including the data of Daemen (1991), was a linear relationship between the wave transmission coefficient C_t and the relative crest height R_c/D_{n50}, which is valid between minimum and maximum values of C_t. In Fig. 23, the basic graph is shown. The linearly increasing curves are presented by

$$C_t = a\frac{R_c}{D_{n50}} + b \tag{40}$$

with

$$a = 0.031\frac{H_i}{D_{n50}} - 0.24 \; . \tag{41}$$

Equation (41) is applicable for conventional and reef-type breakwaters. The coefficient "b" for conventional breakwaters is described by

$$b = -5.42s_{op} + 0.0323\frac{H_i}{D_{n50}} - 0.0017\left(\frac{B}{D_{n50}}\right)^{1.84} + 0.51 \; , \tag{42}$$

and for reef-type breakwaters by

$$b = -2.6s_{op} - 0.05\frac{H_i}{D_{n50}} + 0.85\,. \tag{43}$$

The following minimum and maximum values are derived:
Conventional breakwaters:
Minimum: $C_t = 0.075$; maximum: $C_t = 0.75$. (44)
Reef-type breakwaters:
Minimum: $C_t = 0.15$; maximum: $C_t = 0.60$
for $R_c/D_{n50} < -2$, linearly increasing to $C_t = 0.80$ for $R_c/D_{n50} = -6$. (45)

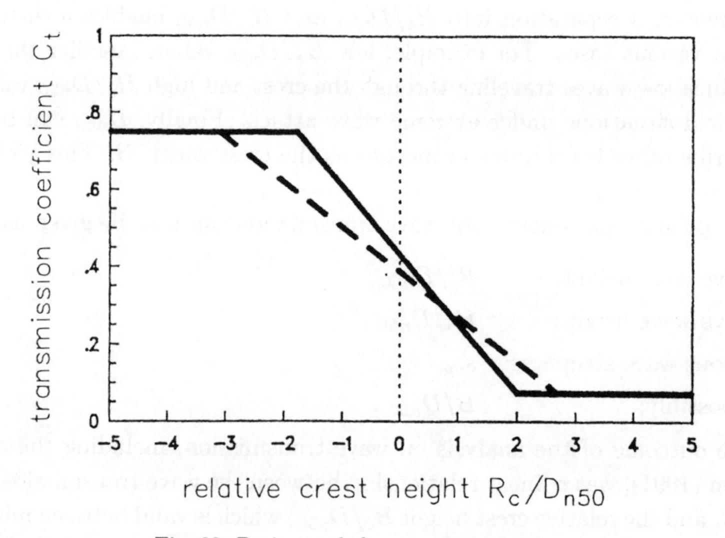

Fig. 23. Basic graph for wave transmission.

The analysis was based on various groups with constant wave steepness and a constant relative wave height. The validity of the wave transmission formula (Eq. 40) corresponds, of course, with the ranges of these groups that were used. The formula is valid for

$$1 < H_s/D_{n50} < 6 \text{ and } 0.01 < s_{op} < 0.05\,.$$

Both upper boundaries can be regarded as physically bound. Values of $H_s/D_{n50} > 6$ will cause instability of the structure and values of $s_{op} > 0.05$

will cause waves breaking on steepness. In fact, boundaries are only given for extremely low wave heights relative to the rock diameter and for very low wave steepnesses (low swell waves).

The formula is applicable outside the range given above, but the reliability is low. Figure 24 shows the measured wave transmission coefficient versus the calculated one from Eq. (40) for various data sets of conventional breakwaters. The reliability of the formula can be described by assuming a normal distribution around the line in Fig. 24. With the restriction of the range of application given above, the standard deviation amounted to $\sigma(C_t) = 0.05$, which means that the 90% confidence levels can be given by $C_t \pm 0.08$. This is a remarkable increase in reliability compared to the simple formula given by Eqs. (37)–(39) and Fig. 22, where a standard deviation of $\sigma(C_t) = 0.09$ is given.

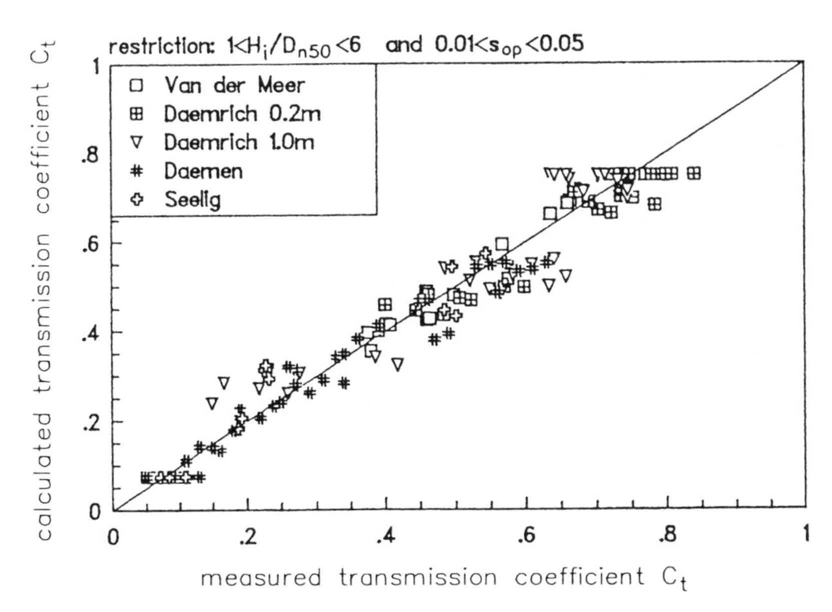

Fig. 24. Calculated (Eq. 40) versus measured wave transmission for conventional breakwaters.

The reliability of the formula for reef-type breakwaters is more difficult to describe. If only tests are taken where the crest height had been lowered to less than 10% of the initial height h'_c, and the test conditions lie within the range of application, the standard deviation amounts to $\sigma(C_t) = 0.031$. If the restriction on the crest height is not taken into account, the standard deviation amounts to $\sigma(C_t) = 0.054$.

3.4. *Reflections*

Waves will reflect from nearly all coastal or shoreline structures. For structures with nonporous and steep faces, approximately 100% of the wave energy incident upon the structure will reflect. Rubble slopes are often used in harbor and coastal engineering to absorb wave action. Such slopes will generally reflect significantly less wave energy than the equivalent nonporous or smooth slope. Although some of the flow processes are different, it has been found convenient to calculate the reflection performance given by C_r using an equation of the same form as for nonporous slopes, but with different values of the empirical coefficients to match the alternative construction. Data for random waves are available for smooth and armored slopes at angles between 1:1.5 and 1:2.5 (smooth), and 1:1.5 and 1:6 (rock).

Data by Allsop and Channell (1988) will be given here, together with data by Van der Meer (1988a), analysed by Postma (1989). Formulas of other references will be used for comparison.

For smooth impermeable slopes, Battjes (1974) gives

$$C_r = 0.1\xi^2 \ . \tag{46}$$

Seelig (1983) gives

$$C_r = \frac{a\xi_p^2}{b + \xi_p^2} \tag{47}$$

with
$a = 1.0, b = 5.5$ for smooth slopes,
$a = 0.6, b = 6.6$ for a conservative estimate of rough permeable slopes.

Equations (46) and (47) are shown in Fig. 25 together with the reflection data of Van der Meer (1988a) for rock slopes. The two curves for smooth slopes are close. The curve of Seelig (1983) for permeable slopes is not a conservative estimate, but it even underestimates the reflection for large ξ_p values.

The best fit curve through all the data points in Fig. 25 is given by Postma (1989) and is also given in the figure:

$$C_r = 0.14\,\xi_p^{0.73} \quad \text{with } \sigma(C_r) = 0.055 \ . \tag{48}$$

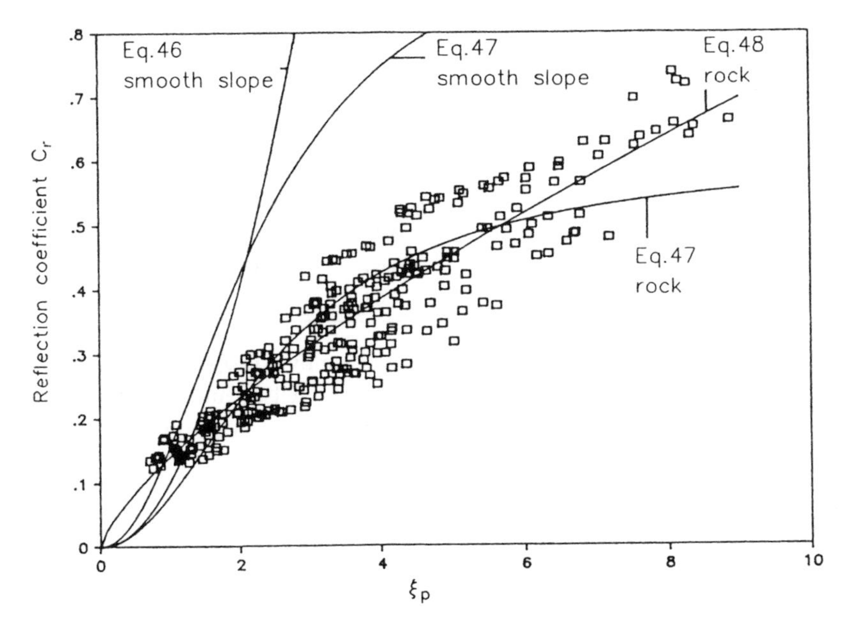

Fig. 25. Comparison of data on rock slopes with other formulas. (Van der Meer 1988a)

The surf similarity parameter did not describe the combined slope angle-wave steepness influence in a sufficient way. Therefore, both the slope angle and wave steepness were treated separately and Postma derived the following relationship

$$C_r = 0.071 P^{-0.082} \cot\alpha^{-0.62} s_{op}^{-0.46}, \tag{49}$$

with

$\sigma(C_r) = 0.036$,

$P =$ notional permeability factor described in Subsec. 2.3.3.

The standard deviation of 0.055 in Eq. (48) reduced to 0.036 in Eq. (49) which is a considerable increase in reliability.

The results of random wave tests by Allsop and Channell (1989), analysed to give values for the coefficients a and b in Eq. (47) (but with ξ_m instead of ξ_p), is presented below. The rock armored slopes used rock in two or one layer, placed on an impermeable slope covered by underlayer stone, is equivalent to

$P = 0.1$. The range of wave conditions for which these results may be used is given by

$$0.004 < s_{om} < 0.052, \text{ and } 0.6 < H_s/\Delta D_{n50} < 1.9\,.$$

Slope type	a in (47)	b in (47)
Smooth	0.96	4.80
Rock, two layers	0.64	8.85
Rock, one layer	0.64	7.22

Postma (1989) also reanalysed the data given by Allsop and Channell which were described above. Figure 26 gives the data by Allsop and Channell together with Eq. (48). The curve is a little higher than the average given in the data. The best fit curve is described by

$$C_r = 0.125\ \xi_p^{0.73} \qquad \text{with } \sigma(C_r) = 0.060\,. \tag{50}$$

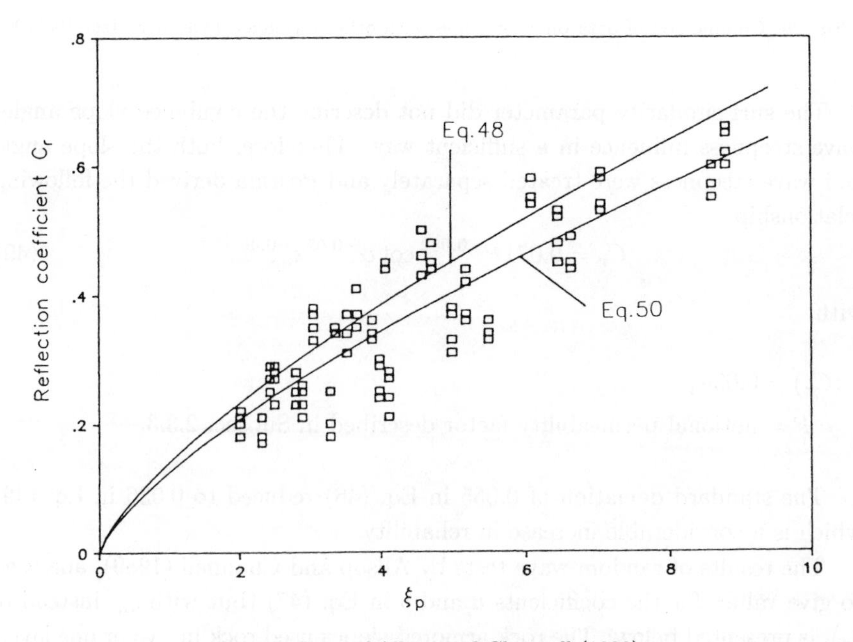

Fig. 26. Data of Allsop and Channell (1989).

4. Structural Response

4.1. *Introduction*

The hydraulic and structural parameters are described in Sec. 2 and the hydraulic responses in Sec. 3. Figure 3 gives an overview of the definitions of the hydraulic parameters and responses as wave run-up, run-down, overtopping, transmission, and reflection, and Fig. 5 shows the structural parameters which are related to the cross-section. The response of the structure under hydraulic loads will be described in this section and design tools will be given.

The design tools described here permit the design of many structure types. Nevertheless, it should be remembered that each design rule has its limitations. For each structure which is important and expensive to build, it is advised to perform physical model studies.

Figure 27 gives the same cross-section as in Fig. 5, but it now shows the various parts of the structure which will be described in the following sections. Some general points and design rules for the geometrical design of the cross-section will be given here. These are

- the minimum crest width,
- the thickness of (armor layers),
- the number of units or rocks per surface area,
- the bottom elevation of the armor layer.

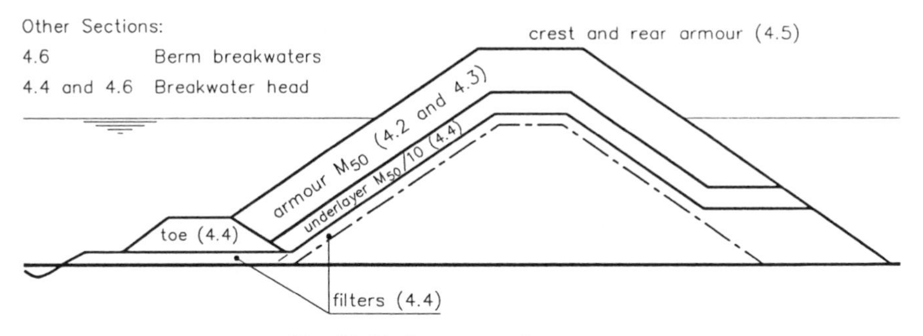

Fig. 27. Various parts of a structure.

The crest width is often determined by construction methods used (access on the core by trucks or crane) or by functional requirements (road/crown wall on the top). Where the width of the crest can be small, a required minimum width $B_{\min}$ should be provided, where (SPM, 1984)

$$B_{\min} = (3 \text{ to } 4)D_{n50}\,. \tag{51}$$

The thickness of layers is given by

$$t_a = t_u = t_f = n\,k_t\,D_{n50}\,. \tag{52}$$

The number of units per m^2 is given by

$$N_a = n\,k_t(1 - n_\nu)D_{n50}^{-2}\,. \tag{53}$$

Where t_a, t_u, t_f = thickness of armor, underlayer, or filter,

n = number of layers,

k_t = layer thickness coefficients,

n_v = volumetric porosity.

Values of k_t and n_v as given in the SPM (1984) are

	k_t	n_v
Smooth rock, $n = 2$	1.02	0.38
Rough rock, $n = 2$	1.00	0.37
Rough rock, $n > 3$	1.00	0.40
Graded rock	–	0.37
Cubes	1.10	0.47
Tetrapods	1.04	0.50
Dolosse	0.94	0.56

Another source for values of k_t and n_v for rock is the CUR/CIRIA Manual (1991):

Shape of rock	Placement	k_t	n_v
Irregular	Random	0.75	0.40
Irregular	Special	1.05–1.20	0.39
Semi-round	Random	0.75	0.37
Semi-round	Special	1.10–1.25	0.36
Equant	Random	0.80	0.38
Equant	Special	1.00–1.15	0.37
Very round	Random	0.80	0.36
Very round	Special	1.05–1.20	0.35

The number of units in a rock layer depends on the grading of the rock. The values of k_t that are given above describe a rather narrow grading (uniform rock). For riprap and even wider graded material the number of stones can not easily be estimated. In that case, the volume of the rock on the structure can be used.

The bottom elevation of the armor layer should be extended downslope to an elevation below minimum still-water level of at least one (significant) wave height, if the wave height is not limited by the water depth. Under depth limited conditions, the armor layer should be extended to the bottom as shown in Fig. 27 and supported by a toe.

4.2. *Rock armor layers*

Many methods for the prediction of rock size of armor units designed for wave attack have been proposed in the last 50 years. Those treated in more detail here are the Hudson formula as used in SPM (1984) and the formulas derived by Van der Meer (1988a).

4.2.1. *Hudson formula*

The original Hudson formula is written as

$$M_{50} = \frac{\rho_r H^3}{K_D \Delta^3 \cot\alpha} \,. \tag{45}$$

K_D is a stability coefficient taking into account all other variables. K_D values suggested for design correspond to a "no damage" condition where up to 5% of the armor units may be displaced. In the 1973 edition of the Shore Protection Manual, the values given for K_D for rough, angular stone in two layers on a breakwater trunk were

$K_D = 3.5$ for breaking waves,

$K_D = 4.0$ for nonbreaking waves.

The definition of breaking and nonbreaking waves is different from plunging and surging waves, which were described in Subsec. 2.1. A breaking wave in Eq. (45) means that the wave breaks due to the foreshore in front of the structure directly on the armor layer. It does not describe the type of breaking due to the slope of the structure itself.

No tests with random waves had been conducted and it was suggested that H_s in Eq. (54) was to be used. By 1984, the advice given was more cautious. The SPM now recommends $H = H_{1/10}$, being the average of the highest 10% of all waves. For the case considered above, the value of K_D for breaking waves was revised downward from 3.5 to 2.0 (for nonbreaking waves, it remained 4.0). The effect of these two changes is equivalent to an increase in the unit stone mass required by a factor of about 3.5!

The main advantages of the Hudson formula are its simplicity, and the wide range of armor units and configurations for which values of K_D have been derived. The Hudson formula also has many limitations. Briefly, they include

- potential scale effects due to the small scales at which most of the tests were conducted,
- the use of regular waves only,
- no account taken in the formula of wave period or storm duration,
- no description of the damage level,
- the use of nonovertopped and permeable core structures only.

The use of $K_D \cot\alpha$ does not always best describe the effect of the slope angle. It may therefore be convenient to define a single stability number without this $K_D \cot\alpha$. Further, it may often be more helpful to work in terms of a linear armor size, such as a typical or nominal diameter. The Hudson formula can be rearranged to

$$H_s/\Delta D_{n50} = (K_D \cot\alpha)^{1/3}. \tag{55}$$

Equation (55) shows that the Hudson formula can be written in terms of the structural parameter $H_s/\Delta D_{n50}$ which was discussed in Subsec. 2.3.1.

4.2.2. *Van der Meer formulas — deep water conditions*

Based on earlier work of Thompson and Shuttler (1975) an extensive series of model tests was conducted at Delft Hydraulics (Van der Meer (1987), (1988a) and (1988b)). These include structures with a wide range of core/underlayer permeabilities and a wider range of wave conditions. Two formulas were derived for plunging and surging waves, respectively, which are now known as the Van der Meer formulas:
for plunging waves,

$$\frac{H_s}{\Delta D_{n50}} = 6.2P^{0.18}\left(\frac{S}{\sqrt{N}}\right)^{0.2} \xi_m^{-0.5}\,, \tag{56}$$

and for surging waves,

$$\frac{H_s}{\Delta D_{n50}} = 1.0P^{-0.13}\left(\frac{S}{\sqrt{N}}\right)^{0.2}\sqrt{\cot\alpha}\,\xi_m^P \,. \tag{57}$$

The transition from plunging to surging waves can be calculated using a critical value of ξ_{mc}:

$$\xi_{mc} = [6.2P^{0.31}\sqrt{\tan\alpha}]^{\frac{1}{P+0.5}} \,. \tag{58}$$

For $\cot\alpha \geq 4.0$, the transition from plunging to surging does not exist and for these slope angles, only Eq. (56) should be used. All parameters used in Eqs. (56)–(58) are described in Sec. 2. The notional permeability factor P is shown in Fig. 6. The factor P should lie between 0.1 and 0.6.

Design values for the damage level S are shown in Table 4. The level "start" of damage, $S = 2 - 3$, is equal to the definition of "no damage" in the Hudson formula (Eq. (54)). The maximum number of waves N which should be used in Eqs. (56) and (57) is 7500. After this number of waves, the structure more or less has reached an equilibrium.

The wave steepness should lie between $0.005 < s_{om} < 0.06$ (almost the complete possible range). The mass density varied in the tests between 2000 kg/m^3 and 3100 kg/m^3, which is also the possible range of application.

The reliability of the formulas depends on the differences due to random behaviour of rock slopes, accuracy of measuring damage and curve fitting of the test results. The reliability of the Eqs. (56) and (57) can be expressed by giving the coefficients 6.2 and 1.0 in the equations a normal distribution with a certain standard deviation. The coefficient 6.2 can be described by a standard deviation of 0.4 (variation coefficient 6.5%) and the coefficient 1.0 by a standard deviation of 0.08 (8%). These values are significantly lower than that for the Hudson formula at 18% for K_D (with mean K_D of 4.5). With these standard deviations it is simple to include 90% or other confidence bands.

Equations (56)–(58) are more complex than the Hudson formula, Eqs. (54) or (55). They also include the effect of the wave period, the storm duration, the permeability of the structure, and a clearly-defined damage level. This may cause differences between the Hudson formula and Eqs. (56)–(58), which are illustrated hereafter.

The $H_s/\Delta D_{n50}$ in the Hudson formula is only related to the slope angle $\cot\alpha$. Therefore a plot of $H_s/\Delta D_{n50}$ or N_s versus $\cot\alpha$ shows one curve for the Hudson formula. Equations (56)–(58) take into account the wave period (or steepness), the permeability of the structure, and the storm duration. The effect of these parameters are shown in Figs. 28 and 29. Figure 28 shows the curves for a permeable structure after a storm duration of 1000 waves (a little more than the number used by Hudson). Figure 29 gives the stability of an impermeable revetment after a wave attack of 5000 waves (equivalent to 5–10 hours in nature). Curves are shown for various wave steepnesses.

The conclusion is clear. The Hudson formula can only be used as a very rough estimate for a particular case. It should be borne in mind that a difference of a factor of two in the N_s-number means a difference in the stone mass of a factor of eight. These figures should be sufficiently convincing for the designer to apply Eqs. (56)–(58) instead of the simple Hudson formula.

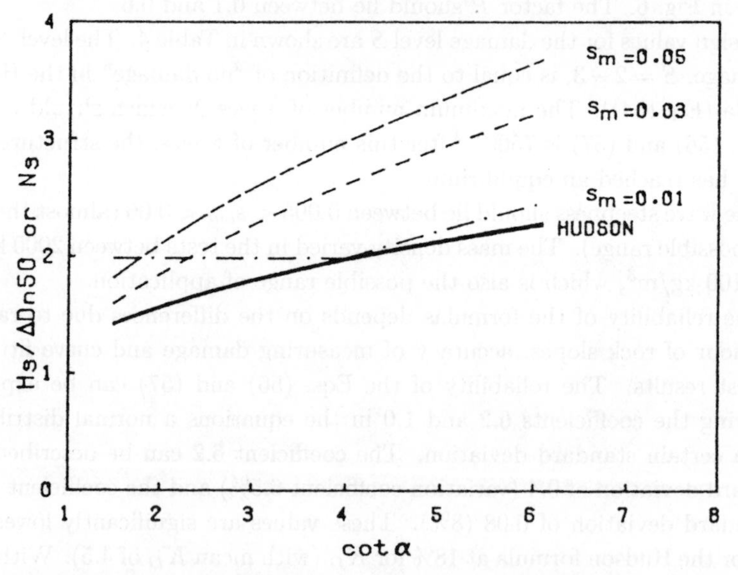

Fig. 28. Comparison between the Hudson formula and Eqs. (56)–(58) for a permeable core and after 1000 waves.

Nevertheless, it is more difficult to work with Eqs. (56)–(58). For a good design it is required to perform a sensitivity analysis for all parameters in the equations.

The deterministic procedure is to make design graphs where one parameter is evaluated. Three examples are shown in Figs. 30–32. Both give a wave height

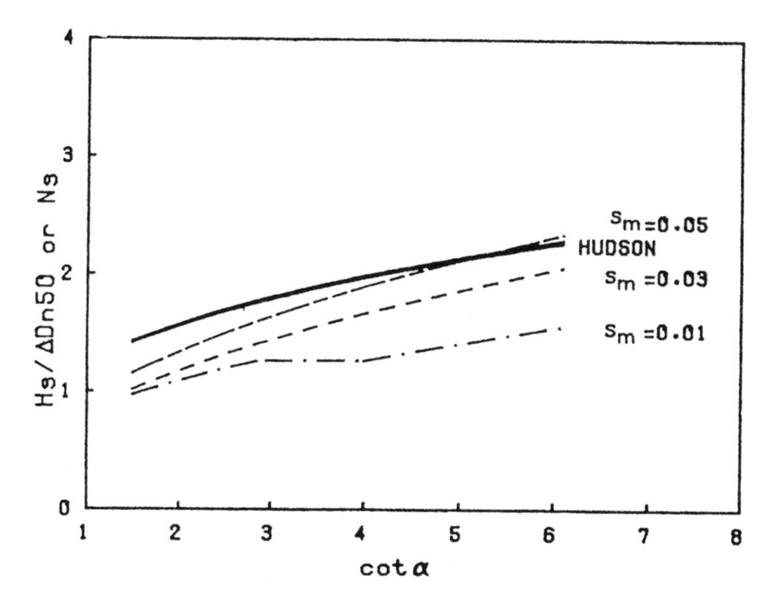

Fig. 29. Comparison between the Hudson formula and Eqs. (56)–(58) for an impermeable core and after 5000 waves.

versus surf similarity plot, which shows the influence of both wave height and wave steepness (the wave climate). The other shows a wave height versus damage plot which is comparable with the conventional way of presenting results of model tests on stability. The same kind of plots can be derived from Eqs. (56)–(58) for other parameters (see Van der Meer, 1988b).

The influence of the damage level S is shown in Fig. 30. Four damage levels are shown: $S = 2$ (start of damage), $S = 5$ and 8 (intermediate damage), and $S = 12$ (filter layer visible). The structure itself is described by $D_{n50} = 1.0\ m$ ($M_{50} = 2.6\,t$), $\Delta = 1.6, \cot\alpha = 3.0, P = 0.5$, and $N = 3000$.

The influence of the notional permeability factor P is shown in Fig. 32. Four values are shown: $P = 0.1$ (impermeable core), $P = 0.3$ (some permeable core), $P = 0.5$ (permeable core), and $P = 0.6$ (homogeneous structure). The structure itself is described by $D_{n50} = 1.0\ m$ ($M_{50} = 2.6\,t$), $\Delta = 1.6, \cot\alpha = 3.0$, $P = 0.5$, and $N = 3000$.

Damage curves are shown in Fig. 32. Two curves are given, one for a slope angle with $\cot\alpha = 2.0$ and a wave steepness of $s_{om} = 0.02$, and one for a

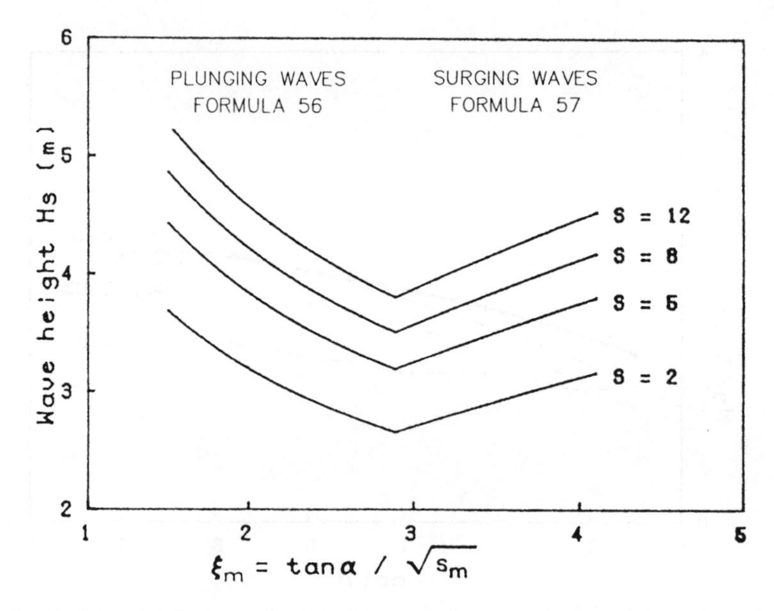

Fig. 30. Wave height versus surf similarity parameter; influence of damage level.

slope angle with $\cot\alpha = 3.0$ and a wave steepness of 0.05. If the extreme wave climate is known, graphs shown in Fig. 32 are very useful to determine the stability of the armor layer of the structure. Figure 32 shows also the 90% confidence levels which give a good idea about the possible variation in stability. This variation should be taken into account by the designer of a structure.

An estimation of the damage profile of a straight rock slope can be made by the use of Eqs. (56) and (57) and some additional relationships for the profile. The profile can be schematised to an erosion area around still-water level, an accretion area below still-water level, and for gentle slopes, a berm or crest above the erosion area. The transitions from erosion to accretion, etc., can be described by heights measured from still-water level (see Fig. 33). The heights are respectively h_r, h_d, h_m and h_b.

The relationships for the height parameters are based on the tests described by Van der Meer (1988a) and will not be given here. The assumption for the profile is a spline through the points given by the heights and with an erosion (and accretion) area according to the stability Eqs. (56) and (57). The method is only applicable for straight slopes.

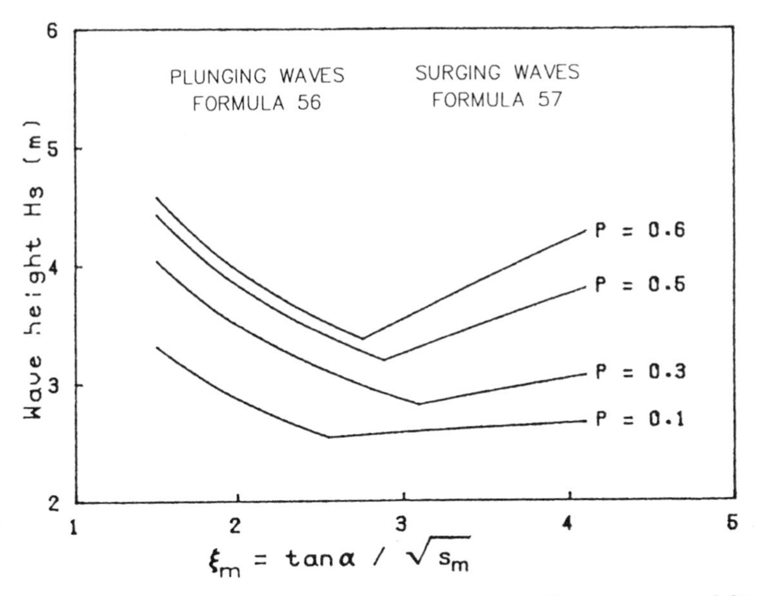

Fig. 31. Wave height versus surf similarity parameter; influence of permeability.

A deterministic design procedure is followed if the stability equations are used to produce design graphs as H_s versus ξ_m and H_s versus damage (see Figs. 30–32) and if a sensitivity analysis is performed. Another design procedure is the probabilistic approach. Equations (56) and (57) can be rewritten to so-called reliability functions and all the parameters can be assumed to be stochastic with an assumed distribution. Here, one example of the approach will be given. A more detailed description can be found in Van der Meer (1988b).

The structure parameters with the mean value, distribution type, and standard deviation are given in Table 8. These values were used in Level II first-order second-moment (FOSM) with the approximate full distribution approach (AFDA) method. With this method, the probability that a certain damage level would be exceeded in one year was calculated. These probabilities were used to estimate the probability that a certain damage level would be exceeded in a certain lifetime of the structure.

The parameter FH_s represents the uncertainty of the wave height at a certain return period. The wave height itself is described by a two-parameter Weibull distribution. The coefficients a and b take into account the reliability of the formulas, including the random behavior of rock slopes.

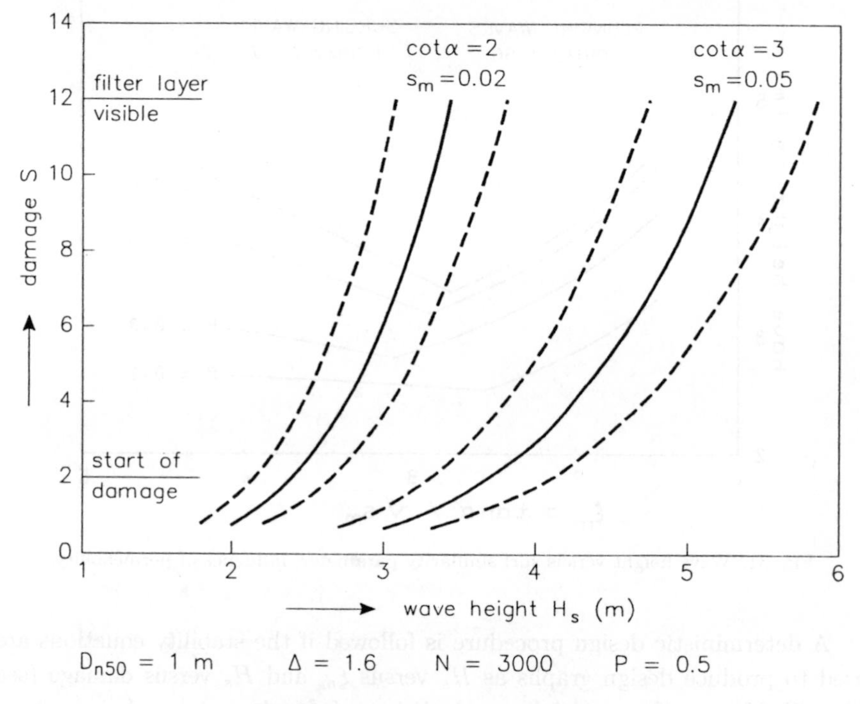

Fig. 32. Wave height versus damage.

Table 8. Parameters used in Level II probabilistic computations.

Parameter	Distribution	Average	Standard deviation
D_{n50}	Normal	1.0m	0.03m
Δ	Normal	1.6	0.05
$\cot\alpha$	Normal	3.0	0.15
P	Normal	0.5	0.05
N	Normal	3,000	1,500
H_s	Weibull	$B = 0.3$	$C = 2.5$
FH_s	Normal	0m	0.25m
s_{om}	Normal	0.04	0.01
a (Eq. 56)	Normal	6.2	0.4
b (Eq. 57)	Normal	1.0	0.08

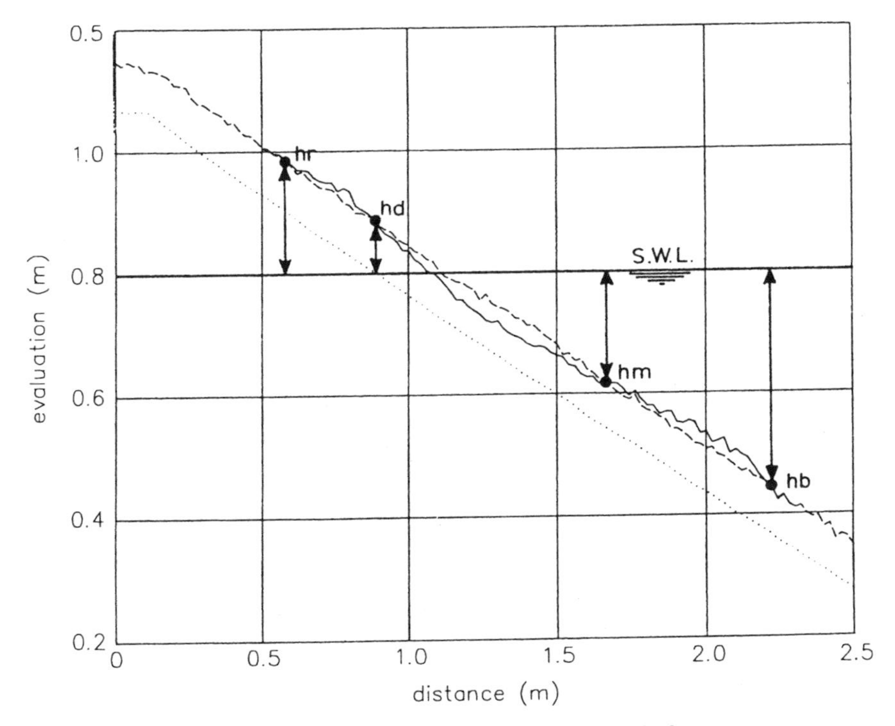

Fig. 33. Damage profile of a statically stable rock slope.

The final results are shown in Fig. 34 where the damage S is plotted versus the probability of exceedance in the lifetime of the structure. From this graph, it follows that the start of damage ($S = 2$) will certainly occur in a lifetime of 50 years. Tolerable damage ($S =$ 5–8) in the same lifetime will occur with a probability of 0.2–0.5. The probability that the filter layer will become visible (failure) is less than 0.1. Probability curves as shown in Fig. 34 can be used to make a cost optimization for the structure during its lifetime, including maintenance and repair at certain damage levels.

Probabilistic Level II calculations as described above can easily be performed if the required computer programs are available. As this is not the case for many designers of breakwaters, it may be easier to use a Level I approach. This means that one applies design formulas with partial safety factors to account for the uncertainty of the relevant parameters. Most guidelines and building codes are based on a Level I approach. The determination of the

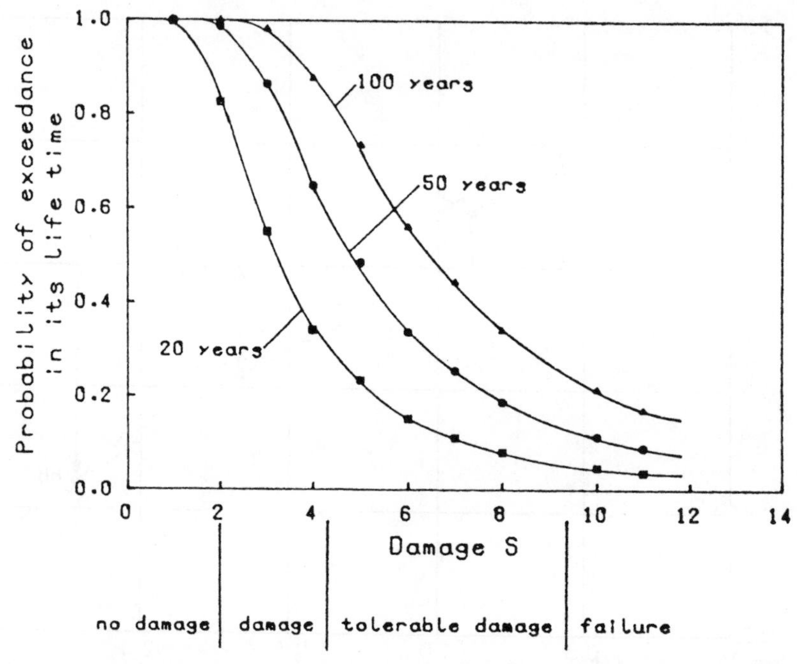

Fig. 34. Probability of exceedance of the damage level S in the lifetime of the structure.

partial safety factors is again based on Level II calculations, but this work has to be done by the composers of the guideline and not by the user.

The safety of rubble mound breakwaters has been described with a Level I approach (partial safety factors) in the recent PIANC report (1993). Two safety factors appeared to be enough to describe the uncertainty: one for the wave height and one for all other parameters together. Partial safety factors for most formulas in this paper are given in PIANC (1993).

4.2.3. *Shallow water conditions*

Up to now the significant wave height H_s has been used in the stability equations. In shallow water conditions, the distribution of the wave heights deviate from the Rayleigh distribution (truncation of the curve due to wave breaking). Further tests on a 1:30 sloping and depth limited foreshore by Van der Meer (1988a) showed that $H_{2\%}$ was a better value for design on depth limited foreshores than the significant wave height H_s, i.e., that the stability of the armor layer in depth limited situations is better described by a higher characteristic value of the wave height distribution $H_{2\%}$ than by H_s.

Equations (56)–(58) can be rearranged with the known ratio of $H_{2\%}/H_s$ for a Rayleigh distribution. The equations become
for plunging waves,

$$\frac{H_{2\%}}{\Delta D_{n50}} = 8.7P^{0.18}\left(\frac{S}{\sqrt{N}}\right)^{0.2}\xi_m^{-0.5}\,, \tag{59}$$

and for surging waves,

$$\frac{H_{2\%}}{\Delta D_{n50}} = 1.4P^{-0.13}\left(\frac{S}{\sqrt{N}}\right)^{0.2}\sqrt{\tan\alpha}\,\xi_m^{P}\,. \tag{60}$$

Equations (59) and (60) take into account the effect of depth limited situations. A safe approach, however, is to use Eqs. (56) and (57) with H_s. In that case, the truncation of the wave height exceedance curve due to wave breaking is not taken into account which can be assumed as a safe approach. If the wave heights are Rayleigh distributed, Eqs. (59) and (60) give the same results as Eqs. (56) and (57), as this is caused by the known ratio of $H_{2\%}/H_s = 1.4$. For depth limited conditions, the ratio of $H_{2\%}/H_s$ will be smaller, and one should obtain information on the actual value of this ratio.

4.2.4. *Effects of armor shape and wide gradings*

The effects of armor shape on stability has been described by Latham *et al.* (1988). They tested five classes of rock with different shape classifications such as fresh, equant, semi-round, very round, and tabular. The damage to the test sections using each of the armor shapes tested was compared with damage calculated using Eqs. (56) and (57). As expected, very round rock suffered more damage than rock of other shapes. The performances of the fresh and equant rock were broadly similar. Surprisingly, the tabular rock exhibited higher stability than other armor shapes.

The coefficient 6.2 in Eq. (56) and 1.0 in Eq. (57) were used to describe the shape effects. These coefficients are summarised in Table 9.

Table 9. Revised coefficients for "nonstandard" armor shapes in Eqs. (56) and (57).

Rock shape class	Plunging waves Alternative for coefficient 6.2	Surging waves Alternative for coefficient 1.0
Fresh	6.32	0.81
Equant	6.24	1.09
Semi-round	5.96	0.99
Very round	5.88	0.81
Tabular	6.72	1.30

The stability of rock armor of (very) wide grading has been investigated by Allsop (1990). Model tests on a 1:2 slope with an impermeable core were conducted to identify whether the use of rock armor of grading wider than $D_{85}/D_{15} = 2.25$ will lead to armor performance substantially different from that predicted by Eqs. (56) and (57). The tests results confirmed the validity of these equations for rock of narrow grading ($D_{85}/D_{15} < 2.25$). Very wide gradings, such as $D_{85}/D_{15} = 4.0$, may in general suffer slightly more damage than predicted for narrower gradings. On any particular structure, there will be greater local variations in the sizes of the individual stones in the armor layer than for narrow gradings. This will increase spatial variations of the damage, giving a higher probability of severe local damage. Considerable difficulties will be encountered in measuring and checking such wide gradings. More information can be found in the above-mentioned references and in Allsop and Jones (1993).

4.3. *Armor layers with concrete units*

The Hudson formula (Eq. 55) was given in Subsec. 4.2 with K_D values for rock. The SPM (1984) gives a table with values for a large number of concrete armor units. The most important ones are $K_D = 6.5$ and 7.5 for cubes, $K_D = 7.0$ and 8.0 for tetrapods, and $K_D = 15.8$ and 31.8 for Dolosse. For other units, one is referred to SPM (1984).

Extended research by Van der Meer (1988c) on breakwaters with concrete armor units was based on the governing variables found for rock stability. The research was limited to only one cross-section (i.e., one slope angle and permeability) for each armor unit. Therefore, the slope angle, $\cot\alpha$, and consequently, the surf similarity parameter, ξ_m, is not present in the stability formulas developed on the results of the research. The same holds for the notional permeability factor, P. This factor was $P = 0.4$.

Breakwaters with armor layers of interlocking units are generally built with steep slopes in the order of 1:1.5. Therefore, this slope angle was chosen for tests on cubes and tetrapods. Accropode are generally built on a slope of 1:1.33, and this slope was used for tests on accropode. Cubes were chosen as these elements are bulky units which have good resistance against impact forces. Tetrapods are widely used all over the world and have a fair degree of interlocking. Accropode were chosen as these units can be regarded as the latest development, showing high interlocking, strong elements, and a one layer system. A uniform 1:30 foreshore was applied for all tests. Only for the highest wave heights which were generated, some waves broke due to depth limited conditions.

Damage to concrete units can be described by the damage number N_{od}, described in Subsec. 2.3.4. N_{od} is the actual number of displaced units related to a width (along the longitudinal axis of the breakwater) of one nominal diameter, $D_n \cdot N_{or}$ and N_{omov} are respectively the number of rocking units and the number of moving units (displaced + rocking).

As only one slope angle was investigated, the influence of the wave period should not be given in formulas including ξ_m, as this parameter includes both wave period (steepness) and slope angle. The influence of wave period, therefore, will be given by the wave steepness s_{om}. Final formulas for stability of concrete units include the relative damage level N_{od}, the number of waves N, and the wave steepness, s_{om}. The formula for cubes is given by

$$\frac{H_s}{\Delta D_n} = \left(6.7\frac{N_{od}^{0.4}}{N^{0.3}} + 1.0\right) s_{om}^{-0.1}, \tag{61}$$

for tetrapods,

$$\frac{H_s}{\Delta D_n} = \left(3.75\frac{N_{od}^{0.5}}{N^{0.25}} + 0.85\right) s_{om}^{-0.2}. \tag{62}$$

For the no-damage criterion $N_{od} = 0$, Eqs. (61) and (62) are reduced to

$$\frac{H_s}{\Delta D_n} = s_{om}^{-0.1}, \tag{63}$$

$$\frac{H_s}{\Delta D_n} = 0.85 s_{om}^{-0.2}. \tag{64}$$

The storm duration and wave period showed no influence on the stability of accropode and the "no damage" and "failure" criterions were very close. The

stability, therefore, can be described by two simple formulas:
start of damage, $N_{od} = 0$,

$$\frac{H_s}{\Delta D_n} = 3.7 \, , \tag{65}$$

failure, $N_{od} > 0.5$,

$$\frac{H_s}{\Delta D_n} = 4.1 \, . \tag{66}$$

A comparison of Eqs. (65) and (66) shows that the start of damage and failure of accropode are very close, although at very high $H_s/\Delta D_n$-numbers. It means that up to a high wave height, accropode are completely stable, but after the initiation of damage at this high wave height, the structure will fail progressively. Therefore, it is recommended that a safety coefficient for design of about 1.5 on the $H_s/\Delta D_n$-value be used. This means that for the design of accropode, one should use the following formula, which is close to design values of cubes and tetrapods:

$$\frac{H_s}{\Delta D_n} = 2.5 \, . \tag{67}$$

The reliability of Eqs. (61)–(66) can be described with a similar procedure as for rock. The coefficients 3.7 and 4.1 in Eqs. (65) and (66) for accropode can be considered as stochastic variables with a standard deviation of 0.2. The procedure for Eqs. (61)–(64) is more complicated. Assume a relationship:

$$\frac{H_s}{\Delta D_n} = af(N_{od}, N, s_{om}) \, . \tag{68}$$

The function $f(N_{od}, N, s_{om})$ is given in Eqs. (61) and (62). The coefficient, a, can be regarded as a stochastic variable with an average value of 1.0 and a standard deviation. From this analysis, it follows that this standard deviation is $\sigma = 0.10$ for both formulas on cubes and tetrapods.

Equations (56) and (57) and (61)–(67) describe the stability of rock, cubes, tetrapods, and accropode. A comparison of stability is made in Fig. 35 where for all units curves are shown for two damage levels: "start of damage" ($S = 2$ for rock and $N_{od} = 0$ for concrete units) and "failure" ($S = 8$ for rock, $N_{od} = 2$ for cubes, $N_{od} = 1.5$ for tetrapods and $N_{od} > 0.5$ for accropode). The curves are drawn for $N = 3000$ and are given as $H_s/\Delta D_n$ versus the wave steepness, s_{om}.

From Fig. 35, the following conclusions can be drawn:

- "Start of damage" for rock and cubes is almost the same. This is partly due to a more stringent definition of "no damage" for cubes ($N_{od} = 0$). The damage level $S = 2$ for rock means that a little displacement is allowed (according to Hudson's criterion of "no damage", however).
- The initial stability of tetrapods is higher than for rock and cubes and the initial stability of accropode is much higher.
- Failure of the slope is reached first for rock, then cubes, tetrapods, and accropode. Stability at the "failure" level (in terms of $H_s/\Delta D_n$-values) is closer for tetrapods and accropode than at the initial damage stage.

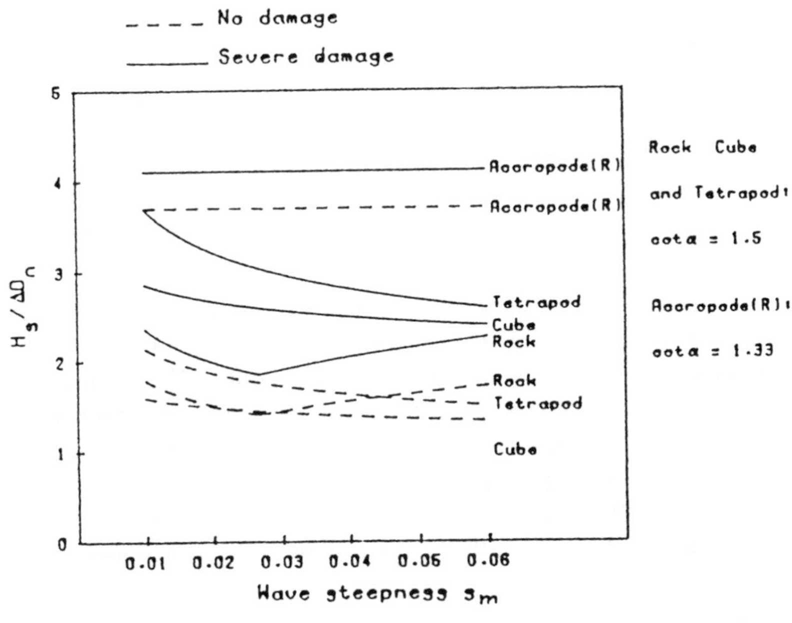

Fig. 35. Comparison of stability of rock, cubes, tetrapods, and accropode.

Another useful graph that can be directly derived from the stability formulas (61) and (62) is a wave height — damage graph. Figure 36 gives an example of cubes and gives the 90% confidence bands too, using the standard deviations described before.

Up to now damage to a concrete armor layer was defined as units displaced out of the layer (N_{od}). Large concrete units, however, can break due to limits in structural strength. After the failures of the large breakwaters in Sines, San Ciprian, Arzew, and Tripoli, a lot of research all over the world was directed towards the strength of concrete armor units. The results of that research will not be described here.

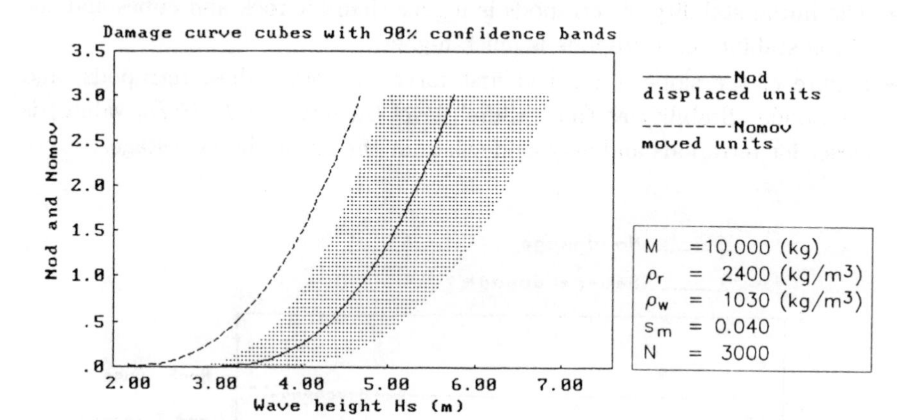

Fig. 36. Wave height — damage curve for cubes with 90% confidence levels.

In cases where the structural strength may play a role, however, it is interesting to know more than only the number of displaced units. The number of rocking units, N_{or}, or the total number of moving units, N_{omov}, may give an indication of the possible number of broken units. A (very) conservative approach is followed when one assumes that each moving unit results in a broken unit. The lower limits (only displaced units) for cubes and tetrapods are given by Eqs. (61) and (62). The upper limits (number of moving units) are derived by Van der Meer and Heydra (1990). The equations for the number of moving units are

for cubes,

$$\frac{H_s}{\Delta D_n} = \left(6.7\frac{N_{omov}^{0.4}}{N^{0.3}} + 1.0\right) s_{om}^{-0.1} - 0.5\,, \tag{69}$$

for tetrapods,

$$\frac{H_s}{\Delta D_n} = \left(3.75\frac{N_{omov}^{0.5}}{N^{0.25}} + 0.85\right) s_{om}^{-0.2} - 0.5\,, \tag{70}$$

The equations are very similar to Eqs. (61) and (62), except for the coefficient −0.5. In a wave height-damage graph, the result is a curve parallel to the one for N_{od}, but which is shifted to the left. For armor layers with large concrete units, the actual number of broken units will probably lie between the curve of N_{od} (Eqs. (61) and (62)) and N_{omov} (Eqs. (69) and (70)).

Holtzhausen and Zwamborn (1992) investigated the stability of dolosse in a basic way, similar to the research on cubes, tetrapods, and accropode described above. Damage was defined as units displaced by more than one diameter and rocking or movements were not taken into account. The aspect of rocking (and breakage) should be considered for heavy dolosse, say, heavier than 10–15 t.

After some rewriting with respect to the damage number N_{od}, which is used in this paper, the stability formula for Dolosse becomes, according to Holtzhausen and Zwamborn (1992):

$$N_{od} = 6250 \left[\frac{H_s}{\Delta^{0.74} D_n}\right]^{5.26} s_{op}^{3} w_r^{20 s_{op}^{0.45}} + E \, , \tag{71}$$

where

w_r = the waist ratio of the dolos,
E = error term.

The waist ratio w_r is a measure to account for possible breakage: a higher waist ratio gives a stronger dolos and should be used for relative severe wave attacks. The applicable range for the waist ratio is 0.33–0.40.

The error term E describes the reliability of the formula. It is assumed to be normally distributed with a mean of zero and a standard deviation of

$$\sigma(E) = 0.01936 \left[\frac{H_s}{\Delta^{0.74} D_n}\right]^{3.32} . \tag{72}$$

As Eq. (71) is a power curve, the no-damage criterion $N_{od} = 0$ cannot be substituted. According to Holtzhausen and Zwamborn (1992) no damage should be described as $N_{od} = 0.1$. The test duration was between 2000 and 3000 waves (one hour in the model). The influence of storm duration was not investigated, which means that Eq. (71) holds for storm durations in the same order as the tests.

Breakage of units has not been treated in this paper. Some relevant references on this topic are Burcharth *et al.* (1991), Burcharth and Liu (1992), Ligteringen *et al.* (1992), Scott *et al.* (1990), Van der Meer and Heydra (1991), and Van Mier and Lenos (1991).

4.4. *Underlayers, filters, toe protection, and head*

4.4.1. *Underlayers and filters*

Rubble mound structures in coastal and shoreline protection are normally constructed with an armor layer and one or more underlayers. Sometimes an underlayer is called a filter. The dimensions of the first underlayer depend on the structure type.

Revetments often have a two-diameter thick armor layer, a thin underlayer or filter, and then an impermeable structure (clay or sand), with or without a geotextile. The underlayer in this case works as a filter. Smaller particles beneath the filter should not be washed through the layer and the filter stones should not be washed through the armor. In this case the geotechnical filter rules are strongly recommended. Roughly, these rules give D_{15}(armor)/ D_{85}(filter) $<$ 4 to 5.

Structures such as breakwaters have one or two underlayers followed by a core of rather fine material (quarry-run). The SPM (1984) recommends for the stone size of the underlayer under the armor a range of 1/10 to 1/15 of the armor mass. This criterion is more strict than the geotechnical filter rules and gives $D_{n50}(\text{armor})/D_{n50}(\text{underlayer}) = 2.2 - 2.3$.

A relatively large underlayer has two advantages. First, the surface of the underlayer is less smooth with bigger stones and gives more interlocking with the armor. This is especially so if the armor layer is constructed of concrete armor units.

Second, a large underlayer results in a more permeable structure and therefore has a large influence on the stability (or required mass) of the armor layer. The influence of the permeability on stability has been described in Subsec. 4.2.

Therefore, it is recommended that we use sizes of $1/10$ to $1/15 M_{50}$ of the armor for the mass of the underlayer.

4.4.2. *Toe protection*

In most cases, the armor layer on the seaside near the bottom is protected by a toe (see Fig. 37). If the rock in the toe has the same dimensions as the armor, the toe will be stable. In most cases, however, one wants to reduce the rock size in the toe. Following the work of Brebner and Donnelly (1962), given in the SPM (1984), who tested toes to vertical-faced composite breakwaters under monochromatic waves, a relationship may be assumed between the ratio h_t/h and the stability number $H/\Delta D_{n50}$ (or N_s), where h_t is the depth of the toe below the water level and h is the water depth (also see Fig. 5). A small ratio of $h_t/h = 0.3$–0.5 means that the toe is relatively high above the bottom. In that case, the toe structure is more a berm structure. A value of $h_t/h = 0.8$ means that the toe is near the bottom.

Sometimes a relationship between $H_s/\Delta D_{n50}$ and h_t/H_s is assumed where a lower value of h_t/H_s should give more damage. Gravesen and Sørensen (1977) describe that a high wave steepness (short wave period) gives more damage to the toe than a low wave steepness. The above-mentioned assumption was based on only a few points. In the CIAD report (1985), this conclusion could not be verified. No relationship was found between $H_s/\Delta D_{n50}$ and h_t/H_s, probably because H_s is present in both parameters.

The CUR/CIRIA Manual (1991) gives a design graph on toe structure stability, based on a collection of site specific tests at Delft Hydraulics and the Danish Hydraulic Institute (see Fig. 37). Three damage classifications were defined: "0–3%" means no movement of stones (or only a few) in the toe; "3–10%" means that the toe flattened out a little, but the function of the toe (supporting the armor layer) is intact and the damage is acceptable. "Failure" means that the toe has lost its function and this damage level is not acceptable.

In almost all cases the structure was attacked by waves in a more or less depth limited situation, which means that H_s/h was fairly close to 0.5. This is also the reason why it is acceptable that the location of the toe, h_t, is related to the water depth, h. It would not be acceptable for breakwaters in very large water depths (more than 20–25 m). Figure 37 is, therefore, applicable for depth limited situations only.

Figure 37 shows that if the toe is high above the bottom (small h_t/h ratio), the stability is much smaller than for the situation where the toe is close to the bottom. The results of DHI (internal paper) are also shown in the graph and correspond well with the 3–10% values of Delft Hydraulics.

A suggested line for design purposes is given in the graph. In general, it means that the depth of the toe below the water level is an important parameter. If the toe is close to the bottom, the diameter of the stones can be more than twice as small as when the toe is half way down the bottom and the water level. A design formula for low and acceptable damage (3–10%) and for more or less depth limited situations is

$$\frac{H_s}{\Delta D_{n50}} = 8.7\left(\frac{h_t}{h}\right)^{1.4}. \tag{73}$$

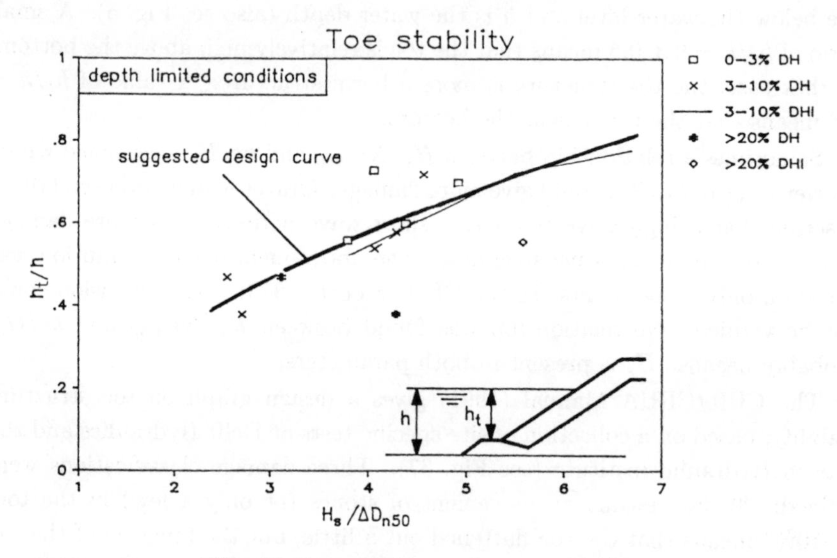

Fig. 37. Toe stability as a function of h_t/h.

Three points are shown in Fig. 37 which indicate failure of the toe. The above given design values are safe for $h_t/h > 0.5$. For lower values of h_t/h, one should use the stability formulas for armor stones described in Subsec. 4.2.

Recent research on toe structure stability was performed by Gerding (1993). His tests were performed in order to establish the influence of wave height, wave steepness, and water depth on toe stability. One of the main conclusions was that the wave steepness had no influence. His analysis resulted in an improved formula with regard to Eq. (73) and included the damage level N_{od} described earlier:

$$\frac{H_s}{\Delta D_{n50}} = \frac{H_{2\%}/1.4}{\Delta D_{n50}} = \left(0.24\frac{h_t}{D_{n50}} + 1.6\right) N_{od}^{0.15} \, . \tag{74}$$

In Eq. (74), $N_{od} = 0.5$ means start of damage, $N_{od} = 2$ means some flattening out, and $N_{od} = 4$ means complete flattening out of the toe. This applies to a "standard" toe size of about 3–5 stones wide and 2–3 stones high. For wider toe structures, a higher damage level can be applied before flattening out occurs. Equation (74) has no restrictions on depth limitation as the water depth h is not used as a parameter. The influence of limited water depth (highest waves break) can be taken into account by using $H_{2\%}$ instead of H_s. A safe approach is followed when H_s is used. Equation (74) can be used in the range:

$0.4 < h_t/h < 0.9$,
$3 < h_t/D_{n50} < 25$.

4.4.3. *Breakwater head*

Breakwater heads represent a special physical process. Jensen (1984) described it as follows:

> "When a wave is forced to break over a roundhead it leads to large velocities and wave forces. For a specific wave direction only a limited area of the head is highly exposed. It is an area around the still-water level where the wave orthogonal is tangent to the surface and on the lee side of this point. It is therefore general procedure in design of heads to increase the weight of the armor to obtain the same stability as for the trunk section. Alternatively, the slope of the roundhead can be made less steep, or a combination of both."

An example of the stability of a breakwater head in comparison with the trunk section and showing the location of the damage, as described in the previous paragraph, is shown in Fig. 38 and was taken from Jensen (1984). The stability coefficient ($H_s/\Delta D_n$ for tetrapods) is related to the stability of the trunk section. Damage is located about 120°–150° from the wave angle. This local damage is clearly found by research with long-crested waves.

Possibly, the actual damage in prototype may be less concentrated as waves in nature are short-crested and multidirectional. Research in multidirectional wave basins should be undertaken to clarify this aspect.

No specific rules are available for the breakwater head. The required increase in weight can be a factor between 1 and 4, depending on the type of armor unit. The factor for rock is closer to 1.

Another aspect of breakwater heads is mentioned by Jensen (1984). The damage curve for a head is often steeper than for a trunk section. A breakwater head may show progressive damage. This means that if both head and trunk were designed on the same (low) damage level, an (unexpected) increase in wave height can cause failure of the head or a part of it, where the trunk still shows acceptable damage. This aspect is less pronounced for heads which are armored by rock.

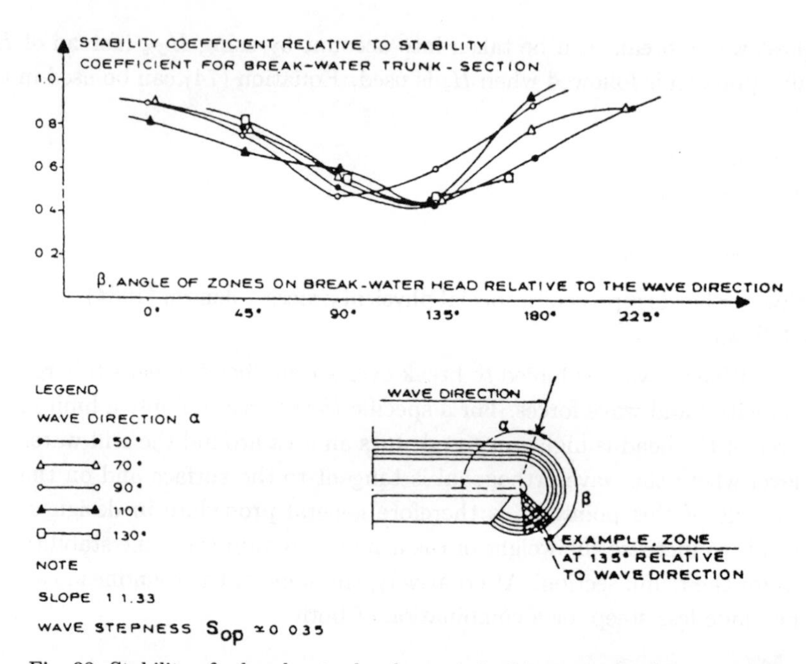

Fig. 38. Stability of a breakwater head armored with tetrapods. (Jensen, 1984)

4.5. *Crest and rear armor (low-crested structures)*

4.5.1 *Classification of low-crested structures*

As long as structures are high enough to prevent overtopping, the armor on the crest and rear can be (much) smaller than on the front face. The dimensions of the rock in that case will be determined by practical matters such as available rock, etc.

Most structures, however, are designed to have some or even severe overtopping under design conditions. Others are so low that under daily conditions, the structure is overtopped. Structures with the crest level around still-water level or below will always have overtopping and transmission.

It is obvious that when the crest level of a structure is low, wave energy can pass over the structure. This has two effects. First, the armor on the front side can be smaller than on a nonovertopped structure, due to the fact that less energy is left on the front side (lower run-down wave forces).

The second effect is that the crest and rear should be armored with rock which can withstand the attack by overtopping waves. For rock structures, the same armor on the front face, crest, and rear is often applied. The methods to establish the armor size for these structures will be given here. They may not hold for structures with an armor layer of concrete units. For those structures, physical model investigations may give an acceptable solution.

Low-crested rock structures can be divided into three categories: dynamically stable reef breakwaters, statically stable low-crested structures (with the crest above still-water level) and statically stable submerged structures.

A reef breakwater is a low-crested homogeneous pile of stones without a filter layer or core and is allowed to be reshaped by wave attack (Fig. 39). The initial crest height is just above the water level. Under severe wave conditions the crest height reshapes to a certain equilibrium crest height. This equilibrium crest height and the corresponding transmission are the main design parameters. The transmission was already described in Subsec. 3.3.

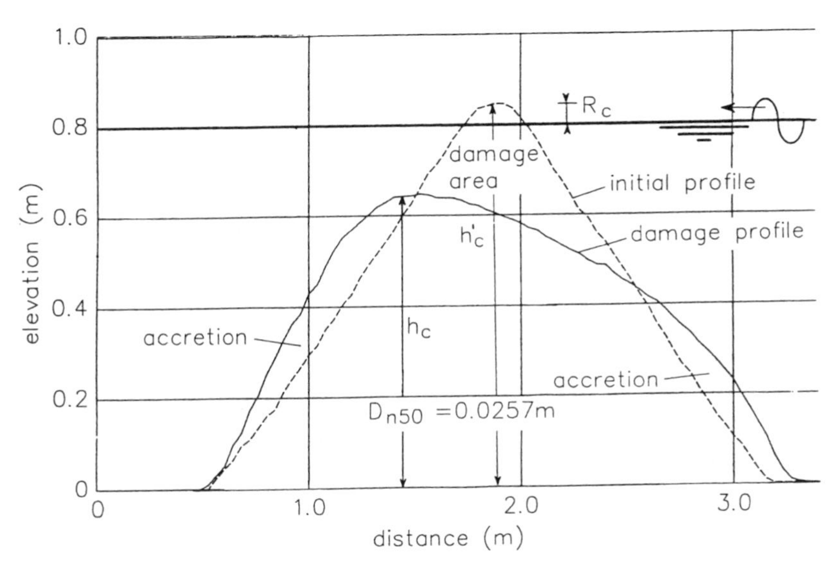

Fig. 39. Dynamically stable reef breakwater.

Statically stable low-crested breakwaters are close to non- or marginally overtopped structures, but are more stable due to the fact that a (large) part of the wave energy can pass over the breakwater (Fig. 40).

All waves overtop statically stable submerged breakwaters and the stability

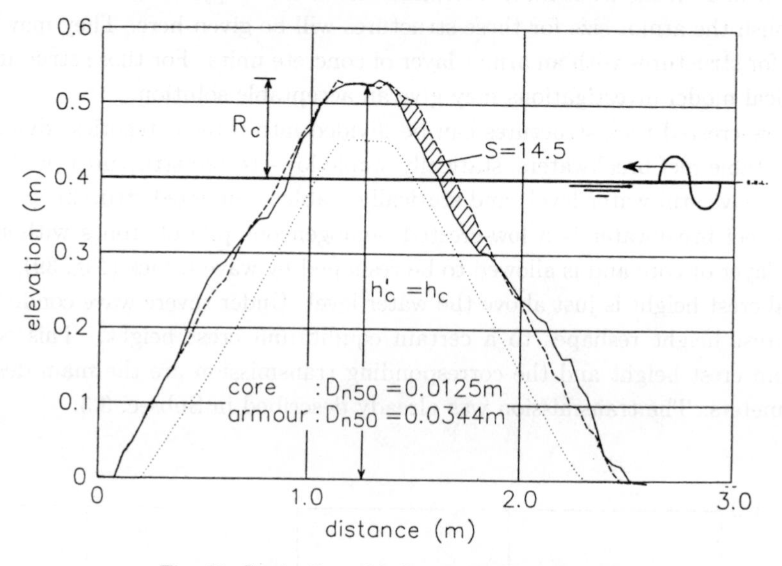

Fig. 40. Statically stable low-crested breakwater.

increases remarkably if the crest height decreases (Fig. 41). It is obvious that the wave transmission is substantial at these structures.

4.5.2. *Reef breakwaters*

The stability analyses conducted by Ahrens (1987, 1989) and Van der Meer (1990a) were concentrated on the change in crest height due to wave attack (see Fig. 39). Ahrens defined a number of dimensionless parameters which described the behavior of the structure. The main one is the relative crest height reduction factor h_c/h_c'. The crest height reduction factor h_c/h_c' is the ratio of the crest height at the completion of a test to the height at the beginning of the test. The natural limiting values of h_c/h_c' are 1.0 (no deformation) and 0.0 (structure not present anymore) respectively.

Ahrens found that for the reef breakwater, a longer wave period caused more displacement of material than a shorter period. Therefore, he introduced the spectral (or modified) stability number, N_s^*, defined by Eq. (9).

That a longer wave period results in more damage than a shorter period is not always true. Ahrens concluded that it is true for reef breakwaters where the crest height lowered substantially during the test. It is, however, not true

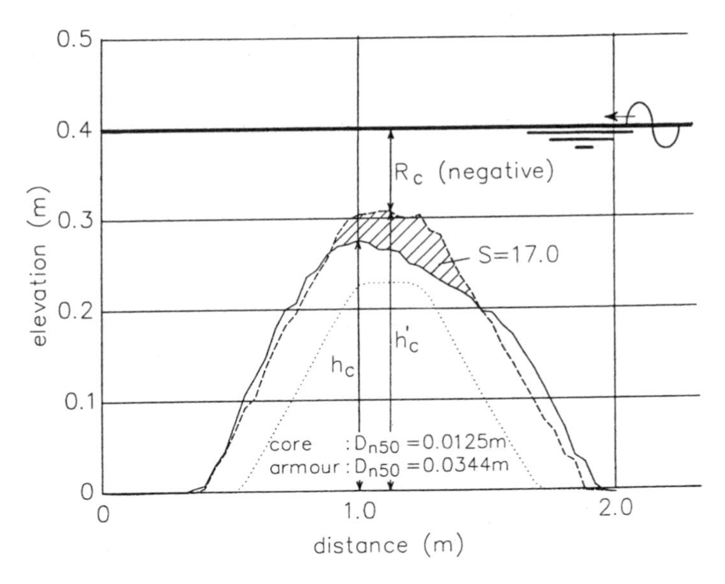

Fig. 41. Submerged breakwater.

for non- or marginally-overtopped breakwaters (Van der Meer, 1987 or 1988). The influence of the wave period in that case is much more complex than suggested by Eq. (9).

The (reduced) crest height, according to Van der Meer (1990a) or Van der Meer and Pilarczyk (1990) including all Ahrens' data, can be described by

$$h_c = \sqrt{\frac{A_t}{e^{aN_s^*}}} \tag{75}$$

with

$$a = -0.028 + 0.045C' + 0.34\frac{h'_c}{h} - 6.10^{-9}B_n^2 \tag{76}$$

and $h_c = h'_c$ if h_c in Eq. (75) $> h'_c$,
where

A_t = area of structure cross-section,

$C' = A_t/h'^2_c$ (structure response slope),

h = water depth at structure toe,

$B_n = A_t/D_{n50}^2$ (bulk number).

The lowering of the crest height of reef-type structures as shown in Fig. 39, can be calculated with Eqs. (75) and (76). It is possible to draw design curves from these equations which give the crest height as a function of N_s or even H_s. An example of h_c versus H_s is shown in Fig. 42. The reliability of Eq. (75) can be described by 90% confidence bands, given by $h_c \pm 5\%$ and is shown in Fig. 42.

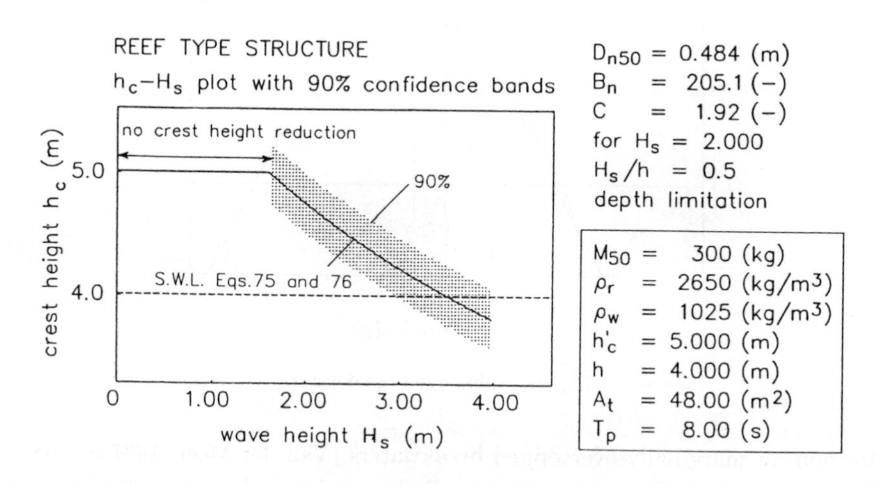

Fig. 42. Design graph of reef-type breakwater.

4.5.3. *Statically stable low-crested breakwaters*

The stability of a conventional low-crested breakwater above still-water level can be related to the stability of a non- or marginally-overtopped structure. For example, stability formulas such as (56) and (57) can be used. The required rock diameter for an overtopping breakwater can then be determined by the application of a reduction factor for the mass of the armor.

The derived equations are based on Van der Meer (1990a):
Reduction factor for

$$D_{n50} = \frac{1}{1.25 - 4.8R_p^*} \tag{77}$$

for $0 < R_p^* < 0.052$,
where

$$R_p^* = \frac{R_c}{H_s}\sqrt{\frac{s_{op}}{2\pi}}\,. \tag{78}$$

The R_p^* parameter is a combination of relative crest height, R_c/H_s, and wave steepness, s_{op}. Equation (77) describes the stability of a statically stable low-crested breakwater with the crest above still-water level simply by the application of a reduction factor on the required diameter of a nonovertopped structure (for example, according to Eqs. (56) and (57)). Equation (77) is shown in Fig. 43 for various wave steepnesses, and can be used as a design graph. The reduction factor to be applied for the nominal diameter can be read from this graph (or calculated by Eq. (77)).

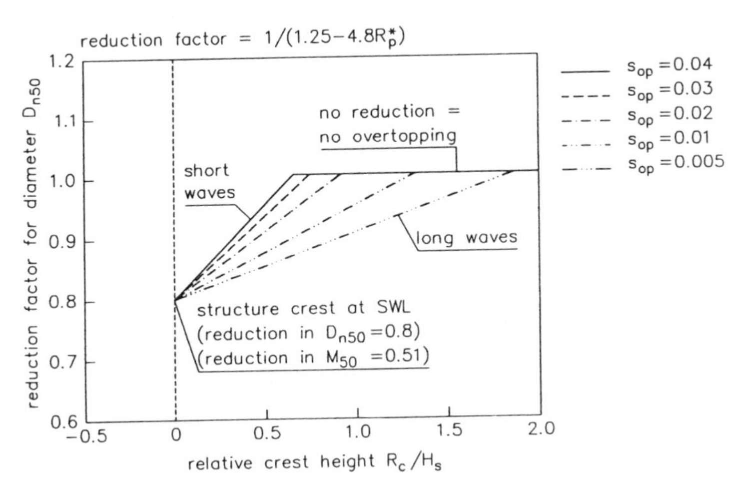

Fig. 43. Design graph for conventional low-crested structures above still-water level.

An average reduction factor of 0.8 in diameter is obtained for a structure with the crest height at the water level. The required mass in that case is a factor $0.8^3 = 0.51$ of that required for a nonovertopped structure.

It is not really required to describe the reliability of the reduction factor in Eq. (77). The reliability of D_{n50} is about the same as for a non- or marginally-overtopped structure, i.e., the reliability depends on the stability formula that is used to calculate the D_{n50} for a nonovertopped structure.

Vidal *et al.* (1992) studied a similar low-crested structure as described above (although including submerged structures), but they divided the armor layer of the structure into three sections: the front slope, the crest, and the back slope. These different sections of the structure have also different stability responses to a sea state condition. The behavior of the total slope protection (as described by Eq. (77)), reflects the stability behavior of each section component. If one wants to optimize the armor weight to obtain a similar security condition in each part of the breakwater, the stability curves in each section should be determined.

Figure 44 gives a comparison of stability between the three sections as a function of the relative crest height and is taken from Vidal *et al.* (1992). The damage level that is described is $S = 2 - 2.5$ for each section (see Eq. (10)).

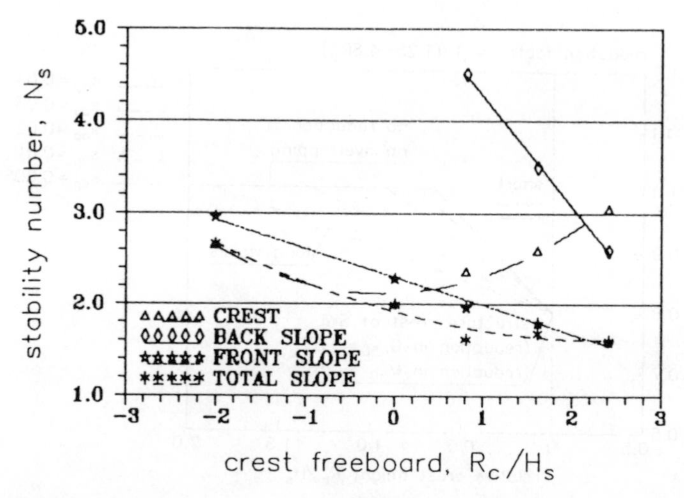

Fig. 44. Stability of front slope, crest, and back slope as a function of relative crest height. (Vidal *et al.*, 1992)

The front slope is least stable for relative crest heights of $R_c/D_{n50} > 0.5$. For $R_c/D_{n50} < 0.5$ the crest section is the least stable. The back slope is the most stable section for $R_c/D_{n50} < 2.0$. For larger values, the crest is more stable (although the stability should be similar for nonovertopped structures). If the freeboard is high (R_c/D_{n50} around 2.5) and the armor on the crest is the same as the back slope, the damage can start at the back slope and desegregate the crest.

The curve corresponding to the total slope reflects the basic principle of Eq. (77) and Fig. 43: the stability increases as the relative freeboard decreases. Moreover, the stability number for non- or marginally-overtopped structures in Fig. 44 amounts to $H_s/\Delta D_{n50} = 1.6$ and for structures with the crest at still-water level to $H_s/\Delta D_{n50} = 2.0$. This gives the same reduction factor of 0.8 for structures with the crest at the water line.

4.5.4. *Submerged breakwaters*

The slope angle has large influence on nonovertopped structures, but in the case of submerged structures the wave attack is concentrated on the crest and less on the seaward slope. Therefore, excluding the slope angle of submerged structures, being a governing parameter for stability may be legitimate.

The stability of submerged breakwaters appeared only to be a function of the relative crest height h_c'/h, the damage level S, and the spectral stability number N_s. The given formulas are based on a reanalysis of the tests of Givler and Sørensen (1986) by Van der Meer (1990a). The stability is described by

$$\frac{h_c'}{h} = (2.1 + 0.1S)e^{-0.14N_s^*} \,. \tag{79}$$

For fixed crest height, water level, damage level, and wave height and period, the required ΔD_{n50} can be calculated from Eq. (79), finally yielding the required rock weight. Also, wave height versus damage curves can be derived from Eq. (79). The equation is shown as a design graph in Fig. 45 for four damage levels. The reliability of Eq. (79) can be described when the factor 2.1 is considered as a stochastic variable with a normal distribution. The data gives a standard deviation of 0.35. With this standard deviation, it is possible to calculate the 90% confidence bands, using $2.1 \pm 1.64 * 0.35$ in Eq. (79). Figure 45 gives the 90% confidence bands for $S = 2$. The scatter is quite large and this should be considered during the design of submerged structures.

4.6. *Berm breakwaters*

4.6.1. *Description of the seaward profile*

Statically stable structures can be described by the damage parameter S, (see Subsec. 2.3.4) and dynamically stable ones by a profile (see Fig. 8). Other typical profiles, but for different initial slopes, are shown in Fig. 46. The main part of the profiles is always the same. The initial slope (gentle or steep) determines whether material is transported upwards to a beach crest or downwards, creating erosion around the still-water level.

Based on extensive model tests (Van der Meer (1988a)), relationships were established between the characteristic profile parameters, as shown in Fig. 8,

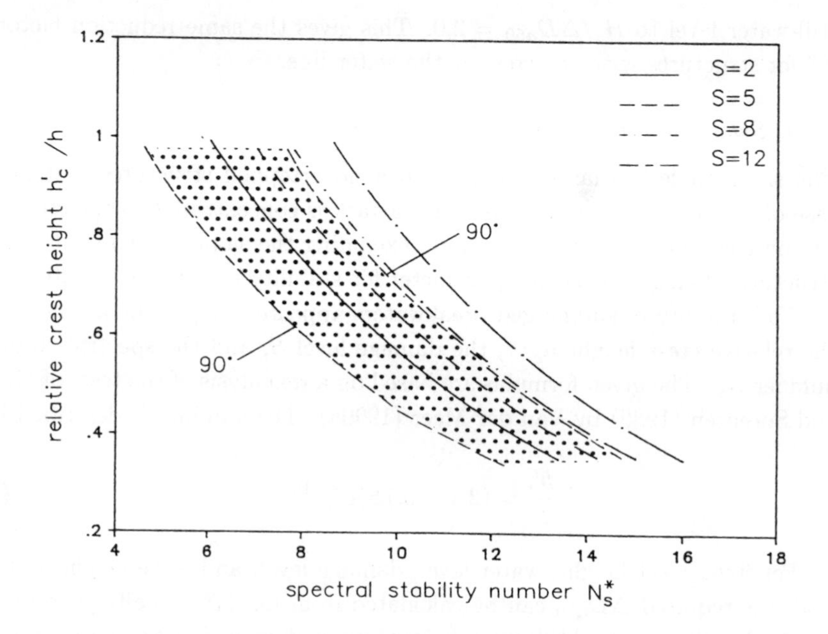

Fig. 45. Design curves for submerged breakwaters.

and the hydraulic and structural parameters. These relationships were used to make the computational model BREAKWAT, which simply gives the profile in a plot together with the initial profile. Boundary conditions for this model are

- $H_s/\Delta D_{n50}$ = 3-500 (berm breakwaters, rock and gravel beaches),
- arbitrary initial slope,
- crest above still-water level,
- computation of a (established or assumed) sequence of storms (or tides) by using the previously computed profile as the initial profile.

The input parameters for the model are the nominal diameter of the stone, D_{n50}, the grading of the stone, D_{85}/D_{15}, the buoyant mass density, Δ, the significant wave height, H_s, the mean wave period, T_m, the number of waves (storm duration), N, the water depth at the toe, h and the angle of wave incidence, β. The (first) initial profile is given by a number of (x, y) points with straight lines in between. A second computation can be made on the same initial profile or on the computed one.

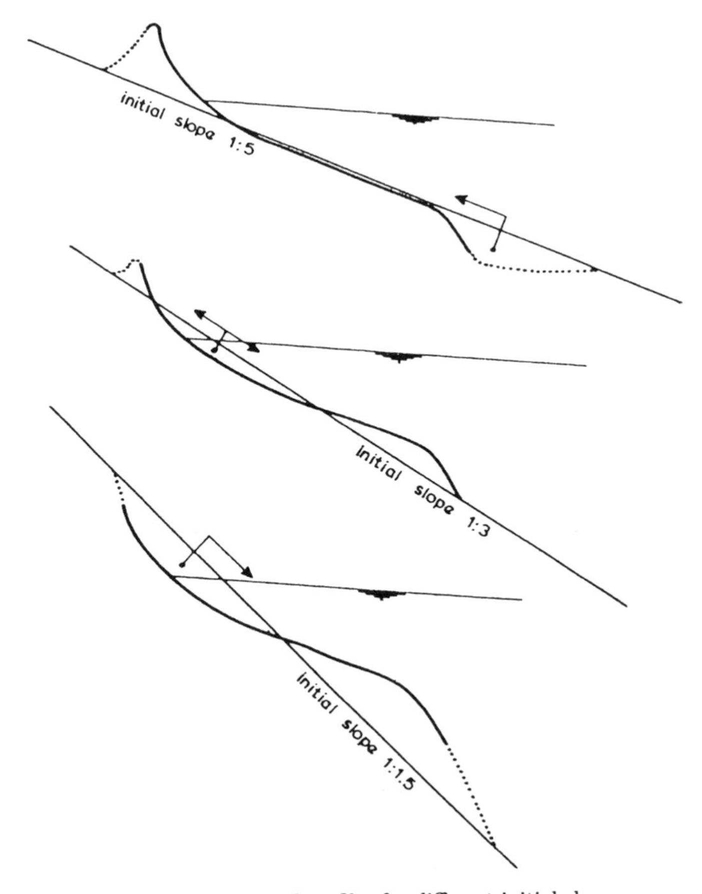

Fig. 46. Examples of profiles for different initial slopes.

The result of a computation on a berm breakwater is shown in Fig. 47, together with a listing of the input parameters. The model can be applied to

- design of rock slopes and gravel beaches,
- design of berm breakwaters,
- behavior of core and filter layers under construction during yearly storm conditions.

The computation model can be used in the same way as the deterministic design approach of statically stable slopes, described in Subsec. 4.2. There

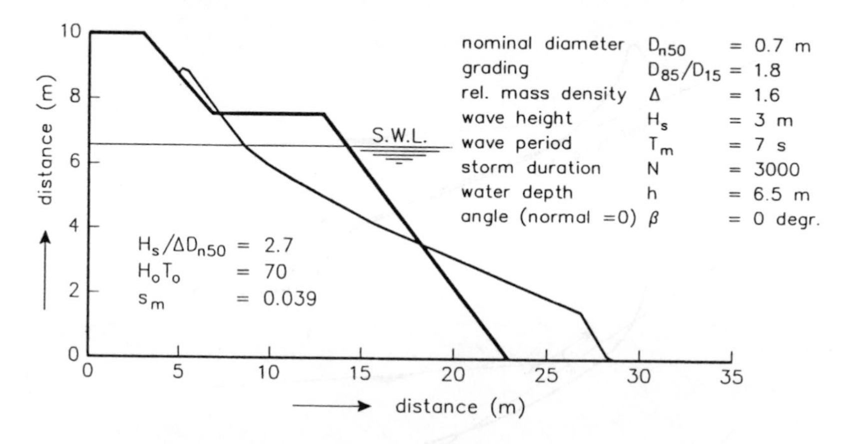

Fig. 47. Example of a computed profile for a berm breakwater.

the rather complicated stability Eqs. (56) and (57) were used to make design graphs such as damage curves, and these graphs were used for a sensitivity analysis. By making a large number of computations with the computational model, the same kind of sensitivity analysis can be performed for dynamically stable structures. Aspects which were considered for the design of a berm breakwater (Van der Meer and Koster (1988)) were, for example

- optimum dimensions of the structure (upper and lower slope, length of berm)
- influence of wave climate, stone class, water depth,
- stability after first storms.

An example to derive optimum dimensions for a berm breakwater will be described in the next section. The influence of the wave climate on a structure is shown in Fig. 48 and shows the difference in behavior of the structures for various wave climates. Stability after first (less severe) storms can possibly be described by the use of Eqs. (56) and (57).

Computations with the computational model can, of course, only be made if the model is available to the user. This is often not the case for the reader of a paper and therefore, a more simple (and less reliable) method should be given which is able to give the user a first impression (but not more than that!) of the profile that can be expected. This method is described below. Other estimates of profiles of gravel beaches are described by Powell (1990) and of berm breakwaters only by Kao and Hall (1990).

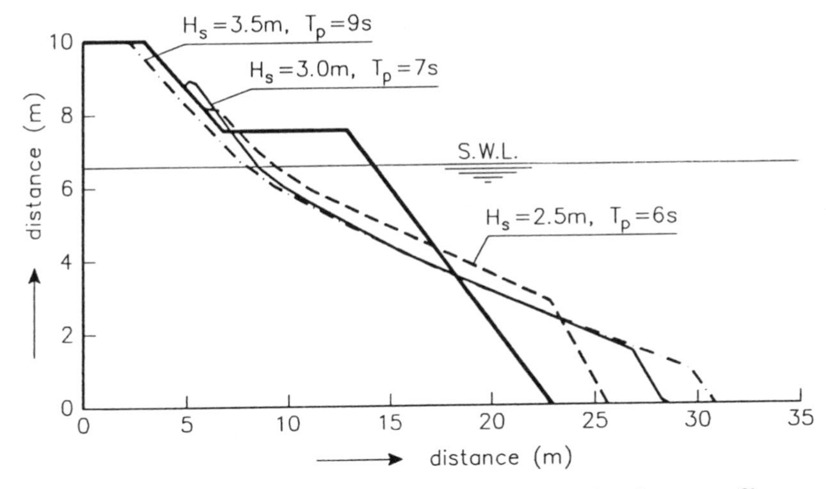

Fig. 48. Example of influence of wave climate on a berm breakwater profile.

Figure 49 gives the schematised profile simplified from the original profiles (see Fig. 8) in the computational model. The connecting point is the intersection of the profile with still-water level. From this point an upper slope is drawn under 1:1.8 and a lower slope under 1:5.5. The crest of the profile is situated on the upper slope and the transition to a steep slope on the lower part. These two points are given by the parameters l_c (length of crest) and l_s (length of step). Of course, a curved line goes through the three points.

The connection with the upper part of the profile and the initial profile is given by l_r (length of run-up). Below the gentle part under still-water level, a steep slope is present, and if the initial profile is gentle ($\cot\alpha > 4$) again there is a gentle slope which gives the "step" in the profile. The transition from a steep to a gentle slope is given by h_t (height of transition). If the initial slope is not a straight line, one should draw a more or less equivalent slope, taking into account the area from $+H_s$ to $-1.5H_s$, which gives $\tan\alpha$. The relationships between the profile parameters and the hydraulic and structural parameters are

$$l_c = 0.041 H_s T_m \sqrt{\frac{g}{D_{n50}}} \,, \tag{80}$$

$$l_s = l_r = 1.8\, l_c \,, \tag{81}$$

$$h_t = 0.6\, l_c \,. \tag{82}$$

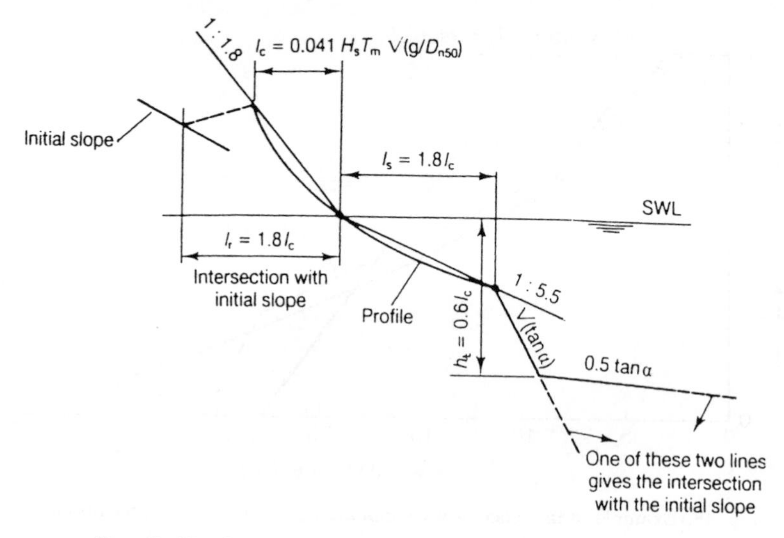

Fig. 49. Simple schematised profile for rock and gravel beaches.

$$\text{Steep slope below still-water level:}\sqrt{\tan\alpha}\ . \tag{83}$$

$$\text{Gentle slope below still-water level:}0.5\tan\alpha\ . \tag{84}$$

Finally, the profile must be shifted along the still-water level until the mass balance is fulfilled. Figure 49 and Eqs. 80–84 give a rough indication of the profile that can be expected. For $H_s/\Delta D_{n50}$ values higher than about 10–15, the prediction is quite reliable. For lower values, the initial profile has a large influence on the profile and therefore, the given method is less reliable. This also applies to berm breakwaters and in that case the method should really be treated as a very rough indication.

4.6.2. *Optimum dimensions for a berm breakwater* (*example*)

A berm breakwater can be regarded as an unconventional design. Displacement of armor stones in the first stage of its lifetime is accepted. After this displacement (profile formation), the structure will be more or less statically stable. The initial cross-section of a berm breakwater can be described by a lower slope 1:m, a horizontal berm with a length b (just above still-water level in this case) and an upper slope 1:n. The lower slope is often steep and close to the natural angle of repose.

The critical design point in the example of Van der Meer and Koster (1988) was that erosion was not allowed on the upper slope above the berm. The minimum required berm length b was established for this criterion with the computational model. The berm length b was determined for various combinations of m and n. Figure 50 shows the final results. Each combination of m, n, and b from this graph gives more or less the same stability (no erosion on the upper slope). It is obvious from Fig. 50 that steep slopes require a longer berm and vice versa.

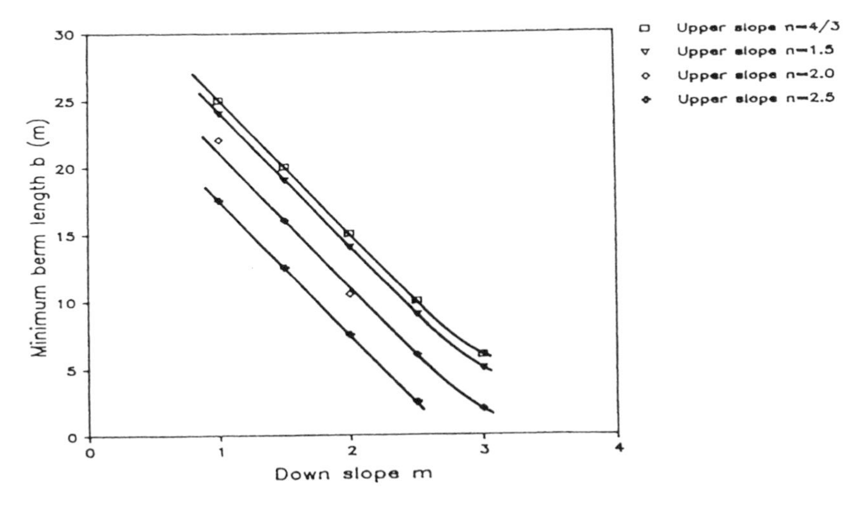

Fig. 50. Minimum berm length as a function of down slope and upper slope for a specific berm breakwater.

Figure 50 gives no information on the optimum values for the slopes. Therefore, another criterion is introduced. The amount of rock required for the construction was calculated for each combination of slopes and berm length. This amount of rock (or cross-sectional area), B, was plotted as a function of the upper and down slope and is given in Fig. 51. It shows that the down slope has minor influence on the required amount of rock (almost horizontal lines) and that a steep upper slope reduces this amount considerably.

It should be noted that actual values given in Figs. 50 and 51 were obtained for a specific structure with specific wave boundary conditions and that they are not generally applicable. But the trends and conclusions are.

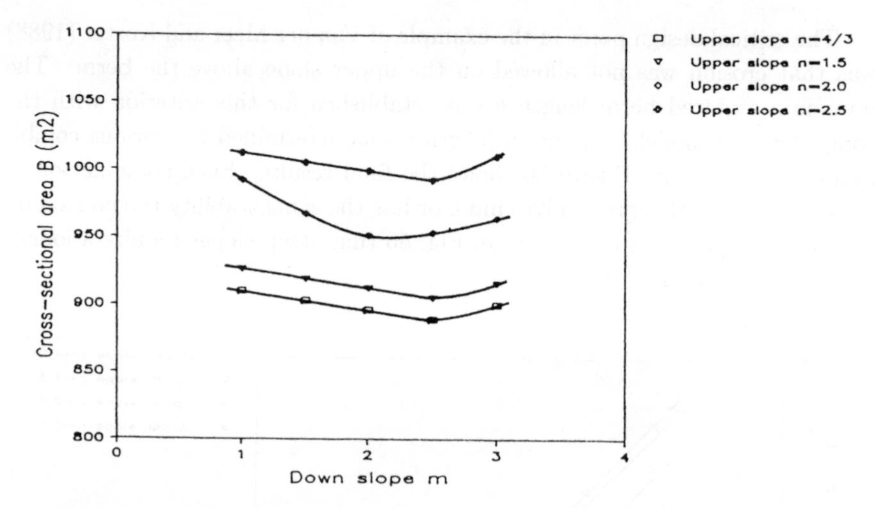

Fig. 51. Cross-sectional area as a function of down and upper slope for a specific berm breakwater.

The relationships for the computational model were based on tests in the range of $H_s/\Delta D_{n50} = 3$–500 (see Van der Meer (1988a)). The model was later verified specifically for berm breakwaters. This is described by Van der Meer (1990c). Tests from various institutes all over the world were used for this verification.

The overall conclusion was that the model never showed large unexpected differences with the test results and that in most cases the calculations and measurements were very close. Compaction of material caused by wave attack and damage to the rear of the structure caused by overtopping are not modeled in the program and this was and is a boundary condition for application of the program.

The combination of the statically stable formulas or model (Fig. 35) with the dynamically stable model showed to be a good tool for the prediction of the behavior of berm breakwaters under all wave conditions.

4.6.3. *Rear stability of berm breakwaters*

Van der Meer and Veldman (1992) performed extensive test series on two different berm breakwater designs. A first design rule was assessed on the relationship between damage to the rear of a berm breakwater and the crest height, wave height, wave steepness, and rock size.

The boundary condition is that the rock at the crest and rear of the berm breakwater has the same dimensions as at the seaward profile. This means that $H_s/\Delta D_{n50}$ is in the order of 3.0–3.5. A further restriction is that the profile at the seaward side has been developed into an S-shape.

The parameter, $R_c/H_s{}^*s_{op}^{1/3}$, showed to be a good combination of relative crest height and wave steepness to describe the stability of the rear of a berm breakwater. The following values of $R_c/H_s{}^*s_{op}^{1/3}$ can be given for various damage levels to the rear of a berm breakwater caused by overtopping waves and can be used for design purposes.

$$\begin{aligned}
&\frac{R_c}{H_s} s_{op}^{1/3} = 0.25 : \text{ start of damage .}\\
&\frac{R_c}{H_s} s_{op}^{1/3} = 0.21 : \text{ moderate damage .}\\
&\frac{R_c}{H_s} s_{op}^{1/3} = 0.17 : \text{ severe damage .}
\end{aligned} \tag{85}$$

A lower value of $R_c/H_s{}^*s_{op}^{1/3}$ means more overtopping and, therefore, more damage. Both a lower relative crest height R_c/H_s and a lower wave steepness give more overtopping and, therefore, more damage.

Andersen *et al.* (1992) performed basic tests on the stability of the rear of a berm breakwater. They included also undeveloped profiles and (very) long berms.

4.6.4. *Head of a berm breakwater*

Burcharth and Frigaard (1987) have studied longshore transport and stability of berm breakwaters in a short basic review. The recession of a breakwater head is shown as an example in Fig. 52, for fairly high wave attack ($H_s/\Delta D_{n50} =$ 5.4). Burcharth and Frigaard (1987) state that, as a first rule of thumb for the stability of a breakwater head, $H_s/\Delta D_{n50}$ should be smaller than 3.

Tests on a berm breakwater head by Van der Meer and Veldman (1992) showed that increasing the height of the berm at this head and, therefore, creating a larger volume of rock, can be seen as a good measure for enlarging the stability of the round head of a berm breakwater, using the same rock as for the trunk.

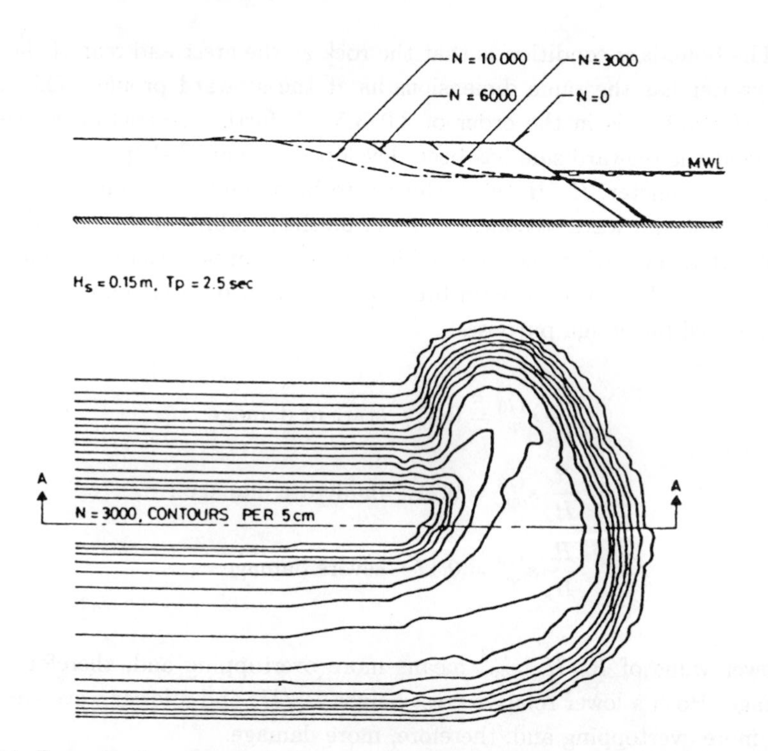

Fig. 52. Example of erosion of a berm breakwater head. (Burcharth and Frigaard, 1987)

4.6.5. *Longshore transport at berm breakwaters*

Statically stable structures as revetments and breakwaters are only allowed to show damage under very severe wave conditions. Even then, the damage can only be described by the displacement of a number of stones from the still-water level to (in most cases) a location downwards. Movement of stones in the direction of the longitudinal axis is not relevant for these types of structures.

The profiles of dynamically stable structures such as gravel/shingle beaches, rock beaches, and sand beaches change according to the wave climate. "Dynamically stable" means that the net cross-shore transport is zero and the profile has reached an equilibrium profile for a certain wave condition. It is possible that during each wave, material is moving up and down the slope (shingle beach).

Oblique wave attack generates wave forces parallel to the alignment of the structure. These forces can cause transport of material along the structure.

This phenomenon is called longshore transport and is well known for sand beaches. Also shingle beaches change due to longshore transport, although the research on this aspect has always been limited.

Rock beaches and berm breakwaters are or can also be dynamically stable under severe wave action. This means that oblique wave attack may induce longshore transport, which can also cause problems for these types of structures. Longshore transport does not occur for statically stable structures, but it will start for conditions where the diameter is small enough in comparison with the wave height. Then the conditions for the start of longshore transport are important.

The start of longshore transport is the most interesting consideration for the berm breakwater where profile development under severe wave attack is allowed, but longshore transport should be avoided. The berm breakwater can roughly be described by $2.5 < H_s/\Delta D_{n50} < 6$. Burcharth and Frigaard (1987) performed model tests to establish the incipient longshore motion for berm breakwaters and their range of tests corresponded to $3.5 < H_s/\Delta D_{n50} < 7.1$. Longshore transport is not allowed at berm breakwaters and, therefore, Burcharth and Frigaard gave the following (somewhat premature) recommendations for the design of berm breakwaters, which are in fact the criterions for incipient motion:

$$\begin{aligned} \text{for trunks exposed to steep waves,} \quad & \frac{H_s}{\Delta D_{n50}} < 4.5 \,, \\ \text{for trunks exposed to oblique waves,} \quad & \frac{H_s}{\Delta D_{n50}} < 3.5 \,, \\ \text{for roundheads,} \quad & \frac{H_s}{\Delta D_{n50}} < 3 \,. \end{aligned} \tag{86}$$

Van der Meer and Veldman (1992) tested a berm breakwater under angles of wave attack of 25 and 50 degrees. Burcharth and Frigaard (1987, 1988) tested their structure under angles of 15 and 30 degrees. Longshore transport was measured by the movement of stones from a colored band. The transport was measured for developed profiles which meant that the longshore transport during the development of the profile of the seaward slope was not taken into account. The measured longshore transport, $S(x)$, was defined as the number of stones that was displaced per wave. Multiplication of $S(x)$ with the storm duration (the number of waves) in practical cases would lead to a transport rate of the total number of stones displaced per storm. Subsequently, the transport rate can be calculated in m^3/storm or m^3/s.

Figures 53 and 54 give all the test results on longshore transport, Both for the tests of Van der Meer and Veldman (1992) and the tests of Burcharth and Frigaard (1988). Both a higher wave height and a longer wave period result in larger transport. According to Eq. (8), a combined wave height-wave period parameter, H_oT_{op}, can be used:

$$H_oT_{op} = \frac{H_s}{\Delta D_{n50}} T_p \sqrt{\frac{g}{D_{n50}}} \,. \tag{87}$$

H_o is defined as the stability number, $H_s/\Delta D_{n50}$, and T_{op} as the dimensionless wave period related to the nominal diameter, $T_{op} = T_p\sqrt{g/D_{n50}}$. With the parameter, H_oT_{op}, it is assumed that wave height and wave period have the same influence on longshore transport. Figures 53 and 54 give the longshore transport $S(x)$ (in number of stones per wave) versus the H_oT_{op}.

Figure 53 gives all the data points. The maximum transport is about 3 stones/wave for $H_oT_{op} = 350$, which is in fact a very high rate for berm breakwaters. The $H_s/\Delta D_{n50}$-value in that case was 7.1, considerably higher than the design value for berm breakwaters. Figure 53 also shows that quite a number of tests had a much smaller transport rate than 0.1 – 0.2 stones/wave.

Therefore, Fig. 54 was drawn with a maximum transport rate of only 0.1 stones/wave. Now, only 4 data points remain of Burcharth and Frigaard (1988), the others are from the tests of Van der Meer and Veldman (1992).

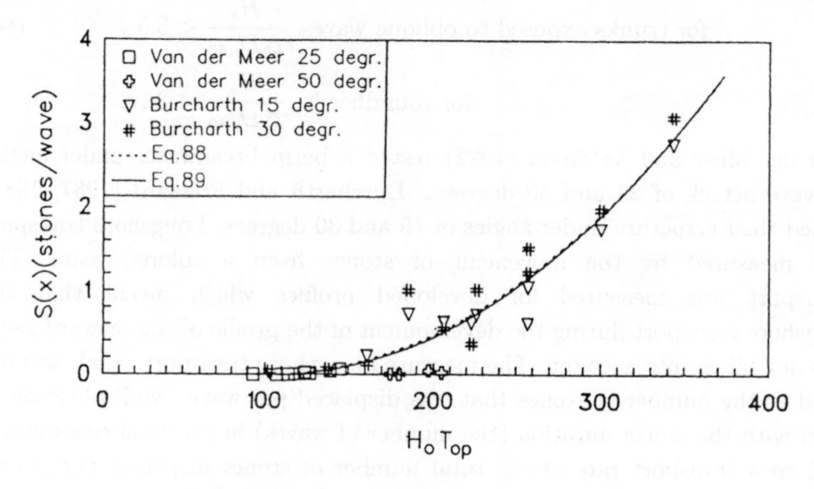

Fig. 53. Longshore transport for berm breakwaters.

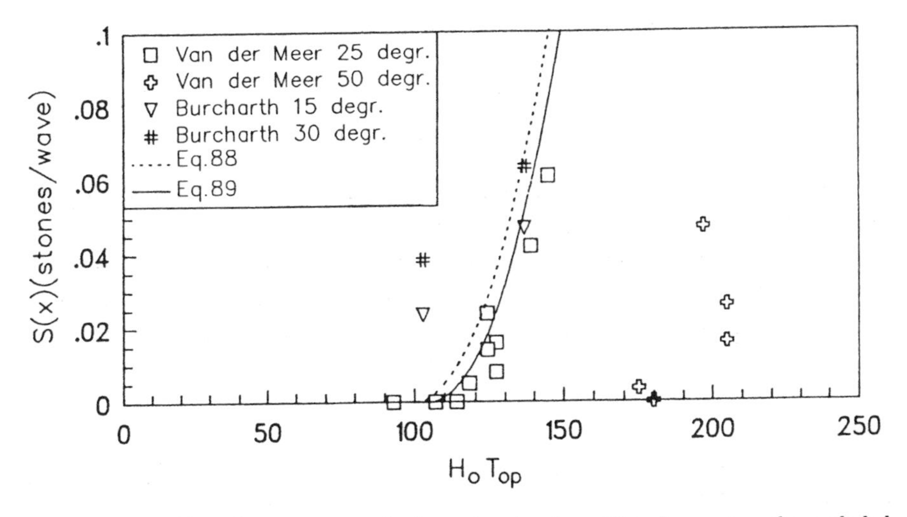

Fig. 54. Onset of longshore transport for berm breakwaters. This figure gives the exploded view of the part in Fig. 53 with $S(x) < 0.1$.

Figure 54 shows that the transport for large wave angles of 50 degrees is much smaller than for the other angles of 15–30 degrees. The two lowest points of Burcharth and Frigaard show transport for $H_oT_{op} = 100$, where the tests of Van der Meer and Veldman do not give longshore transport up to $H_oT_{op} = 117$.

Vrijling *et al.* (1991) use a probabilistic approach to calculate the longshore transport at a berm breakwater over its total lifetime. In that case, the start or onset of longshore transport is extremely important. They use the data of Van der Meer and Veldman (1992) and the data of Burcharth and Frigaard (1987), but not the extended series described in Burcharth and Frigaard (1988). Based on all data points, (except for some missing data points, these are similar to those in Fig. 53), they come to a formula for longshore transport:

$$\begin{aligned} S(x) &= 0 \text{ for } H_oT_{op} < 100 \ , \\ S(x) &= 0.000048(H_oT_{op} - 100)^2 \ . \end{aligned} \tag{88}$$

Equation (88) is shown in Figs. 53 and 54 with the dotted line. The equation fits nicely in Fig. 53, but does not fit the average trend for the low H_oT_{op}-region (see Fig. 54). The equation overestimates the start of longshore transport a little (except for 2 points of Burcharth and Frigaard). Therefore, Eq. (88) was changed a little in order to describe the start of longshore transport better:

$$
\begin{aligned}
S(x) &= 0 \text{ for } H_oT_{op} < 105 \ , \\
S(x) &= 0.00005(H_oT_{op} - 105)^2 \ .
\end{aligned}
\tag{89}
$$

This equation holds for wave angles roughly between 15 and 35 degrees. For smaller or larger wave angles, the transport will be (substantially) less. Equation (89) is shown in Figs. 53 and 54 with the solid line and fits better in the low H_oT_{op}-region. The upper limit for Eq. (89) is chosen as $H_s/\Delta D_{n50} < 10$. With Eq. (89), the longshore transport for berm breakwaters has been established.

References

Ahrens, J. P. (1987). Characteristics of reef breakwaters. CERC, Vicksburg, Technical Report CERC-87-17.

Ahrens, J. P. (1989). Stability of reef breakwaters. *J. Wtrwy., Port, Coast. and Oc. Eng. ASCE.* **115**(2):221–234.

Allsop, N. W. H. and A. R. Channell (1989). Wave reflections in harbours: Reflection performance of rock armored slopes in random waves. Hydraulics Research, Wallingford, U.K., Report OD 102.

Allsop, N. W. H. (1990). Rock armoring for coastal and shoreline structures: Hydraulic model studies on the effects of armor grading. Hydraulics Research, Wallingford, U.K., Report EX 1989.

Allsop, N. W. H. and R. J. Jones (1993). Stability of rock armor and riprap on coastal structures. *Proc. Int. Riprap Workshop*, Fort Collins, Colorado, USA. 99–119.

Aminthi, P. and L. Franco (1988). Wave overtopping on rubble mound breakwaters. *Proc. 21st ICCE. ASCE.* Malaga, Spain.

Andersen O. H., J. Juhl, and P. Sloth (1992). Rear side stability of berm breakwater. *Proc. Final Overall Workshop of MAST G6S Coastal Structures*, Lisbon, Portugal.

Battjes, J. A. (1974). Computation of set-up, longshore currents, run-up, and overtopping due to wind-generated waves. *Comm. on Hydraulics*, Department of Civil Engineering, Delft Univeristy of Technology, Report 74–2.

Bradbury, A. P., N. W. H. Allsop, and R. V. Stephens (1988). Hydraulic performance of breakwater crown wall. Hydraulics Research, Wallingford, U.K., Report SR 146.

Brebner, A. and P. Donnelly (1962). Laboratory study of rubble foundations for vertical breakwater. Queen's University Kingston, Ontario, Canada, Engineer Report No. 23.

Burcharth, H. F. and P. Frigaard (1987). On the stability of berm breakwater roundheads and trunk erosion in oblique waves. *Seminar on Unconventional Rubble-Mound Breakwater. ASCE.* Ottawa, Canada.

Burcharth, H. F. and P. Frigaard (1988). On 3-dimensional stability of reshaping berm breakwaters. *Proc. 21st ICCE. ASCE.* Malaga, Spain. Ch. 169.

Burcharth, H. F., G. L. Howell, and Z. Liu (1991). On the determination of concrete armor unit stresses including specific results related to Dolosse. *J. Coast. Eng.* **15**:107–165.

Burcharth, H. F. and Z. Liu (1992). Design of Dolos armor units. *23rd ICCE. ASCE.* Venice, Italy. 1053–1066.

CIAD, Project group breakwaters (1985). Computer-aided evaluation of the reliability of a breakwater design, Zoetermeer, The Netherlands.

CUR/CIRIA Manual (1991). Manual on the use of rock in coastal and shoreline engineering. Gouda, The Netherlands, CUR Report 154. CIRIA special publication 83, London, U.K.

Daemen, I. F. R. (1991). Wave transmission at low-crested breakwaters. M.Sc. thesis. Delft University of Technology, Faculty of Civil Engineering, Delft, The Netherlands.

Daemrich, K. F. and W. Kahle (1985). Shutzwirkung von Unterwasserwellen brechern unter dem einfluss unregelmässiger Seegangswellen. Eigenverlag des Franzius-Instituts für Wasserbau und Küsteningenieurswesen, Heft 61 (in German).

De Gerloni, M., L. Franco, and G. Passoni (1991). The safety of breakwaters against wave overtopping. *Proc. ICE Conf. Breakwaters and Coastal Structures*, Thomas Telford, London, U.K.

De Waal, J. P. and J. W. Van der Meer (1992). Wave run-up and overtopping at coastal structures. *Proc. 23rd ICCE. ASCE.* Venice, Italy. 1758–1771.

Engering, F. P. H. and S. E. J. Spierenburg (1993). MBREAK: Computer model for the water motion on and inside a rubble and mound breakwater, Delft Geotechnics, MAST-G6S Report.

Franco, L. (1993). Overtopping of vertical face breakwaters: Results of model tests and admissible overtopping safes. *MAST 2-MCS-project Monolithic (vertical) Coastal Structures; Proc. of final workshop*, Madrid, Spain.

Führböter, A., U. Sparboom, and H. H. Witte (1989). Großer Wellenkanal Hannover: Versuchsergebnisse über den Wellenauflauf auf glatten und rauhen Deichböschungen mit de Neigung 1:6. Die Küßte. Archive for Research and Technology on the North Sea and Baltic Coast (in German).

Givler, L. D. and R. M. Sørensen (1986). An investigation of the stability of submerged homogeneous rubble-mound structures under wave attack. Lehigh University, H. R. IMBT Hydraulics, Report #IHL-110-86.

Gravesen, H. and T. Sørensen (1977). Stability of rubble mound breakwaters. *Proc. 24th Int. Navigation Congress*, Leningrad, Russia.

Holtzhausen, A. H. and J. A. Zwamborn (1992). New stability formula for dolosse. *23rd ICCE. ASCE.* Venice, Italy. 1231–1244.

Jensen, O. J. (1984). A monograph on rubble mound breakwaters. Danish Hydraulic Institute, Denmark.

Kao, J. S. and K. R. Hall (1990). Trends in stability of dynamically stable breakwaters. *Proc. 22nd ICCE. ASCE.* Delft, The Netherlands. Ch. 129.

Kobayashi, N. and A. Wurjanto (1989). Numerical model for design of impermeable coastal structures. University of Delaware, USA, Research Report No. CE-89-75.

Kobayashi, N. and A. Wurjanto (1990). Numerical model for waves on rough permeable slopes. *J. Coast. Res.*, Special issue 7:149–166.

Latham, J.-P., M. B. Mannion, A. B. Poole, A. P. Bradbury, and N. W. H. Allsop (1988). The influence of armorstone shape and rounding on the stability of breakwater armor layers. Queen Mary College, University of London, U.K.

Ligteringen, H., J. C. Van der Lem, and T. Silveira Ramos (1992). Ponta Delgada Breakwater Rehabilitation. Risk assessment with respect to breakage of armor units. *23rd ICCE. ASCE.* Venice, Italy. 1341–1353.

Owen, M. W. (1980). Design of seawalls allowing for wave overtopping. Hydraulics Research, Wallingford, U.K., Report No. EX 924.

PIANC (1993). Analysis of rubble mound breakwaters. Report of Working Group no. 12 of the Permanent Technical Committee II. Supplement to Bulletin No. 78/79. Brussels, Belgium.

Postma, G. M. (1989). Wave reflection from rock slopes under random wave attack. M.Sc. thesis. Faculty of Civil Engineering, Delft University of Technology, Delft, The Netherlands.

Powell, K. A. and N. W. H. Allsop (1985). Low-crest breakwaters, hydraulic performance and stability. Hydraulics Research, Wallingford, U.K., Report SR 57.

Powell, K. (1990). Predicting short term profile response for shingle beaches. Wallingford, U.K., Research Report SR 219.

Scott, R. D., D. J. Turcke, C. D. Anglin, and M. A. Turcke (1990). Static loads in Dolos armor units. *J. Coast. Res.*, Special issue 7:19–28.

Seelig, W. N. (1980). Two-dimensional tests of wave transmission and reflection characteristics of laboratory breakwaters. Vicksburg, USA, CERC Technical Report No. 80-1.

Seelig, W. N. (1983). Wave reflection from coastal structures. *Proc. Conf. Coastal Structures '83. ASCE.* Arlington, USA.

SPM (1984). Shore Protection Manual. Coastal Engineering Research Center. U.S. Army Corps of Engineers.

Steetzel, H. J. (1993). Cross-shore transport during storm surges. Ph.D. thesis. Delft University of Technology, Delft, The Netherlands.

TAW (1974). Technical Advisory Committee on Protection against Inundation. Wave run-up and overtopping. Government Publishing Office, The Hague, The Netherlands.

Thompson, D. M. and R. M. Shuttler (1975). Riprap design for wind wave attack. A laboratory study in random waves. HRS, Wallingford, U.K., Report EX 707.

Van der Meer, J. W. (1987). Stability of breakwater armor layers – Design formulas. Elsevier *J. Coast. Eng.* **11**:(3)219–239.

Van der Meer, J. W. (1988a). Rock slopes and gravel beaches under wave attack. Ph.D. thesis. Delft University of Technology, The Netherlands. Also, Delft Hydraulics Communication No. 396.

Van der Meer, J. W. (1988b). Deterministic and probabilistic design of breakwater armor layers. *Proc. ASCE, Wtrwy. Port, Coast. and Oc. Eng.* **114**(1):66–80.

Van der Meer, J. W. (1988c). Stability of cubes, tetrapods and accropode. *Proc. Breakwaters '88*, Eastbourne. Thomas Telford.

Van der Meer, J. W. and M. J. Koster (1988). Application of computational model on dynamic stability. *Proc. Breakwaters '88*, Eastbourne. Thomas Telford.

Van der Meer, J. W. and K. W. Pilarczyk (1990). Stability of low-crested and reef breakwaters. *Proc. 22th ICCE ASCE.* Delft, The Netherlands. 1375–1388.

Van der Meer, J. W. (1990a). Low-crested and reef breakwaters. Delft Hydraulics Report H 198/Q 638.

Van der Meer, J. W. (1990b). Data on wave transmission due to overtopping. Delft Hydraulics Report H 986.

Van der Meer, J. W. (1990c). Verification of BREAKWAT for berm breakwaters and low-crested structures. Delft Hydraulics Report H 986.

Van der Meer, J. W. and G. Heydra (1991). Rocking armor units: Number, location and impact velocity. Elsevier *J. Coast. Eng.* **15**(1 & 2):21–40.

Van der Meer, J. W. and K. d'Angremond (1991). Wave transmission at low-crested structures, in *Coastal Structures and Breakwaters Proc. ICE*, London, U.K. Thomas Telford.

Van der Meer, J. W. and J. J. Veldman (1992). Stability of the seaward slope of berm breakwaters. Elsevier *J. Coast. Eng.* **16**(2):205–234.

Van der Meer, J. W., H. A. H. Petit, P. Van den Bosch, G. Klopman, and R. D. Broekens (1992). Numerical simulation of wave motion on and in coastal structures. *Proc. 23rd ICCE. ASCE.* Venice, Italy. 1772–1784.

Van der Meer, J. W. and C. J. M. Stam (1992). Wave run-up on smooth and rock slopes of coastal structures. *J. Wtrwy., Port, Coast. and Oc. Eng. ASCE.* **118**(5):534–550.

Van der Meer, J. W. and I. F. R. Daemen (1994). Stability and wave transmission at low-crested rubble mound structures. *J. Wtrwy., Port, Coast. and Oc. Eng. ASCE.* **120**(1):1–19.

Van Gent, M. R. A. (1994). The modeling of wave action on and in coastal structures. Elsevier *J. Coast. Eng.* **22**(3 & 4):311–339.

Van Mier, J. G. M. and S. Lenos (1991). Experimental analysis of the load-time histories of concrete to concrete impact. Elsevier *J. Coast. Eng.* **15**(1 & 2):87–106.

Vellinga, P. (1986). Beach and dune erosion during storm surges. Ph.D thesis. Delft University of Technology, The Netherlands.

Vidal, C., M. A. Losada, R. Medina, E. P. D. Mansard, and G. Gomez-Pina (1992). A universal analysis for the stability of both low-crested and submerged breakwaters. *23rd ICCE. ASCE.* Venice, Italy. 1679–1692.

Vrijling, J. K., E. S. P. Smit, and P. F. De Swart (1991). Berm breakwater design; the longshore transport case: A probabilistic approach. *Proc. Coastal Structures and Breakwaters. ICE.* London, U.K. Thomas Telford.

Symbols

A_c	Armor crest freeboard, relative to still-water level
A_e	Erosion area on profile around still-water level
B	Structure width, in horizontal direction normal to face
C_r	Coefficient of wave reflection
C_t	Coefficient of total transmission, by overtopping or transmission through
D	Particle size, or typical dimension
D_n	Nominal block diameter $= (M/\rho_r)^{1/3}$
D_{n50}	Nominal diameter $(M_{50}/\rho_r)^{1/3}$
D	Sieve diameter
D_{50}	Sieve diameter, diameter of stone which exceeds the 50% value of sieve curve
D_{85}	85% value of sieve curve
D_{15}	15% value of sieve curve
D_{85}/D_{15}	Armor grading parameter
E_i	Incident wave energy
E_r	Reflected wave energy
E_t	Transmitted wave energy
F_c	Difference of level between crown wall and armor crest $= R_c - A_c$
G_c	Width of armor berm at crest
g	Gravitational acceleration
H	Wave height, from trough to crest
H_i	Incident wave height
H_{m0}	Significant wave height calculated from the spectrum $= 4\sqrt{m_0}$
H_r	Reflected wave height
H_t	Transmitted wave height
H_s	Significant wave height, average of highest one-third of wave height
$H_{2\%}$	Wave height exceeded by 2% of waves
$H_{1/10}$	Mean height of highest one-tenth of waves
H_oT_o	Dimensionless combined wave height-period parameter
h	Water depth
h_c, h'_c	Armor crest level relative to seabed, after and before exposure to waves
h_t	Depth of toe below still-water level
k_t	Layer thickness coefficient

L	Wave length, in the direction of propagation
L_o	Deep water or offshore wave length, $gT^2/2\pi$
M	Mass of an armor unit
M_{50}, M_i	Mass of unit given by 50%, i%, on mass distribution curve
m	Seabed slope
m_0	Zeroth moment of wave spectrum
$m\text{-}nth$	nth moment of spectrum
N	Number of waves in a storm, record or test
N_{od}, N_{od}, N_{omov}	Number of displaced, rocking, units per width D_n across armor face
N_s	Stability number $= H_s/\Delta D_{n50}$
N_s^*	Spectral stability number
P	Notional permeability factor, defined by van der Meer
P	Probability that x will not exceed a certain value; often known as cumulative probability density of x
p	Probability density of x
q	Overtopping discharge, per unit length of seawall
Q	Dimensionless overtopping discharge; various definitions
R	Strength descriptor in probabilistic calculations
R	Dimensionless freeboard parameters; various definitions
R_c	Crest freeboard, level of crest relative to still-water level
R_d	Run-down level, relative to still-water
$R_{d2\%}$	Run-down level, below which only 2% pass
R_u	Run-up level, relative to still-water level
$R_{u2\%}$	Run-level exceeded by only 2% of the incident waves
S	Loading descriptor in probabilistic design
S	Dimensionless damage, A_e/D_{n50}^2; may be calculated from mean profiles or separately for each profile line, then averaged
$S(x)$	Number of rocks displaced per wave in longshore direction
s	Wave steepness, H/L
s_{om}	Wave steepness for mean period, $2\pi H_s/gT_m^2$
s_{op}	Wave steepness for peak period, $2\pi H_s/gT_p^2$
s_p	Wave steepness with local wave length
T	Wave period
T_m	Mean wave period
T_p	Spectral peak period, inverse of peak frequency
t_a, t_u, t_f	Thickness of armor, underlayer or other layer

V	Variation coefficient $= \sigma/\mu$
Z	Reliability function in probabilistic design; $Z = R - S$
α	Structure front face angle
β	Angle of wave attack with respect to the structure
Δ	Relative buoyant density of material considered, e.g., for rock $= \rho_r/\rho_w - 1$
$\mu(x)$	Mean of x
ξ	Surf similarity parameter, or Iribarren number, $= \tan\alpha/\sqrt{s}$
ρ_r	Mass density (saturated surface dry density)
ρ_r, ρ_c, ρ_a	Mass density of rock, concrete, armor
ρ_w	Mass density of water
γ	Total reduction factor
γ_b	Reduction factor for a berm
γ_f	Reduction factor for rough slopes
γ_h	Reduction factor for depth limited waves
γ_β	Reduction factor for oblique wave attack
$\sigma(x)$	Standard deviation of x
ξ_m	Surf similarity parameter based on T_m
ξ_p	Surf similarity parameter based on T_p
ξ_{mc}	Critical surf similarly parameter